Starman Boo

HANDBOOK of BINOCULAR ASTRONOMY

By Michael Poxon, B.A. (Hons)

Dedicated to the memory of my parents
Who selflessly ensured I had the education
they could never have

Handbook of Binocular Astronomy

Published by Starman Books

ISBN: 978-0-9562394-0-2

CONTENTS

Appendices

INTRODUCTION

To the general public, an astronomer is always thought of as somebody with a telescope, and there are a fair number of folks who are of the opinion that an ordinary pair of binoculars are only useful for maybe birdwatching, or a day at the races. That this is simply not true in the field of astronomy is what I aim to demonstrate in this book. Binoculars are easily accessible, often inexpensive and are useful in many fields of astronomy, whether it be the making of scientifically useful observations or the simple pleasure of gazing up into the sky, taking in the beauty, the interest, and the sheer magnificence of it all. The chief attractions of binoculars from an astronomical perspective are their large fields of view (that is, you can see a much larger area of the sky than you can with a telescope) and their relatively good light-grasp - meaning that you can see many more stars that are well below the limit of your eyes alone.

Of course, for some fields of astronomy, binoculars are of limited use. For instance, unless you are using large (and probably expensive) instruments, you will not be able to see very much detail on the planets. However, the dice rolls both ways - there are *variable stars* for instance; these are stars that change in brightness over periods of time, and you can follow hundreds of these fascinating objects with an ordinary pair of binoculars. Also, there are imposing groups of stars and large clusters which lose their interest and beauty when seen with a telescope, because telescopes show only a small part of the sky at any one time. This is a book that covers the whole field of astronomy, from some of the basic concepts through to a more advanced stage where you will be able to make a real contribution to our knowledge of the denizens of the sky.

Above all, this is a *practical* book, designed to get you out into the night air. Virtual reality is all very fine, but real reality is better!

In the eighteenth and nineteenth centuries, the heyday of amateur science, brilliant observers typified by the Herschels, Struves and the Rev. T.W. Webb surveyed the night sky with their telescopes in order to discover more about the universe, which they felt to be a beautiful and awe-inspiring thing. Coming nearer to our own times, we have had observers like Leslie Peltier in the USA and George Alcock in Britain who bestowed the same sort of dedication upon their beloved skies. What I try and demonstrate in this book is that this voyage of discovery can be shared in the same spirit by today's amateur astronomers too - and contrary to popular belief, it does not have to be a pastime for the well-off.

Michael Poxon

Norfolk, UK

April 2009

CHAPTER 1. BINOCULARS

This is a book for both the beginner as well as the seasoned observer. If you are an absolute beginner, and don't know your asteroids from your adenoids, it is probably a good idea to stay with this chapter. If, on the other hand, you spend every clear night God sends out of doors no matter what the thermometer has to say, and know all about the Step and fractional comparison methods, you can happily go to the final chapter without serious loss. That chapter is an exhaustive guide to the sights that can be seen with binoculars in each of the 88 constellations in the heavens. Then again, maybe you are one of those many people who have developed an interest in the sky and would like to progress a bit further but are reluctant to dispense with large wads of paper (or of course plastic) to pay for that impressive, gleaming new electronic telescope with all its mysterious knobs and controls. This chapter is for you.

The first thought of somebody with more than just a nodding interest in the heavens is usually "what sort of telescope will I need?" To which I answer STOP! Maybe you don't need a telescope at all - at least, not yet, and what you certainly don't need is one of those awfully-tempting black-and-white jobs on a wooden tripod to be found in City stores, camera shops and mail-order catalogues everywhere. These things are really toys, so I will call them Toy Telescopes or TT's for short. Though they look like what the public expect a telescope to look like, i.e., a long tube on a tripod, they are nearly always disappointingly useless as an introduction to serious astronomy, and could well frustrate you or put you off for life. Buy one and you will probably be both out of pocket and out of luck. The chief problems with TT's are their small fields (i.e., the amount of sky you can see through them) and the fact that the finders (the little telescope attached to the main one, which *do* show a large amount of sky, so that you can find the object you are interested in, hence the name) are only sufficiently powerful to find the brighter stars. The engineering of the stands may also be found wanting. In fact, early in 2002 I was speaking to a friend whose young son had been given one of these things for Christmas by well-meaning relatives. It was utterly useless - and a waste of money to boot. A pair of binoculars, on the other hand, can be about a third to a quarter of the price of your average TT, will have a large, bright field, and will be easily portable to a dark sky site (even by bicycle!) and of course useful for those other activities I skated over in the introduction such as bird-watching. This is no flippant remark, as I have found that an interest in one branch of natural science is usually coupled with an interest in others. However, in this connection, it may be useful to tell any neighbours you may have about your nightly activities, or they could well become concerned at seeing you peering upwards with a pair of binoculars! This is, though, not the only reason to tell them; you may also develop their interest in the sky as well - or encourage them to reduce light pollution by drawing curtains or turning off unwanted lights. Such light pollution is a highly-relevant facet of modem astronomy.

Binoculars come in various powers and sizes, and this is how they are sold - for instance 7 x 50, 8 x 30 and so on. The first number tells you how many times the binoculars magnify, while the second gives the diameter (in millimetres) of the objective. The objective is the large lens that faces the sky and collects light from whatever you happen to be looking at. The bigger the objective is, the more light it collects, and the fainter you can see. Binoculars, of course, have a pair of objectives. Each half of a binocular is thus a telescope - a strange-looking, modified tele-

Fig.1: Schematic diagram of Standard Binoculars

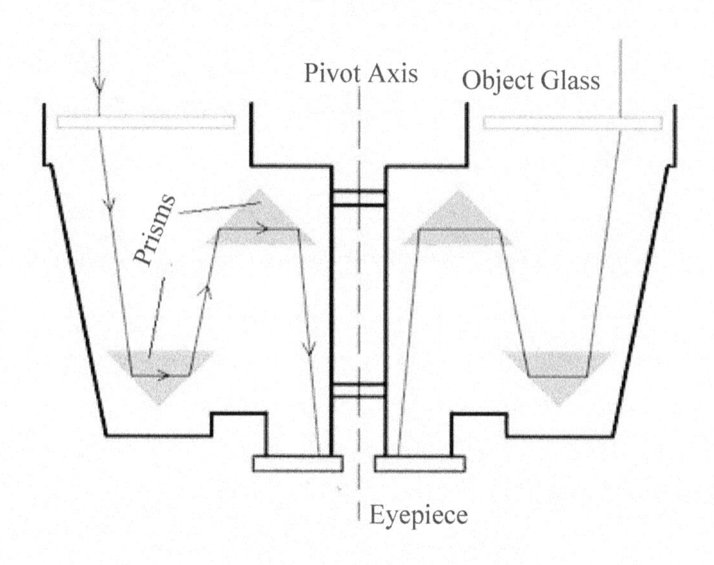

scope undoubtedly, but still basically a telescope. It collects the light from a distant object (bird in a tree, star, etc.) and magnifies it. The original patent submitted by Zeiss is for a "double telescope with increased objective separation". So there you are.

Between each objective and eyepiece, light is folded by a pair of prisms (Fig.1). This folding means that the instruments are much more compact and able to be held in the hands, so that manipulation tends to be steadier than with a long-tubed telescope. The glass of which the prisms are made, as well as the prism design itself, has a bearing on the optical image. The diagram shows a schematic view; in practice, one prism is rotated vertically through 90° with respect to the other, thus saving space. The cheaper varieties use a type of right-angled prism called a porro prism made of a type of glass called BK7, whereas the better-quality models employ glass with a component of the element Barium and are known as BaK4. Prism misalignment due to accidents, rough treatment, and so on is probably the main cause of serious defects, and again the better-quality binoculars, while not immune altogether, will feature stronger mountings for the prisms. It is not necessary to know the science behind these facts, but bear in mind the above differences when you see this sometimes daunting terminology used in advertisements. Advertisers sometimes cannot resist blinding the consumer with science! In this regard, never buy zoom binoculars for astronomy. The extra optical components they inevitably involve causes too much light loss. As an example, 7 x 50's magnify an object 7 times and its objectives, also called object glasses, are each 50 mm in diameter. A pair of 7 x 50's are ideal as a good, all-round instrument with which to view the sky for the beginner, for as the magnification increases, the amount of sky seen through the lens decreases, and you will tend to become 'lost in space' more easily. For the moment, much of what follows relates to average-sized binoculars: say, everything from 6 x 30's up to 16 x 50's.

Larger binoculars than this are easily available, especially from the companies listed in Appendix I. Even in your everyday TT outlet, instruments such as 20 x 70s can be readily found. I bought a pair of this very size from this very type of store a few years ago, and they were fine. Bear in mind, however, that they will be much heavier than their more modestly-sized cousins, and prolonged hand-holding will lead to aching arms and a complete inability to support the instruments steadily. Also to the fore recently we have seen the introduction of Image-Stabilised Binoculars, which use sensors to detect motion and adjust the optics accordingly through small internal motors. ISB's could prove useful for astronomical purposes, but bear the following facts in mind: firstly, they weigh over a kilogram, and all the image stabilisation in the world will not prevent your arms from aching. Secondly, they are not cheap, the above models retailed at nearly £900 when they first appeared, though prices are coming down. Thirdly, you can get perfectly stable images by devising or buying a decent stand, as I describe in the next chapter under "Observing Aids". Finally, many have small 'exit pupils' so you are going to get some wasted light. If after all this they are still a must-have, then several of the outlets in Appendix I stock them. Absolutely finally, the chip that does the smart work is battery-powered, and the batteries' life (at about three hours) may not be long enough for protracted observational sessions. You do not want all that expensive high-tech wizardry proving useless because the power is low! I have tried a pair of these out (on Mauna Kea in Hawai'i, no less!) and was not overly impressed. Another way of sparing your arm-strain is to employ Roof-Prism Binoculars, which use a different optical design to the standard right-angled Porro Prism configuration and are consequently lighter and less chunky. However, the manufacturing standards involved need to be much more precise and this has to be paid for. Furthermore, the geometry of the prisms used has other optical effects, which must be counteracted. Again this comes at a price. The cost of roof prism binoculars is also moderating but, as in the case of the Image-Stabilised models, be sure to buy them from an established optical or astronomical dealer who will be able to advise you further. Also light in weight are *monoculars* which, as the name suggests, is simply one half of a pair of binoculars. They are readily available and often inexpensive, but obviously lack the ease of use and depth of image that binoculars possess, and have the disadvantage that, since both eyes are not being used, you need to do something with the inactive eye, and this can introduce eye strain. A monocular can be very useful as a "quick check" resource, I have found, and is not to be despised. Again, they can be bought from the suppliers listed in the appendices.

If you are a beginner, I strongly advise you to start off with something conventional; like everything else, you ought to know what the rules are before you attempt to bend them! Rather like the person who buys all the expensive golfing equipment but has little idea of how to play the game, you may come to feel that you have parted with a large amount of money for a limited reward and will feel yourself having to justify it. You will soon become disillusioned if those large, powerful binoculars show you such a small area of star-packed sky that you have no idea what you are looking at!

TESTING

Having had some thoughts on the model, there remains another important, not to say crucial, process to go through before you actually part permanently with your money. We need to make sure the optical quality is good enough for astronomy. Don't forget, we will be pushing these things to their limits, and what may look fine in the middle of the race-track or the woods may not be quite so impressive when you are trying to see a tiny, faint point of light in the sky, with your neck at a disagreeable angle. Both *Astronomy Now* and *Sky and Telescope* carry regular advertisements for established and experienced suppliers. There are, after all, bound to be binocular equivalents of the dreaded TT, though often the average camera shop sells perfectly adequate, if not top-of-the-range, binoculars. One such dealer in my home town nevertheless assured me that binoculars were no use for astronomy!

When you go to inspect and purchase the binoculars (especially from somewhere like a high street shop) first establish with the owner that you will need them for astronomy and would like to test them at night. However, by day, you can still get some idea of the state of their optics. Focus on a distant brick wall and see how sharp the image is, and try each of these criteria.

1. The layers of mortar in the wall should be equally well-defined, whether they are horizontal or vertical.
2. All parts of the field of view should be equally bright and distinct, though a very small amount of false colour around the circumference of the field can be tolerated.
3. There should be no double vision evident, and the distance between the eyepieces should be smoothly adjustable so that each eye can be brought to the centre of the field of view, which should be circular. The 'figure 8' outline so beloved of Spy movies serves only to indicate that our hero is using binoculars!
4. At least one eyepiece must be independently focusable.

If the binoculars fail any of these tests, ask to try another pair. If they pass, hand over the money, get a receipt, and ask if you can test them at home with a proviso to return them if they do not meet your expectations. If the reply is doubtful, think about asking for your money back or try somewhere else.

Okay - so you have the binoculars at home and it looks like a clear night. If you have no stand for them, support the binoculars firmly in your hands (still with the strap attached just in case - you don't want to drop them!) and rest your elbows on something sturdy such as a wall. Incidentally, it is possible to buy shoulder- or neck-cradles which are convenient, and will also leave your hands free. A prominent, but not brilliant, white star makes a good test. It needs to be nice and high in the sky to minimise the effects of colour and distortion sometimes produced by the effects of low altitude, and should show no false colour or spikiness when the binoculars are properly focused with the star near the centre of the field. One of the eyepieces may have a scale scribed with numbers around its barrel. This is called the *dioptre* ring and indicates the degree of focus for that eye, in which connection I should add that when you do arrive at the sharpest view, any further change of focus should be immediately evident. Again, if the star test proves disappointing, return the binoculars as soon as possible, and try again. If you have acquired them from a reputable astronomical dealer, you should not be dismayed; but test them anyway.

MAINTENANCE

As I pointed out earlier, you will need one of the eyepieces to be independently adjustable for focus, and in this regard a very occasional (i.e.,yearly rather than weekly), small amount of grease on the metal axle of the central focusing wheel may nol go amiss. Do not under any circumstances get grease, oil, or anything similar remotely near any optical surface - don't even think about it. Bear in mind that your binoculars have a pair of compound eyepieces, two prisms (i.e., four altogether) and compound object glasses; all the surfaces of these components have to be accurately polished, fixed and aligned minutely. The wonder, after considering all this, is that binoculars are so inexpensive! So treat them with care.

You can see that binoculars are potentially fragile affairs, so a few words on the subject of treatment may be in order. There are three broad reasons for declining performance - dirt, optical misalignment and mechanical misalignment. Treat your binoculars with care and do not become complacent. Contrary to the opinion of small hands, they are not playthings. If for some reason you do suspect misalignment has taken place, on no account should you poke around inside to try and rectify the situation. According to chaos theory, small causes can lead to great effects, and unskilled - however intelligent - hands manipulating unknown components will readily ensure that theory becomes fact, and chaos will ensue!

Dirtwise, a few specks of dust are unavoidable, will make very little difference to image degradation, and can be ignored. When, however, your binoculars do get to the stage where some cleaning is necessary, try to avoid physical contact wherever possible. Photographic shops sell little air-puffers which are very useful in this regard, though avoid extended use of the attached brush. It may look and feel soft, but remember that it has been designed for photographic lenses, not binocular objectives. If you use the brush, apply a light stroking motion, and never, *ever*, rub. If you use cotton wool, always make sure it is the 'pure' variety: the common type is often treated with oils or other substances which are perfect for babies' bottoms but which will have unwanted effects on optical glass. Before using cleaners, please consult an astronomical specialist first, and never use anything intended for household cleaning. Any excessive abrasion will not only cause possible damage to the anti-reflection coatings which usually protect the lenses and decrease the amount of light wasted, but will also produce optical scratching. These will combine to produce an awful, diffuse image which really is worse than useless.

Diffuseness in the short term, however, is caused by dewing of the objectives under the influence of damp night air. In order to prevent this from happening, you can make a pair of simple 'dewcaps" by constructing two cylinders of card, lined on the insides if you wish with matt black paper. The length of the caps should be at least double the diameter of the Object Glasses to be effective. This is a method I use to good effect on my own 'binoculars' which actually take the form of a 10 x 50 finder perched on my large (36 cm) telescope.

Give your objectives no protection and the following will invariably happen... a beautiful clear night full of promise opens above you. After half an hour, your lenses start to mist up, so you take them into the warm of the house for a few minutes, ruining your delicate night vision in the process. When you finally return outside, you find that the stars are being devoured by a giant bank of cloud which has magically appeared out of nowhere during the time you were indoors.

CHAPTER 2. A FEW BASICS

Let's now spend a little while skating over a few basic astronomical concepts, since they will be encountered again and again throughout the course of the book, and arc sufficiently important to spend no little time on. In my experience, what confuses beginners most are those factors having to do with appearances - scale, brightness, orientation and so forth. Strangely enough, a frequent question that gets asked is "how do you know what the names of the stars are?" to which one can only give the reply "the same way you know what your friends' names are!" We know what the stars' names are because we named them - but it has to be said that star designations can look forbidding when you first meet them!

MAGNITUDE

The first term we need to understand is magnitude, which is simply the astronomer's word for 'brightness'. The magnitude set-up works backwards rather like an awards system: so, just as first prize is more valuable than second prize, a first magnitude star is brighter than a second magnitude one, and so on. There is, however, one big difference; you can't get better than first prize, but you can get brighter than first magnitude!

Very bright sky objects therefore can have zero or even negative magnitudes; the Full Moon has a magnitude of -12 and the Sun of -27. Sirius, the brightest star in the night sky, is of magnitude -1.5, and several of the other bright stars have zero magnitude - though confusingly they are still colloquially called 'first-magnitude' stars; the ghastly word *zeroth* has thankfully never really caught on. At the other end of the scale, the naked eye can just detect stars of the sixth magnitude, and av-

Fig.2: Graphical representation of the Magnitude system

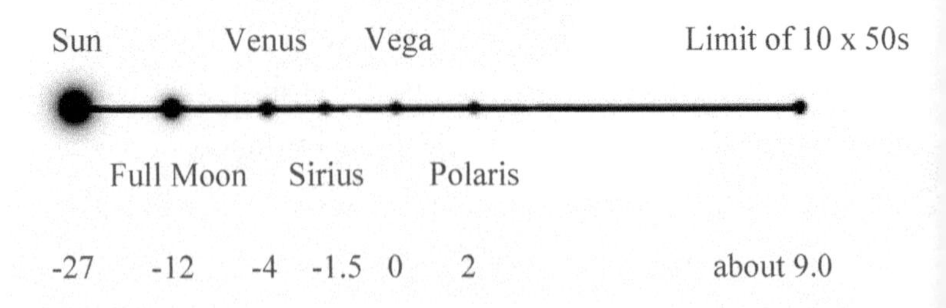

erage binoculars will go down to magnitude 9. We sometimes abbreviate this to -27m, -1.5m, 9m and so on. My own telescope - with an objective diameter of 36 cm (just over 14 inches) has revealed stars as faint as 16.5m.

You will notice that within the general magnitude groups, we further subdivide into decimals of a magnitude, so that a star of, say, 4.5m is very slightly brighter than one of 4.6m. (The smaller the number, the brighter the star, don't forget!) This difference is, in point of fact, about the smallest that an experienced eye can detect, and I will use no greater accuracy than this throughout the book for the excellent reason that it is not necessary for our purposes. Fig.2 gives an easy graphical representation of magnitudes on a 'number line' where you can see that sometimes it is more appropriate to give accurate values (0.0 as distinct from 0.9) or at other times more general ones (i.e., 'magnitude 6' as being at the naked-eye limit of visibility). It is just a matter of context, or 'horses for courses'.

THE CONSTELLATIONS

When you first looked into the starry sky, what did you see? A jumble of glittering points of light dotted, so it seemed, haphazardly above you? Maybe you kept looking up; after a time, that formlessness would have started to take on shape, for the human mind is exceptionally skilled at making any sort of order out of seeming chaos. Perhaps you noticed a regularly-spaced line, or a bright triangle, or something that reminded you of an everyday object. For thousands of years mankind has been busily making itself at home in the sky by arranging patterns of stars into recognisable groups, and over the course of centuries tales were woven around the shapes we fancied that we saw. These are the *constellations* - such as the great hunter Orion who strides across the Winter sky, or the Great and Little Bears swinging around the heavens' North Pole. Most of the familiar constellation figures, together with the tales that surround them come down to us by way of classical Greece, but at a much later date when Europeans, both navigators and those with less laudable aims, sailed South of the equator, they encountered stars and constellations they had never seen before - since the solid Earth obstructs our view of the stars on the other side from where we are - and for which there were no classical myths, though of course the people already living in the Southern countries such as the Australian Aborigines and the Bushmen of Southern Africa had woven their own rich mythological systems around their own familiar Southern stars.

It must be borne in mind that a constellation is only a unit because we have, by virtue of our pattern-making tendencies, deemed it to be so. The individual stars that make up a constellation pattern are not, at least in the great majority of cases, actually connected with each other - they merely happen to lie in the same sort of direction as we look at them. If the Earth revolved around some other star, the familiar stars we see would appear in completely different patterns, for not only would we be at a different viewpoint, but the individual stars would also all be at different distances from this viewpoint, and would appear of different brightnesses than they appear to us currently. (In fact, I have written software which will do this for you!) Maybe the stars in the pattern we call the Plough or Big Dipper would look like a ladder instead of a ladle, and would be called the "Stairway to Heaven" or something similar. There are 88 constellations that the sky is subdivided into, each with their own well-established and unique boundaries, rather like a vast jigsaw puzzle, so that there can be no part of the sky which does not belong to a given constellation. These boundaries were drawn up in 1930 and superseded the old and unscientific idea of placing stars by saying, for instance, that a given star was in the "left elbow of Orion" or whatever. Some of the old star atlases are primarily works of art rather than anything else. When I was about twelve years old, I actually owned one of these, a beautiful volume dating from 1822 by one Alexander Jamieson and dedicated to King George III (I think it was him; history isn't my strong point). I thought it would be interesting to (tastefully, of course) colour in the figures. When I came to try and sell it to an antiquarian book dealer a few years later, he was not amused! Many of these old atlases show constellation figures that no longer exist, such as *Sceptrum Brandenburgicum* (the Sceptre of Brandenburg, now included in the large constellation of Eridanus, the River), or the Hot-Air Balloon *Globus Aerostaticus*, whose stars are now part of the constellation Cetus, the Whale. Cetus, in fact, is one of a large group of constellations involved in one of the best-known stories of the celestial stage, that of Perseus and Andromeda. Perseus was the dashing hero, Andromeda the damsel in distress. Her mother Cassiopeia and her father Cepheus are nearby in the sky, as is Pegasus, the winged horse ridden by Perseus when he saved Andromeda from Cetus, the sea monster (Yes, I know Cetus means "whale" but you have to remember that whales were seen as fearsome creatures before Greenpeace el al came on the scene). In the sky, the main stars of Pegasus form a large square, and for a time the star at the top left-hand comer of the square was thought of as being in both Pegasus as well as the neighbouring Andromeda. This is one reason why the constellation boundaries needed to be scientifically determined, and the star, whose name is Alpheratz, is now officially in Andromeda, though it is always shown as a star in the Square of Pegasus just to aid pattern recognition. Appendix 4 lists all the constellations, together with their official 3-letter abbreviations, which are used throughout the book.

NAMING THE STARS

On the subject of names, whilst the well-known stars like Polaris, Betelgeuse and so on have their own personal names if you will, most of the fainter ones do not, and so some method needed to be devised to identify them. In 1603 the astronomer Bayer came up with the Greek letter system that we find is still in use today, whereby the brightest star in a constellation is accorded the first letter of the Greek alphabet, alpha (α); the second brightest is beta (β) and so on through the 24-letter Greek alphabet to omega (ω). Since Alpheratz is the brightest star in Andromeda, it is also known as alpha Andromedae. Sorry; because the constellation names are Latin, we have to use Latin grammar to match, but I promise there isn't any more! *Andromedae* is the 'genitive' or 'apostrophe-s' form of Andromeda, so means 'of Andromeda'. These forms will soon become second nature as you meet them, and who knows - you may even develop an interest in Latin, as I did. Most people have heard of "alpha Centauri" - well, that's a genitive, from "Centaurus"!

Returning now to another classical language, the Greek alphabet is given in Appendix 2, though in some cases the strict order has become somewhat less than strict. In Sagittarius, for instance, alpha and beta are quite faint, while the two actual brightest stars in the constellation are in fact epsilon and sigma (ε and σ)! The Greek alphabet, with its 24 letters, provides for only this number of stars in each constellation, so when we run out of these, we use a system devised by *Flamsteed*, the first Astronomer Royal. He numbered stars (including those which also have Greek letter names or proper names) sequentially from West to East, thus the most Westerly star in Perseus, for example, is 1 Persei, our friend Alpheratz can also be named as 21 Andromedae and the brilliant blue star overhead during Summer nights in Europe or North America can be called Vega, α Lyrae or 3 Lyrae. In practice of course, everyone calls it Vega! In many constellations, chiefly in the Southern skies, stars fairly close together will be given the same letter and are distinguished by superscripts, such as α^1, α^2 and so on. Another practice prevalent in the Southern skies is to use ordinary letters (such as a, b, A, B, etc) instead of numbers when the Greek alphabet has been used up. Finally, in some rare cases, a star name has even been given to an object that is not a star! Both 47 Tucanae and omega Centauri are objects with conventional star designations, a Flamsteed number and a Greek letter - though they are not actually stars at all, but in fact are giant balls of thousands of stars called Globular Clusters.

Before we change the subject unduly, let's visit the Summer sky for a while and look at the constellation of Cygnus, the swan, sometimes called the Northern Cross. (Fig. 3) Its brightest star is α Cygni or Deneb, a star of magnitude 1. The stars in the cross-beam (or the wings of the swan if you like) are all of the second magnitude, while at the foot of the cross is a 3rd-magnitude star called Albireo. Halfway between Albireo and the central member of the cross is a still-fainter star called eta (η) Cygni, which is of magnitude 4. When you have found Cygnus, go and take a look at these stars, noting the progression from bright to faint both within the whole group, as well as within the stars of the cross-beam; though these latter are all of round about magnitude 2, they are not equal

Fig.3: The Constellation of Cygnus, showing stars of various magnitudes

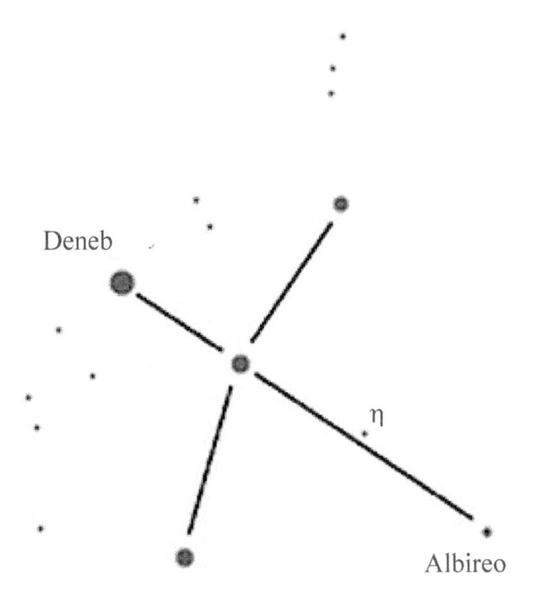

in brightness. This is where the decimals of a magnitude come into play. If we were talking generally, as above, we would just call them "second magnitude stars" whereas if we wanted to stress their individual bright-

Fig.4: Using the Plough and the Pole Star as degree measurers

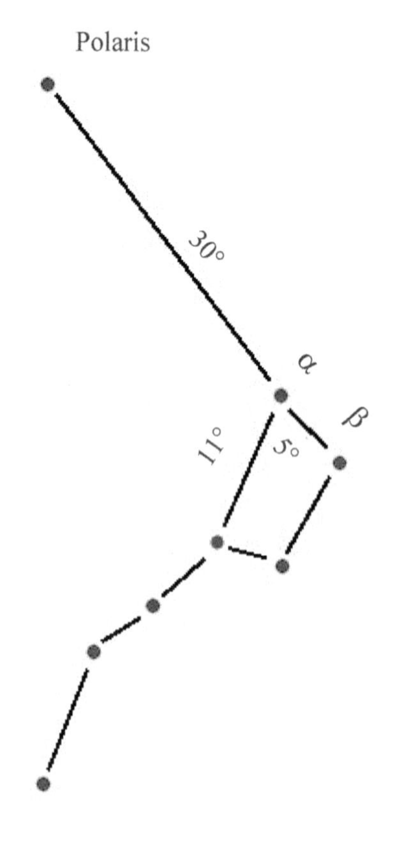

Fig.5: The Celestial Sphere - not to scale. Instead, the sphere should be thought of as having an indefinite size.

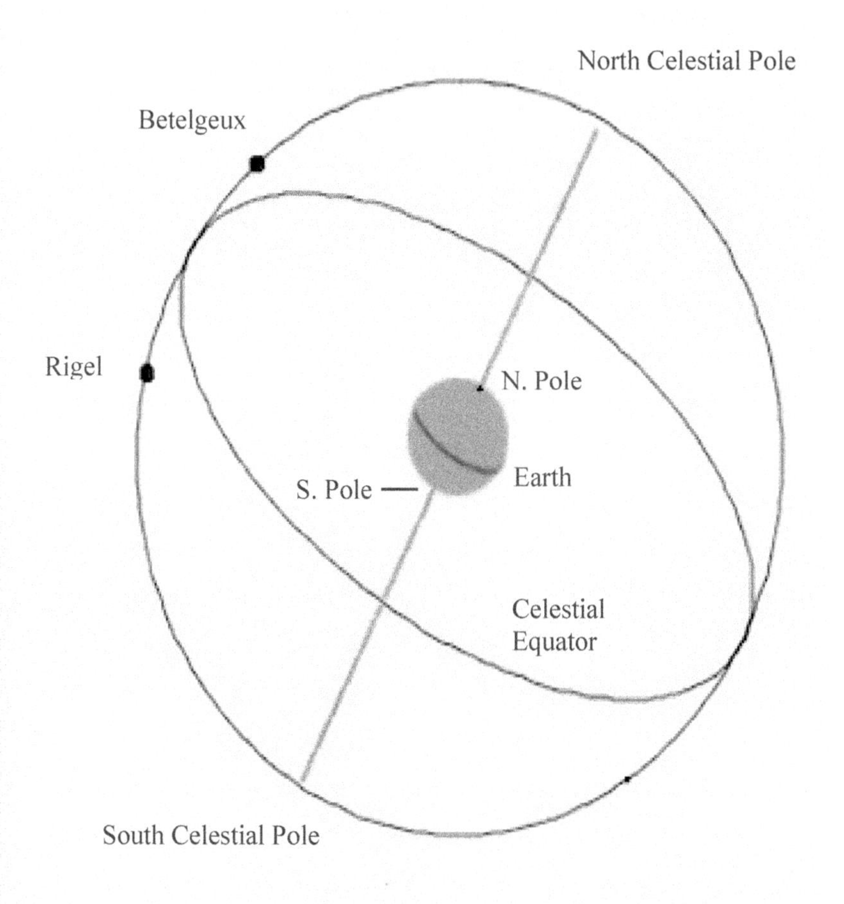

nesses, we would quote their more accurate values such as 2.1, 2.4 or whatever. It is again a question of "horses for courses".

DISTANCES IN THE SKY

Let us now leave magnitude for a while to look at the concept of distance in the sky. I do not mean the actual physical distance between the stars, which is truly immense, rather the apparent distance from one to another as we look at them in the sky. For this we use *angular measure*, and our units are degrees. We all know there are 360° in a circle, so if you find any point on the horizon, describe a line through the point right above you (the *zenith*) and hit the opposite horizon, you will have gone through 180°, from the horizon to the zenith is 90°, and so on. A spread hand at arm's length covers about 25 to 30 degrees, about the width of the Plough or Big Dipper as you look at it in the sky.

So let's have a look now at some angular distances in the night sky. Find the Plough and use the diagram here (Fig.4) to get some idea of what various angular (or "arc") distances look like in reality. The pointers, α and β Ursae Majoris, are 5 degrees apart, so dividing this distance by 5 will give you some idea of what one degree looks like. It is actually twice the apparent diameter of the Moon in the sky - but it is no good telling people that! The fact is, because the Moon is so bright, it actually looks much larger than this. When we start using optical aid to magnify, we home in on smaller portions of the sky than we see with the naked eye, and one degree now appears as a fairly large unit. So we need to subdivide it - into 60 minutes (written 60') although a minute can in turn be subdivided into 60 seconds (60"). Now, a second is a *very* small distance in the sky, but even so, an ordinary pair of binoculars can discern two stars about 25" apart under average conditions. Such *double stars* are interesting to look at and provide good tests for the quality of your optics as well as of your own experience and sometimes, persistence. There are hundreds of them in this book, and you are sure to find at least one that makes a favourite object for you. Note, by the way, that the "double-apostrophes" - more properly called a *double-prime* symbol, used as the contracted form for 'seconds of arc' have also been used to mean inches, but unless otherwise noted, this book always intends the meaning 'seconds of arc'.

Still on the subject of angles, we need to acquaint ourselves with the sky's co-ordinate system, since that is how we find and properly identify many objects. On the Earth, we can locate any spot accurately by means of its Latitude (North or South of the equator) and its Longitude (East or West of Greenwich). The sky system is very similar but actually easier, since we only measure in one direction - Eastward - from the celestial analogy of the Greenwich meridian, so there is no astronomical International Date Line halfway round the celestial sphere to confuse you! This sky version of the Greenwich line is called the *First Point of Aries*, and we measure positions from it in an Easterly direction, that is, leftwards. We do not normally use degrees, however, but hours, minutes and seconds of time. This is because the Earth rotates in about 24 hours, and the effect is for the sky to seem to rotate around a stationary 'us'. Our measurement here is called *Right Ascension* (RA) instead of Longitude. A star whose RA is (say) 3 hours 20 minutes, written 3h 20m, culminates - or reaches its highest point in the sky - 3 hours and 20 minutes after the 'Aries meridian' has done so. Think of it as being 3 hours and 20 minutes behind 'Aries'. The RA of the Aries meridian is, of course, zero, and just as with lines of longitude on the Earth, lines of RA extend from one pole to the other. For the moment, don't worry about what the first point of

Aries actually is - just think of it as the fact that if we are going to measure, then we have to start measuring from somewhere, and this is where we happen to start from!

The sky version of Latitude is *Declination*, and it determines which stars we can see and which ones we can't. Just like latitude on the Earth, we measure it in degrees North or South of the equator - though in this case it is the celestial, rather than the Earthly equator, of course. As you may see from Figure 5, the Poles and Equator of the sky are merely projections of those of the Earth; they have no physical reality of their own, though it just so happens that Polaris, the Pole Star, lies close to that part of the sky that our Earth's North Pole currently points to - Its declination is +90° just as the latitude of the North Pole is. You can see that, if you were to stand at the North Pole itself, the Pole Star would be right above you. Therefore, the altitude of the celestial pole, whether North or South, is equal to your own latitude on the Earth. If you lived on the equator (zero latitude!) the altitude of the sky poles would also be zero, and they would in fact lie on opposite sides of your horizon, with the celestial equator passing overhead. You may also see from this that the declination of the overhead point is the same figure as your latitude. This needs to be borne in mind when observing with ordinary binoculars. For, although objects are easier to see when near the zenith as there is less mist and atmosphere through which a star's light has to pass, you are also quite likely to assume an uncomfortable position in looking straight up for any length of time.

So you can see that, unless you do live on the equator, there will always be some stars that you can never see - as you move away from the equator, one sky pole climbs higher in the sky, while the other will fall below the horizon. If you go to latitude 30° North, the Pole Star will be at an altitude of 30° too - and the stars within a 30-degree radius of it, that is, between 60° and 90°North declination - will in fact never set. To you, they will be *circumpolar*, visible all year round. Conversely, the South celestial pole will now be 30 degrees *below* the horizon, and stars within a thirty-degree radius of that (i.e., with Declinations between 60 and 90 degrees South) will never climb above the horizon at all, and so will be permanently invisible from your latitude. Fig. 5 shows declination in action using the two chief stars in Orion. The declination of Betelgeuse is about +10° and that of Rigel about -10°, That of the Pole Star, of course, is about +90°.

So - when you know your latitude, you can work out which stars will be circumpolar and which will be forever hidden from you; merely subtract your latitude from 90°. Stars with declinations greater than this figure will never set. Stars with declinations less than the same figure *but with the opposite + or - sign* will never be visible. For example, I live in the UK at a latitude of about +52°. Subtracting this from 90 gives 38, so stars whose declinations are greater than +38° will be circumpolar; those with declinations of -38° or less are permanently invisible to me, so if I want to see (say) the Southern Cross, at a declination of about 60° South, I will have to go to a part of the world that has a latitude of less than 30 degrees. I can confirm as of 2002 that the Southern Cross is indeed visible from Hawai'i, which is at a latitude of 20 degrees! As for the stars in between your 'extremes', these will rise and set normally, though obviously the nearer they are to your visible sky pole, the shorter their period below the horizon will be.

THEORY INTO PRACTICE

We saw earlier how optical aid allows us to magnify, and see objects that are close together in the sky. The ability of an optical system to separate close objects is called its *resolving power*. The nineteenth-century British observer W. R. Dawes determined a formula to calculate this, and by applying it, we emerge with a resolving limit of about 10" for a 50mm lens - though other optical effects come into play with this theoretical limit, and in practice standard binoculars will actually only resolve to about twice that distance for persons with normal eyesight, as a rule.

We also need to know how much sky, or *field*, we can see when

Binocular	Mag. Limit	Field (°)	Res. Power (sec)
8 x 30	7.5	4	40
7 x 50	8.5	7	30
10 x 50	9.3	5	25
12 x 60	10.00	5	20
20 x 70	10.5	3.5	15

we look through the eyepieces. Field size also depends upon magnification - the higher the power, the smaller the field. When you are first beginning a detailed study of the sky with binoculars, a large field is more useful, as you will then usually be able to navigate your way around using prominent naked-eye stars.

The table here sums up what we have seen so far. Whenever possible, I have used my own findings to construct it, and these should be taken as guides rather than fixed values. For instance, you could be using expensive, high quality mounted binoculars in the middle of the Arizona

Desert - on the other hand, you could be stuck in the middle of a large town with a winter wind blowing around your fifteenth-floor dwelling!

One more value remains to be discussed, and that is what is known as the *exit pupil.* If you point your bins at the daylight sky (obviously avoiding the Sun!) you will be able to see a little disc of light a few millimetres from the eyepiece. This is the real image your eyes physically see when you look through the eyepiece, and ideally it should match the diameter of the pupil of your eye, so that no light is wasted. The diameter of the dark-adapted eye pupil is 7mm, though this tends to decrease with age, reaching about 5mm at age 40. Dividing the objective diameter by the magnification will give you the diameter of the exit pupil, so you can see that 7 x 50 glasses, with an exit pupil of 7mm, will give the best return on your optical investment. For the same reason, so will 10 x 70's - but they will be heavier and more expensive!

THE OBSERVER...

Few people, including some astronomers, stress the importance of the observer, who is after all the final link in the observational chain. In these days of Black Holes, Dark Matter, Space Telescopes and the rest, the person at the eye-end can sometimes be relegated to the status of a mere appendage to the technology, rather than the reverse. In order to make the most of the technology, you need to make the most of yourself. No less an authority than Sir William Herschel pointed out that an object is easier to see when you know it is there, and the instances of a small telescope being able to catch an object that has been discovered with a larger one are many. What has changed here is not usually a physical quantity, but the experience of the observer, a strangely metaphysical quality indeed, but none the less real for all that. Be ready, then, to look at the same object under different conditions: get fully used to the dark; reduce strain (this may indeed become a subconscious effort eventually), relax the eyes by taking a break every now and again; keep your hands warm by using thermal gloves - you will soon become used to handling flimsy paper charts and suchlike; I speak from experience. As you get to know the night sky better, you will absorb more comfort from its inhabitants and its lore and will feel more at home there, and less anxious about getting lost. Finally, if you are having difficulty seeing faint objects, use *averted vision*, which consists in using the outer edge of the retina, looking "out of the comer of the eye" in fact. Just why this works as it does needs an explanation as to how your eyes work. So here we go!

Fig.6: Schematic diagram of the Eye

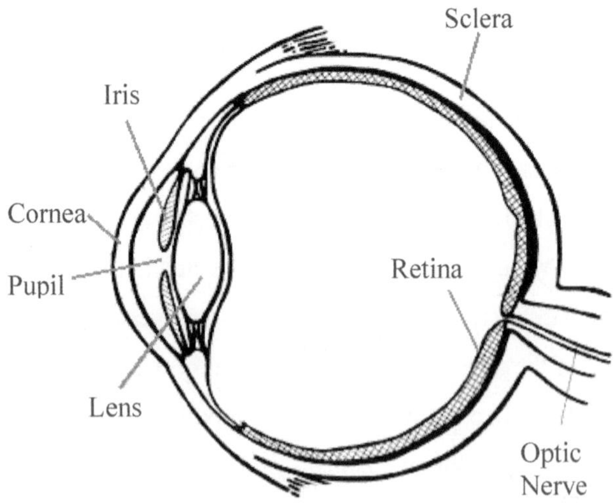

...AND HIS EYES

The eye - what a truly amazing work of nature it is - basically unchanged throughout the entire vertebrate kingdom for many millions of years, suggesting a design that cannot be improved upon. Considering all the beauty we fathom through the eyes, we take them largely for granted. This is probably why many works on amateur astronomy do the same - but this book is different!

There are three basic parts to the eye - the physical eye itself, the optic nerve, and the visual centre in the brain. For our purposes, the first of these is the only one we are concerned with. It takes the form of a fairly spherical chamber that we call the eyeball, the clear outer coating being the *sclera* from the Greek word for 'hard'. This covers the *cornea*, which together with the *lens* focuses incoming light onto the retina at the back of the eye. If you think of the retina as the film in an old-fashioned camera, you will get the idea. It is not coated with chemical crystals however, but backed with a layer of cells of two kinds called *rods* and *cones*. Each of these carries out a different function; the cones are sensitive to differences in wavelengths of light and so distinguish colour, while the rods are sensitive to degrees of light and so are responsible for the general perception of light and dark. The central area of the retina comprises only cones, but rods become more plentiful as we go toward the edge, which is where the rods overwhelmingly predominate. That is why we use averted vision to see faint objects, although by doing so we sacrifice some colour vision in the process. The ends of the rod cells contain a substance called *visual purple*, responsible for optimising light-sensitivity, which is white in normal light, but which becomes renewed as the environment darkens, a process which can take up to twenty minutes. This is why it is not a good idea to spoil all that hard-won dark adaptation! The amount of light which falls upon the retina is regulated by the *iris*, an opening in the eye controlled by two sets of muscles around it. Using again the analogy of a camera, think of the iris as the aperture, but one which is completely instinctive and automatic. Fig. 6 shows a basic diagram of the human eye.

Some eye defects are a function of ageing, though there are some good yoga exercises for strengthening the eyes that I have found extremely effective. Short or long sight, if it is not excessive, can be compensated for by focusing adjustments in the eyepieces, though sufferers of astigmatism (the result of an irregularly-shaped cornea, though on rare occasions it can be internal) will need to keep their glasses on. Vitamin A helps to maintain the eyes, and nicotine has harmful effects upon dark-adaptation, so avoid the evil weed and eat carrots for maximum benefit!

OBSERVING AIDS

Let's now have a peer at matters which are rather less definable, such as comfort. It will be evident that an observation made when you are warm, comfortable and looking at an easily-visible object will be more reliable than one made when you are knee-deep in stinging nettles, cringing up at the zenith in the vain hope of reaching some obscure ninth-magnitude star at the limit of vision (believe me, I've been there!) At these times you will need some sort of stand to rest or hold your binoculars steady. One method is to use a wall or fence as a rest for the arms; another is to sit back in a deck chair or lay on a large board such as a railway sleeper. All these have the glaring disadvantage that you have no hands free to examine books or charts, and is for this reason that you should seriously think about some sort of stand. I have tried a few solutions but have to say that I prefer above all others the common or garden camera tripod - they may look flimsy when compared with some telescope stands you might see, but provided the tripod is well-maintained and the extended legs properly splayed, there is nothing better - I have used an ordinary camera stand to support binoculars as large as 40 x 70 without any trouble. A camera stand will not only enable you to point the bins to any point in the sky, but more importantly, to lock the position once found. The heads of camera stands are provided with holes and slits for attachments, and even an engineering muppet like myself has no difficulty in fashioning an effective clamp using 'Meccano' (erector sets in the USA) or something similar. It is possible to buy dedicated, inexpensive binocular clamps from camera stores, but in my experience they are not always of high enough strength or quality. Those obtained from specialist suppliers are more likely to last. Some of these suppliers (see the appendices) provide dedicated mounts for binoculars sometimes including seats, counterbalances, and so on, most of which work on a parallelogram design principle. If you get serious about observing, you may like to consider one of these. To sum up; unmounted bins permit quick freedom of movement and are easily portable, which is fine when you are just ranging across the sky. Mounted binoculars increase comfort by decreasing arm strain, and will give you up to 1½ magnitudes gain over hand-held ones. But you have probably settled on a solution that suits you, so now it is time to get down to to the reason why you bought the things in the first place - observing.

CHAPTER 3. THE SOLAR SYSTEM

THE SUN

The Solar System is our own small corner of the vast Galaxy and is, as the name suggests, dependent on and centred around the Sun, which is a good a place as any to start. The Sun is a star, and all stars are, like the Sun, vast globes of gas which radiate heat and light by nuclear processes deep within their interiors; though compared with most of the stars you can see at night, our Sun is not very large, luminous, young, nor old - but like baby bear's porridge it is just right - thankfully for us. The only reason the Sun is so bright is because it is far, far nearer than any of the night-time stars. It is 150,000,000 km distant, nearly one and a half million kilometres across, and our entire world could fit inside it a million times over, with room to spare. The titanic pressure of the gases at the Sun's core forces up the temperature there to fifteen million degrees, so that nuclear reactions then take place. This is what powered the Sun into full life some five billion years ago, keeps it shining today, and will indeed ensure we will see its cheerful yellow face for another five billion years to come - though by the end of that time its face will have become rather larger and redder as its evolution progresses.

Astronomers are interested in the Sun for many reasons. For instance, it is the only star we can study in such detail. Learning more about the Sun, as a representative star, can tell us a great deal about stars in general. The second reason is more down-to-Earth (literally!) Clearly the Sun is absolutely crucial to events here on Earth, chiefly as regards weather and climatic phenomena. As we understand more about the Sun, we can get a better understanding of its effects upon us such as Global

Warming. Bear in mind here that both Solar and weather phenomena are vast and complicated interrelated and chaotic systems, and there will virtually never be a simple cause-and-effect relationship between the parameters involved. However, the more accurately these many parameters can be determined, the more likely our physical models of the 'big picture' are to approach some sort of reality.

Ever since he began to walk upright, mankind has been interested in the Sun, and despite the battery of high-tech instruments we now possess, there is still much that we do not understand about it, even though it is our home star. And while it is extremely unlikely that one single observer will make a solo impact upon our store of Solar knowledge, the pooled contributions of many such people may well be significant at some point. This is, in fact, a general truth in many fields of astronomy (and indeed, most science) today.

From the outset two things need to be said; firstly, that the Sun is not the ideal object for the binocular owner, though observing it makes a good introduction if you wish to interest yourself in it further. Secondly and more importantly, **NEVER, UNDER ANY CONDITIONS *AT ALL*, LOOK AT THE SUN THROUGH YOUR BINOCULARS**. You have only to see what happens to a piece of paper when you focus the Sun's light upon it with a cheap plastic magnifying glass to get some idea of what will happen to your retina. In fact, it will be fried - and you will be left permanently blind.

There are two methods you can use to look at the Sun safely. The first is to use special reflective filters (silvered on the Sunward side, obviously!) which fit over the objectives. Even here, you need to ensure that these are proper, dedicated astronomical filters. Never trust your local camera shop or TT outlet (remember the legendary TTs?) Once again, I cannot stress the importance of using filters that are the Real Mc-Coy. Never play chicken with the Sun. If it were me, I would want to wear Sunglasses too, not so much to look cool, but as a final check. You can never be too pernickety where the future of your vision is concerned. The second method is in my view much to be preferred, at least by the beginner. It takes the form of projecting the magnified image of the Sun upon a sheet of white card, held in such a position so as to give a sharp view. This is the method I used to observe the Total Eclipse in 1999 which, from my home in Eastern England, was actually about 97% total.

Fig.7: Mechanics of a Solar Eclipse

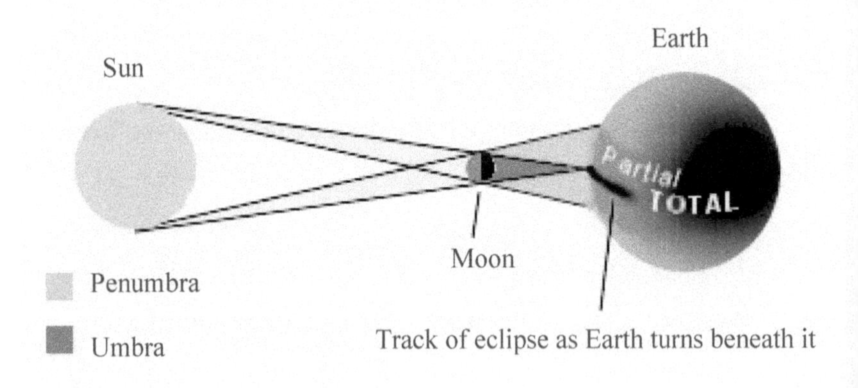

My 11-year-old son was quite impressed at seeing a few Sunspots too. In order to actually sweep up the image of the Sun on the card, do not be tempted to look, even fleetingly, through the eyepiece; instead, point the bins in the general direction of the Sun until its image appears on your card. You will, of course, get a pair of images - which is also a good check as to whether that small dark spot you can see is a bona fide Sun-spot and not just a speck of dirt on the card.

Sunspots are dark areas on the Sun which, although they are called spots, can be extremely large. Bear in mind that our star is about a million and a half kilometres in diameter - and that little spot you can see may actually be larger than the entire Earth! Sunspots were once thought to be holes in the Sun's surface, but we now know that they are areas that are slightly cooler than the surrounding *Photosphere* - the surface of the Sun, if you can use the word 'surface' to describe an object which is completely gaseous. The temperature of the photosphere is nearly 6,000° C, and sunspots have temperatures of about 4,500 degrees; by everyday standards searingly hot, but cool in comparison with the rest of the Sun's surface, and so seeming dark by contrast only. Spots are produced as a result of the way the Sun's magnetic field interrupts the flow of hot gas just below the surface, so although we saw that the old idea of spots being holes in the Sun was inaccurate, there is something of the truth about it. The magnetic disturbances with which spots are associated extend not only over the area of a spot, but also upward into space for a great distance as well. Often associated with them are *Solar Flares*, not a new type of hip streetwear, but titanic eruptions of billions of tons of super-hot gas vomited into space. This ejected gas is highly charged magnetically, and has on occasion knocked out electric power stations and communications facilities when the streams of particles hit the Earth. Our knowledge of flares is still not as complete as we would like, and they remain slightly mysterious.

It is an interesting exercise to follow a spot, or group of them, as they march serenely across the face of the Sun, sometimes changing their shape and position as they go, sometimes disappearing altogether. If you are lucky enough to find a really large specimen, you may even see something of its structure, that of a main central darkness or umbra surrounded by a lighter area or penumbra. Make one observation a day and see what you come up with. Try and estimate the time it takes for a spot to get from one edge (or 'limb') of the Sun to the other. Note that as a spot - you will only notice this with the large ones - approaches the limb, it may appear foreshortened due to the effects of perspective.

When observing, note the diameter of the image on your card at which you get the best view, and stick to this subsequently. That way, you can standardise your procedure and develop a more scientific attitude toward the recording of your observations, and will also notice any changes more readily. Even such well-known phenomena as sunspots retain some mysteries; spots do not appear at random, but are only observed within definite zones North and South of the Sun's equator, and follow what is known as the Sunspot Cycle. The spot-generation process begins at the far extremities of the sunspot zones and slowly migrates toward the equator. When the generation cycle reaches the equator it is at its minimum, by which time, however, a new cycle has just begun again at the extremities. The sunspot cycle has a rough period of 11 years, but in practice has fluctuated between about 7 and 14 years. There is much about the process that still remains uncertain, and while we have a general scenario in place as regards the cycle, its nature occupies the same sort of grey area as does long-range weather forecasting. The effects of the cycle upon our world may be many and various. Some have even suggested links to the hem-lengths of skirts!

More spectacular than Sunspots, though not an everyday occurrence, is a *Solar Eclipse*, when the Moon passes between us and the Sun. Coincidentally, the Moon and Sun appear the same size as seen from Earth, so the occasions when one disc exactly covers the other (i.e., a Total eclipse) are rare. More frequent are partial eclipses. These result from the exact same cause but are seen much more often, as the conditions which determine their appearance operate over a much wider geographical area. As figure 7 shows, the Moon's shadow takes the form of a cone, with the zone of totality being the small tip that touches the Earth. The *penumbra*, or region inside which the entire Sun is not completely covered by the disc of the Moon, is much larger, and so the ensuing partial eclipses are seen by more people. Another type of Solar obscuration is an *annular* eclipse. This is just a total eclipse in which the Moon is slightly farther away in its orbit from the Earth, and so looks a little smaller; thus the Sun appears as a bright ring of light surrounding the dark disc of the Moon. You can see that there is no great difference between all these eclipse types - it is really more a question of where you happen to be when the eclipse is going on. If you are outside even the penumbra, of course, you see no eclipse at all and it is just a normal day for you. For instance, the total eclipse of August 11 1999 was visible as total from parts of Britain. From the South-West of the country the eclipse was total, although - being Britain of course - the darn thing was largely clouded out, in fact with rain in Cornwall. At my home in South-Eastern England, however, we had much less cloud, which actually cleared right on time to reveal a brilliant blue sky for the not-quite-total eclipse. The light at totality was decidedly eerie, and I followed the progress of the eclipse with 10 x 50 binoculars, projecting twin images onto white paper. My large telescope which I use to observe very faint stars with was useless for this, so it sadly missed the event! If you are lucky enough to witness totality, you will be able to catch the *Corona*, the normally-invisible outer atmosphere of the Sun, which takes the form of a beautiful, streaming pearly radiance (though this is purely a naked-eye phenomenon). Even here, we are not completely au fait with what is happening, since the temperature of the Corona is about a million degrees - far hotter than the surface of the Sun itself! Just why this should be so is unknown. Less mysterious but no less imposing are *Baily's Beads* and the *Diamond Ring Effect*. These are caused by shafts of Sunlight streaming through mountain valleys on the limb of the Moon - to which object we now turn.

THE MOON

There is an inn, a merry old inn
beneath an old grey hill,
And there they brew a beer so brown
That the Man in the Moon himself came down
one night to drink his fill.

J. R. R. Tolkien, *The Lord of the Rings*, I, 9

The Moon, at an average distance of 384,000 km, is our closest neighbour in space by a long chalk, and is usually the first thing that any budding astronomer will want to look at. If the Sun may have sounded a little disappointing from the purely practical viewpoint, the Moon is anything but, since even the naked eye will show a lot of details on the surface of our satellite, which is some 3,450 km in diameter. Who does not know the Man in the Moon after all? Why, the traditional version of the rhyme above even had him visiting my home city of Norwich! His round face is made up of large, vaguely circular grey areas which the old skywatchers fancied may have been seas like our own, and were so named. These strange, often beautifully wistful names belie the stark reality of the Moon, but are still retained today; thus we have the *Mare Tranquillitatis* (Sea of Tranquillity), *Sinus Roris* (Bay of Dews) and many others.

Certainly the Moon is as full of tranquillity as it as empty of dew, both due to the fact that since it has no atmosphere, there is no sound to be heard there, nor any type of weather either.

The *Seas* (Mare - two syllables, plural Maria) are the lowlands of the Moon, vast plains of solidified lava flows. The presently accepted theory of their origin takes us back to a time so deep that the Earth itself had not yet completely solidified, in a Solar System still swarming with the unused leftovers of creation, rocky bodies by the million, some of them more or less planet-sized. One of these mammoth objects, probably a third the size of the fledgling Earth itself, smashed into our forming world and spewed trillions of tons of the still-molten matter away into space. At the time of the impact, the heavy elements such as iron had begun to sink towards the Earth's core, so the less-dense elements only were ejected to form the embryonic Moon, which is why it contains far less iron than does the Earth, and so is only three-fifths as dense. This ejected molten matter gradually formed itself into a large sphere under the mutual gravitational attraction of its component parts, and started to settle down into planetary (or at least satellitic) respectability. But of course the neighbourhood of this infant Moon was still a dangerous place to be; not only because of the already-existing swarms of rocky material, but also now because of the remaining innards of the young Earth which had not yet decided to complete the shaping of our satellite-to-be. But the new sphere was too large, its gravity too strong. Thousands of chunks of rock rained down on the new Lunar surface, causing the underlying molten rock to well up and flood the craters the impacts had just formed. The heavily-cratered uplands of the Moon date from this time. The seas, however, were formed later. We know this because they are far less cratered, and date from a time when much of the meteoritic material had by now been used up to produce the already-existing craters. In essence, however, they are the same; the sites of mighty meteor impacts, which then filled up with the molten bowels of the Moon shortly after the main phase of meteorite bombardment (by 'shortly', I mean only a hundred million years or so). The Solar System at this time was a much more dynamic - and hazardous - place than it is now!

The *craters* are certainly the Moon's best-known and most often-observed features, and many are visible in binoculars. Even through a humble pair of 10 x 50's, your first view of a large, well-placed crater is a rewarding experience, its rugged walls casting ink-black shadows onto the grey floor below. But don't leave things there; try and make a drawing. As a rough guide to scale, make the feature twice the diameter of your objectives, so that if you are using 10 x 50s, the drawing will be about 100 mm across. Pick an object that is large, but not too complicated, since as you draw, you will start to notice finer detail than you saw at first. The next clear night (if it is not too far in the future) draw the same object, noting how the appearance has changed under a slightly different angle of sunlight. Try again the next month. You may already find that you are seeing yet more new detail; things are becoming easier to see because you are becoming more used to looking at them. You are becoming more experienced, in fact.

Craters range in size from tiny pits at the limit of vision up to giant superbowls over 250 km across, the so-called walled plains; they are traditionally named after scientists, poets, and other worthy folk. Of late, there has also been a tendency to name features after astronauts, space probes and so on, and the Moon holds a strange assortment of bedfellows - where else would you find Julius Caesar, Newton and the Bay of the Astronauts? A special type of crater is the *ray crater*, where bright streaks like the spokes of a wheel radiate out from the crater which was the impact site. The ray material is believed to be composed of newly-ejected minerals which have not yet had time to darken through chemical reaction to the airlessness of space. They are extremely bright, those from Tycho for instance dominating the naked-eye view of the Full Moon. Other bodies in the Solar System also possess ray craters. The Moon also has numerous mountain ranges just like the Earth does. In fact, in terms of sheer size they are even more imposing, and several ranges are visible with binoculars. I do not suppose we shall have too long to wait before some daredevil mountaineer runs out of Earthly Munros to bag and heads for the crags of the Moon instead. There are also a few solitary peaks to be found, but certainly the ranges are easier to find at first. All mountain-building activity on the Moon has of course long since ceased.

In terms of observing, we first need to see how the Moon moves. It revolves around us once a month, always keeping the same face towards us, a phenomenon known as *Captured Rotation.* Fig. 8 shows the scheme of things, but its two-dimensional rendering does not demonstrate the fact that the Moon's orbit is inclined to that of the Earth's own path around the Sun, an important point to bear in mind. As can be seen, when the Moon is between us and the Sun, sunlight falls on the side farthest from us, and we cannot see the Moon at all; this is the time of New Moon. By the time our satellite has travelled another ninety degrees around its orbit, its phase increases, or waxes, until we see half of its facing side - but only a quarter of the whole Moon, since we do not obviously see the corresponding half of the far side too! This is why we call this phase 'first quarter' although most people, including me, still call it a 'half Moon'. Another ninety degrees of travel now sees the whole Earth-

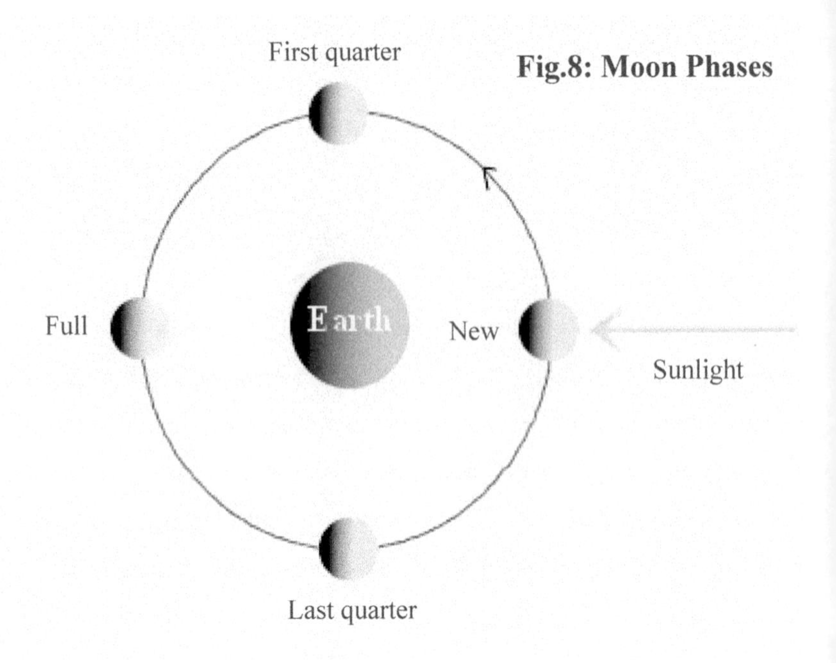

Fig.8: Moon Phases

facing side of the Moon in full sunlight, and we are at Full Moon; after this, the phase shrinks (or wanes) and another 90° brings us to a position where the left-hand half of the Moon is illuminated, or Last Quarter. Obviously, at intermediate positions is it at intermediate phases, and it has long been a source of friendly competition among astronomers as to who can see as thin a crescent after New Moon as possible. The record is said to be fifteen hours.

Observing the Moon can be a lovely pastime. You don't necessarily have to draw or record anything if you don't want to. Just take a little time to look at it - if it is only a thin crescent soon after New, you may well see the dark side faintly illuminated by Sunlight reflecting off our own world and onto the Moon. This we call Earthshine and at these times you may even be able to see features on the dark side. Do not forget that the Earth is larger and more reflective than the Moon, so throws much more light upon it than it throws upon us. A few days later, at First Quarter, you will have a good opportunity to draw some features, since now there are several large craters on view as well as imposing shadows, which are long enough to be interesting, though not so long as to wipe out everything in their path. Since there is no atmosphere on the Moon, there is no subtle shading. It is what computer graphics people call grey-scale - black and white, with only different shades of grey in between. As the *lunation*, or cycle of Lunar phases wears on, shadows become shorter and shorter until at Full Moon, with the Sun shining directly down, they disappear altogether. This is definitely the worst time to observe not only the Moon, but everything else as well (except possibly any Planets that may be about). One type of object that can be seen well at this Lunar midday are the rays, as mentioned above. On the whole, however, my opinion of the Full Moon is that it is good for one thing only - and that has little to do with astronomy!

THE MOON MAPS (FIGS. 9 AND 10)

These I have divided into two, broadly to be thought of as being for the waxing and waning Moon. As far as detail goes, I have made no attempt at pinpoint accuracy or exact draughtsmanship, since the binocular owner will not only find this unnecessary with his relatively small magnification, but may actually be misled. Note that, as with all charts in this book, the image is - contrary to works on telescopic astronomy - erect, as seen by the naked eye (or of course binoculars). Telescopes, in order to eliminate needless lenses which absorb light, give an inverted view. The erect charts used in this book also help in relating what you see through your binoculars with the view as seen with your eyes.

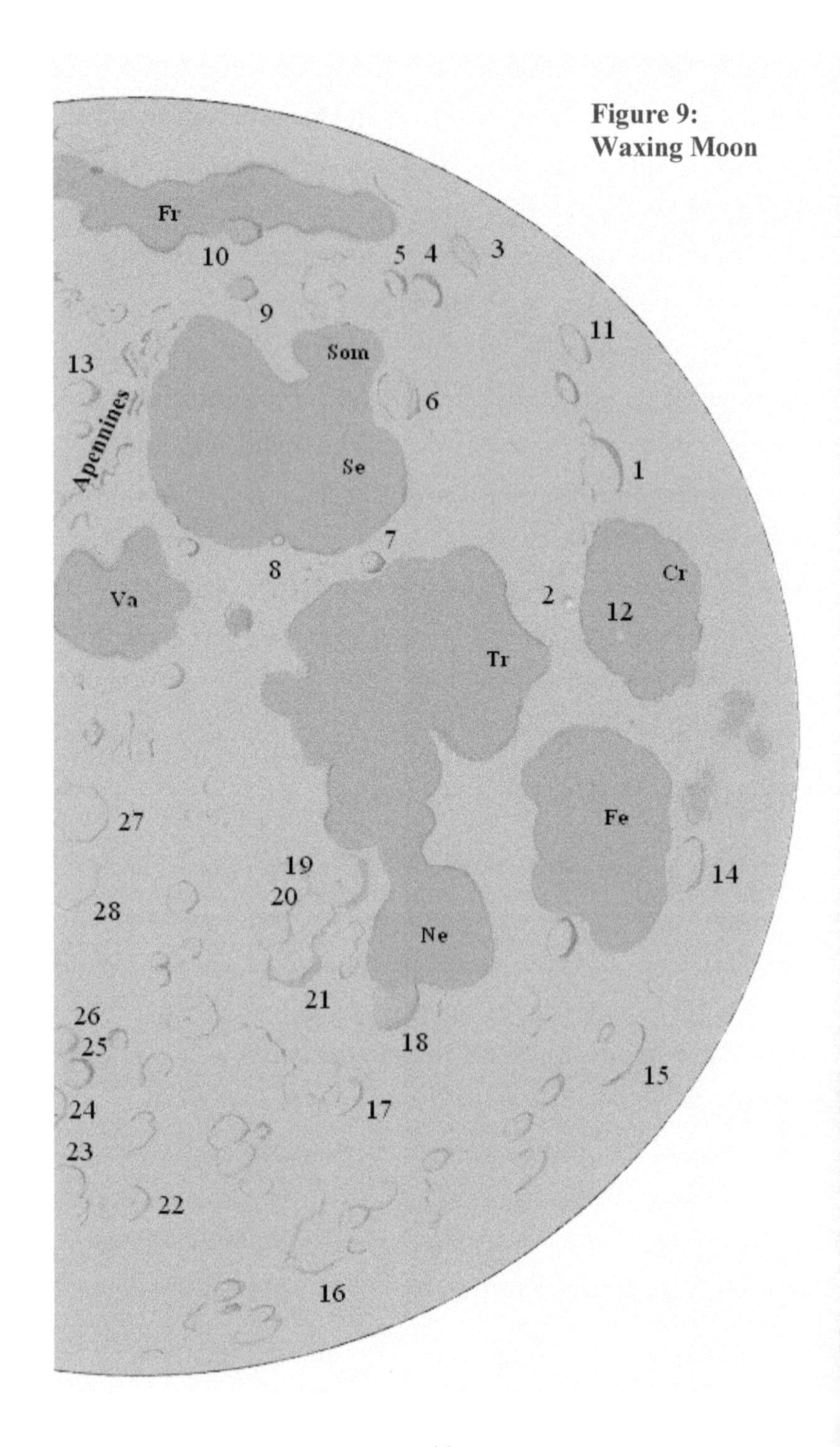
Figure 9:
Waxing Moon
Fr
10
5 4
3
9
11
Som
13
Apennines
6
Se
1
7
8
Cr
2
Va
12
Tr
Fe
27
19
14
20
28
Ne
21
26
25
18
15
24
17
23
22
16

Figure 10:
Waning Moon

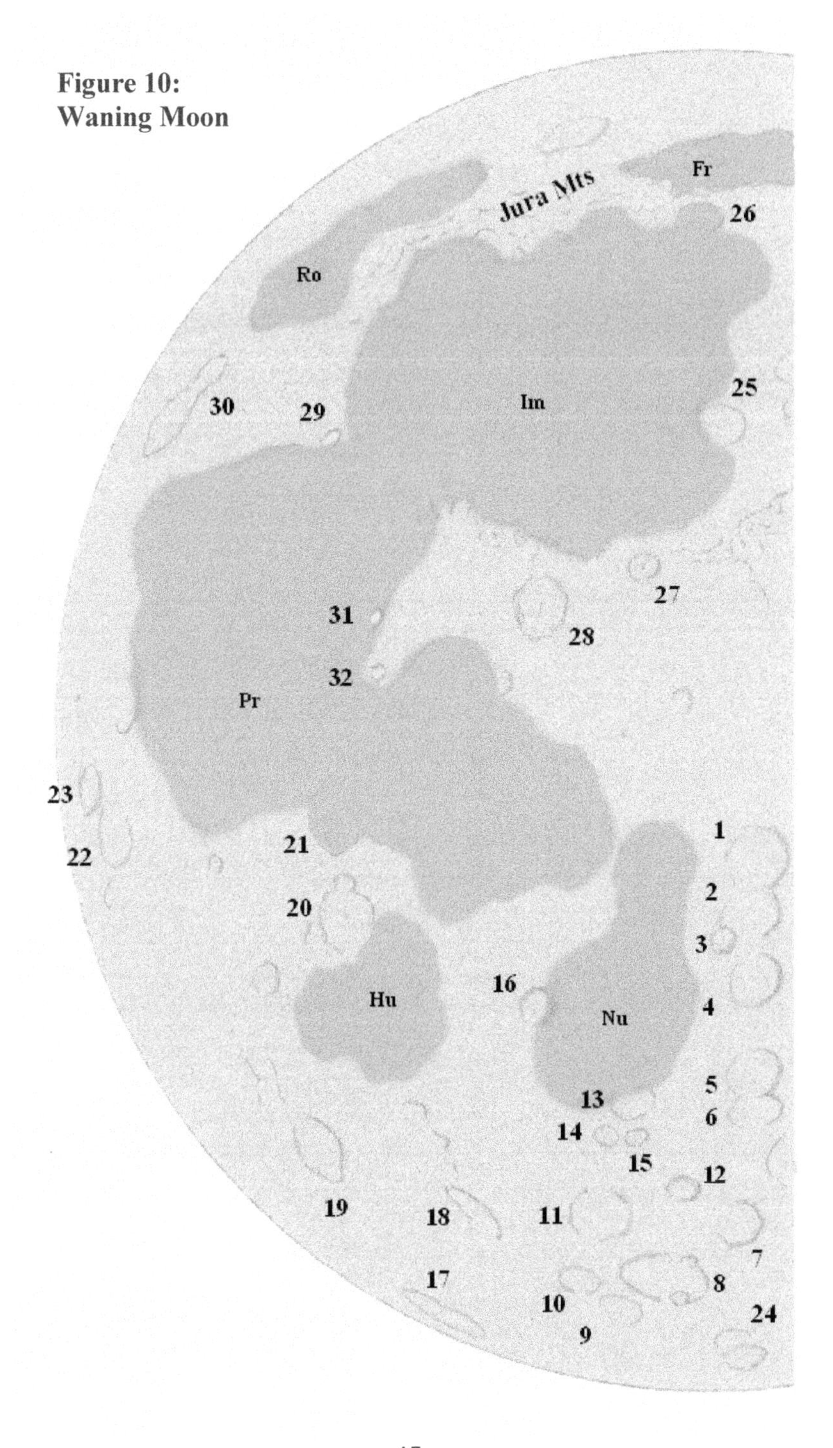

Waxing Moon

1. Cleomedes	A crater 130 km across, with walls over 2,000 metres high.
2. Proclus	Although only 30 km across, its great brightness renders it easily visible in binoculars. It is the centre of a promi nent ray system extending out into Mare Crisium.
3. Endymion	A 130 km formation with a very dark floor
4. Atlas	A noble feature over 80 km across that forms a twin with its slightly smaller neighbour Hercules (5).
6. Posidonius	A large and rather ghostly-looking walled plain 96 km across
7. Plinius	On the edge of the Mare Serenitatis, this is rather smaller than the preced ing crater.
8. Menelaus	This is only 16 km across, but is very bright at or near full Moon

9. Eudoxus	Named after a Greek mathematician, this is 72 km across, with a dark floor and walls 3,350 metres in height.
10. Aristoteles	A slightly larger crater, again with high walls
11. Messala	Over 120 km wide, this has low walls. *Messala* is just a Latin version of the name of an eighth-century astronomer, Mashallah Ibn Athari who, despite his Arab name and the fact that he lived in Basra, was in fact Jewish!
12. Picard	Named after a French scientist rather than the captain of the starship Enter prise, this is a small but very bright spot in the Mare Crisium.
13. Aristillus	A fine formation 56 km in diameter
14. Langrenus	A large, fine crater with terraced inner walls and a prominent central moun tain massif.
15. Petavius	This fine formation is 160 km across, with a famous cleft in the floor, which is unfortunately not visible with bin oculars.

16. Janssen	This vast, ruined hexagonal formation shows a great wealth of detail both inside and around it.
17. Piccolomini	A large ring 112 kilometres across, with a central peak
18. Fracastorius	This even sounds broken up, as it is! It intrudes into the Mare Nectaris, with the Pyrenees Mountains close by, which reminds us that many of the Lu nar mountain chains have been named for Earthly ones.
19. Theophilus	A 103 km crater with a central eleva tion. It is one of a well-known group near the middle of the Lunar disc, the others being Cyrillus (20) and Catha rina (21).
22. Maurolycus	A very large, ruined object in a rugged part of the landscape
23. Stöfler	A large, craggy formation 240 km across, one of the largest on the Moon.

24. Walter	A fine, circular crater with a bright peak
25. Aliacensis	This, together with its neighbour Werner (26) has high walls and a dark floor.
27. Hipparchus	A circular feature nearly 150 km across
28. Albategnius	A Latinised name again, this time of Al-Battani, who was a famous Arab astronomer from the golden age of Is lamic science. The formation itself is slightly smaller than the previous ob ject.

Waning Moon

1. Ptolemaeus	A fine, regular feature some 185 km in diameter
2. Alphonsus	Another fine feature, with Alpetragius (3) intruding into its walls.
4. Arzachel	The third of a chain with the two pre vious craters, this is slightly smaller, but with high walls rising to 3,660 me tres above the floor.

5. Purbach	Like Regiomontanus (6) nearby, this is ruined and irregular in form
7. Maginus	Another large and rugged feature, 160 km across
8. Clavius	This huge plain, 240 km across, has high and detailed walls and floor. It is visible with the naked eye when well-placed and makes a good first object to draw.
9. Blancanus	A crater nearly 100 km wide, this has a twin in the nearby Scheiner (10).
11. Longomontanus	A large and clearly hexagonal forma tion
12. Tycho	A great astronomer with a great crater to commemorate him, Tycho is the centre of the largest ray system on the Moon.
13. Pitatus	An elongated feature with low walls and a dark floor. The twin craters Gau ricus (14) and Wurzelbauer (15) lie to the South.

16. Bullialdus	A smaller but well-marked crater with fine walls and a notable central peak.
17. Bailly	Although technically the largest crater of all, this 260 km formation is so close to the limb that it is difficult to see.
18. Schiller	A similar formation, named after the author of the 'Ode to Joy'. This crater is nearly 180 km across.
19. Schickard	A dark-floored feature 220 km across, again a markedly hexagonal crater.
20. Gassendi	A similar object, with a dark floor
21. Letronne	A rugged plain some eighty km across
22. Grimaldi	This great walled plain is 225 km wide, and has a dark floor, unlike its neighbour...
23. Riccioli	Somewhat smaller and lighter in hue than Grimaldi
24. Moretus	Another dark object with massive walls

25. Archimedes	You've found it! This 80 km sized cra ter has low walls.
26. Plato	One of the most distinctive of all cra ters, this has a smooth, dark grey floor and low walls.
27. Eratosthenes	A small but prominent feature, due to its high surrounding walls and central mountain.
28. Copernicus	One of the most beautiful of the Lunar craters as well as being the centre of a prominent ray system, its detailed and terraced walls are visible with binocu lars. It lies in a mountainous area and has a diameter of 90 km.
29. Aristarchus	This is said to be the brightest spot on the Moon's surface, and is of course readily visible in binoculars. It has long been suspected of harbouring some sort of volcanic activity, al though some think that this may have something to do with the brightness of the object rather than actual vulcan ism.

30. Otto Struve — Named after the famous observer of double stars, this is most fittingly a double crater! Though it is large at 240 km, it is too close to the limb to be seen well.

31. Kepler — This, together with its neighbour Encke (32) is an extremely bright ray crater.

Mountain Ranges

Apennines — A fine, extensive range contiguous with the Caucasus Mountains. Its peaks are best seen near the boundary between the lit and unlit parts of the Moon (called, rather frighteningly, the *Terminator* by astronomers!)

Jura Mts — Another fine set of peaks, again deriving its name from an Earthly range and bordering the area known as Sinus Iridum (bay of rainbows). When astride the terminator, the bay and its surrounding mountains produce an effect of great beauty as the sunlight catches the tops of the tallest peaks.

'Seas'

Cr = Mare Crisium (sea of crises): Tr = M. Tranquillitatis (sea of tranquillity): Se = M. Serenitatis (sea of serenity): Fr = M. Frigoris (sea of cold): Som = Lacus Somniorum (lake of sleep): Fe = M. Foecunditatis (sea of fertility): Ne = M. Nectaris (sea of nectar): Pr = Oceanus Procellarum (ocean of storms): Nu = M. Nubium (sea of clouds): Hu = M. Humorum (sea of humours): Va = M. Vaporum (sea of vapours): Im = M. Imbrium (sea of rains): Ae = Sinus Aestuum (bay of emptiness): Ro = Sinus Roris (bay of dews): Ir = Sinus Iridum (bay of rainbows).

THE PLANETS

Of the nine known planets which orbit the Sun, few can be profitably observed with average binoculars, so this will of necessity be a fairly short section, The table here gives some facts and figures relating to the planets. To the table we should add the following comments: no figures are given for the angular diameters of Uranus and Neptune because they appear as stars through binoculars and even small telescopes. Pluto has been omitted altogether, as it is far too faint, burning up the sky at magnitude fourteen! In this regard, the magnitudes given are, of course, the brightest that each planet attains.

	Distance from Sun (km)	**Max Magn.**	**Diam (")**
Mercury	58,000,000	-1.2	12
Venus	108,000,000	-4.4	64
Mars	228,000,000	-2.7	25
Jupiter	778,000,000	-2.5	50
Saturn	1,426,000,000	-0.4	21
Uranus	2,870,000,000	5.5	
Neptune	4,497,000,000	7.5	

The words *Planet* and *Plankton* come from the same Greek word meaning 'to wander' - plankton wanders through the sea, and planets wander through the heavens, or at least through the Zodiac, that band in the sky which represents the orbital plane of the chief inhabitants of the Solar System as viewed from Earth. Mercury and Venus are closer to the Sun than we are, and they are called the *Inferior* planets. Both Mercury and Venus show phases like the Moon, though those of Mercury are not visible with binoculars. *Figure 11* demonstrates this; in position A Ve-

Fig.11: Aspects of an Inferior Planet

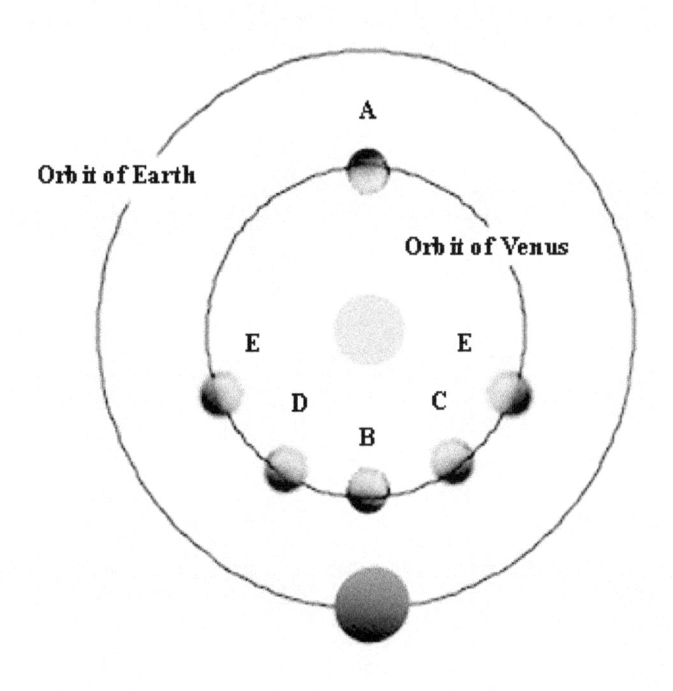

nus, our inferior planet in this case, is on the far side of the Sun from us, and so appears "full" but at the same time smaller. In position B we have "new Venus" and it cannot be seen at all except on those rare occasions when it passes in front of, or *transits* the Sun. Near C and D the planet is at its best for observation, and we see the phase as a crescent. Some peo-

Fig. 12: Retrograde motion of a superior planet

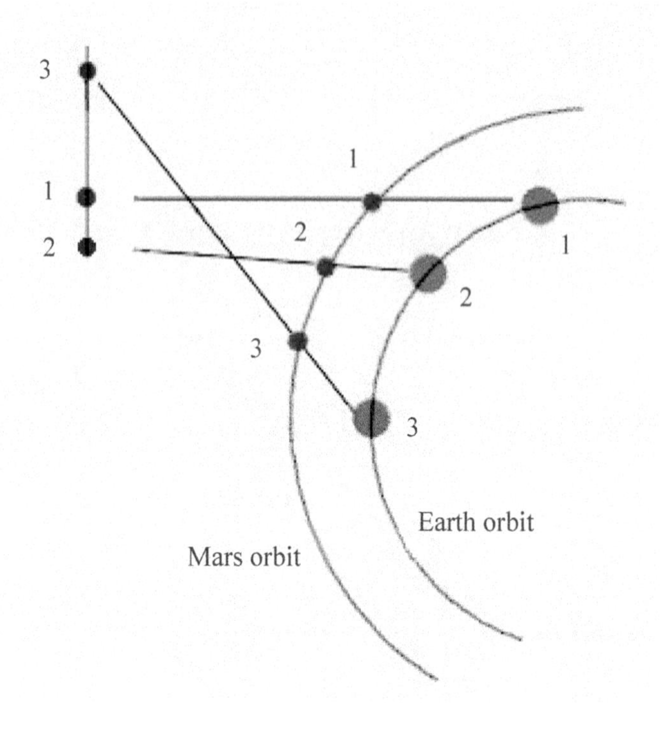

ple even claim to have seen this shape with no optical aid at all, and with binoculars the best time is undoubtedly in a less than dark sky, for otherwise the brightness of the planet will be so great, none but the most perfect of lenses will show it as anything other than a searingly brilliant star. The points at E represent the planet at its greatest possible angular distance from the Sun and thus those times when it is visible the longest. The planet is then said to be at *elongation* (western or eastern).

The planets whose orbits lie farther from the Sun than ours (called the *superior* planets), because of their slower motion, lead to them appearing to double back on themselves as seen against the sky background as we first approach and then recede from them over the course of a few weeks or months. Figure 12 illustrates this. From positions 1 to 2 Mars seems to be moving conventionally, but as Earth overtakes it, it seems to lag behind and appears to move backwards against the starry background, from 2 to 3. This is called 'retrograding'.

MERCURY

This is a small world much like the Moon both in size and geology, with many craters and mountain ranges across its surface. Orbiting so close to the Sun, it takes just 88 days to make its 'yearly' revolution, only 1½ times longer than its axial rotation period, or 'day' which is 59 earth days. Due to its proximity to the Sun, it is never the easiest object to try and locate, and here a word of warning is in order. If the Sun is still visible, position yourself so that it is hidden behind a wall or something similar, or otherwise in trying to sweep up Mercury, you may accidentally locate the Sun instead. Mercury has an eccentric orbit, and at the widest possible separation from the Sun of 28°, it is always best-placed for observers in the Southern hemisphere at such times. High-power glasses may help to find it easier, since they let in less sky-light with the result that the field background appears darker. Being an inferior planet, it runs through a full cycle of phases like the Moon, and although these are not visible with everyday binoculars, owners of good-quality larger models may like to try and note them. Mercury is actually at its brightest about two weeks before the precise date of elongation, so this may be a better time to try and glimpse it. You will then have the satisfaction of seeing a planet that neither Copernicus nor your humble author has ever seen.

Another way to see Mercury is to catch it in *transit* across the face of the Sun. These are fairly rare; those in the first decade of this century were 7 May 2003 and 8 November 2006, both best visible from the Southern hemisphere. Use the projection method to view transits, and try not to confuse the small dark disc of Mercury with any sunspots that may be visible.

VENUS

It is easy to see why this planet was named after the Goddess of love, when on a clear evening she hangs in the sky like a silver-yellow lamp, at times bright enough to cast shadows. It is said that beauty is but skin deep, and that is certainly the case with Venus. Her rich radiance hides a stormy reality of sulphuric acid rain falling in torrents from clouds so thick that the cosmos outside the planet is never visible. If, in the words of the old song "On a clear day you can see forever", then you can see for no distance at all on Venus, since the weather forecast is extremely dull, forever hot, rainy and very cloudy, outlook similar. Add to that the fact that the surface temperature is several hundred degrees above that of Earth, courtesy of a runaway Greenhouse Effect, and you have a world that is hardly top choice for a romantic holiday. The planet is also peculiar in that it takes longer to spin on its axis than it does to orbit the Sun so that its 'day' of 243 days is longer than its 'year' of 225 days! Oh yes - and just to be difficult it rotates backwards as well (East to West, as distinct from the other planets). It used to be thought that Venus could be some sort of twin to the Earth, complete with water and even life, but in fact the only way the two planets are obviously alike is in size - Venus is very nearly as large as our own world, and has a rather similar geology.

Finding Venus is no problem. I once saw it myself, aged nine, on a hot Summer afternoon. This is probably the best time to turn your bins on it if you can, since when seen against a dark sky, its very brilliance will show up every tiny imperfection and irregularity in your binoculars. The gentle pressures of a light sky are altogether more beneficial. This is especially true if you want to see the phases of the planet. It is not that these are hard to see in and of themselves - they have in fact been seen with the naked eye - rather because the sheer brilliancy of the object will redound upon your optics as mentioned above. Indeed, the very brightness of the disc has been the basis for several anomalies associated with the planet, and many observers in the eighteenth and nineteenth centuries fancied that they saw vague grey markings on both the bright and dark sides of the planet. Such anomalies are, with little doubt, appearances brought on by effects in our own atmosphere, the condition of the observers' eyes and optics, or more likely, a combination of all of these. Venus does, after all, come complete with her mirror as befits the Goddess of love, and this mirror appears to be of the distorting variety!

Even so, it is still instructive to follow the changing phases of Venus, noting the time (in order to later assess if the colour of the sky could be relevant) and any other factors. You may see the dark side apparently faintly illuminated, or grey markings. Note or draw what you see at all times, not what you *expect* to see. Even if they are phantoms, their cause may one day hold some significance. Nobody can tell in science just what use some hitherto-neglected phenomenon or observation may one day be put to.

As with Mercury, Venus transits the disc of the Sun, though in her case the transits are much rarer, and occur in pairs a few years apart, with each pair being separated by a two-hundred year gap. Luckily, we are approaching a transit pair at the moment. One occurred on June 8, 2004 and the other is due on June 6, 2012. The same observation techniques should be used, though of course the disc of Venus will be much larger than that of Mercury, and since it takes almost a day to cross the face of the Sun, you should have no problems following it.

MARS

For many years of the last century Mars was thought of as the world on which life might have evolved outside our own sphere. Who has not heard of the bug-eyed Martians and their canals? It is a great shame that the Martians are ever bit as non-existent as their canals, and we can happily put aside any visions of gaily-painted boats or jaunty hoop-jerseyed Martian gondoliers.

However, in many ways it is a world like our own, though much smaller. It rotates in the same period, and has the same axial inclination, though of course it takes longer to orbit the Sun, doing so in 687 days. It also has polar caps where water might well lurk, and surface details sent back by recent spacecraft show gullies and other geological features which strongly suggest the presence of running water in the past - and possibly at present, maybe running even as you read these words. The main problem with Mars as an Earthlike planet is its small size. Because of that, it has less mass and so cannot retain a very thick Earthlike atmosphere.

Through binoculars, Mars is best viewed near opposition, and owners of large instruments may be able to make out some of the main surface features. Using 20 x 70's I have seen the large, well-known triangular dark patch known as the *Syrtis Major,* an observation later confirmed by telescopic observers. It is also possible for the planet to show a slight amount of phasing when appropriately placed, though this it not noticeable with conventional binoculars, and even if you do have large glasses, the optics will have to be outstanding in their definition. Mars is also like the Earth in that it possesses large Polar Caps of ice; these grow and shrink with the changing of the Martian seasons, and they may also be seen when large and prominent with binoculars. The smaller glasses will not show the polar caps, but will at least give a good demonstration of why Mars is called the Red Planet, provided that there is not a dust storm going on. These can be extremely widespread; a storm in June 2000 covered half the entire planet. Dust storms are inaugurated by the rapid change in atmospheric pressure as Mars warms up in its approach to the Sun on its rather eccentric orbit. Even if you do not have binoculars powerful enough to show detail on Mars, you can make a note of the colour. It is probably best to use some sort of quantitative record rather than using colour terms, since these are highly personal. Maybe you could have a "redness index", with 0 being pure white and 10 deep red. Stellar observers sometimes use a similar system. Finally, Mars has two tiny satellites called Phobos and Deimos which are beyond all but very large telescopes.

JUPITER

From these small, rocky worlds we now approach the gas giants, so called because though they are not entirely gaseous, their atmospheres make up the majority of their bulk and do not allow us to see down to any of their surfaces. Jupiter is much the largest of the gas giants. A thousand Earths could fit inside it - with room to spare! Its large size (it is over 100,000 km in diameter) and light colouring make it a brilliant object at all times, and it has, like Venus, been seen in broad daylight. I remember one occasion some years back when, for a few confusing seconds, I thought I was watching the Moon rise through sparse Autumnal trees; I had just moved into a rural situation and was not used to seeing Jupiter as bright as this! When it reaches opposition in the Autumn it is best-placed for all purposes, for then it is highest in the sky and visible for a long time at a stretch. Why, even yours truly, a fanatical avoider of planets, will give it an occasional visit, to look at the brownish belts on its surface and its entrancing family of four bright satellites.

Probably the first thing you will notice about the planet, especially if you are using larger than average binoculars, is its shape. Jupiter's rotation period, its day if you will, is only about ten hours long (or should that be 'short'?) and so it bulges toward the equator due to centrifugal force. The binocular astronomer may also be able to see two or three of Jupiter's belts. All the gas giants have belts as a result of their predominantly gaseous constitutions and rapid rotation periods, but those of Jupiter are the best-known and the easiest to see. They were first recorded in 1630, not long after the invention of the telescope. The two main ones lie either side of Jupiter's equatorial region - the planet is only tilted from the plane of its orbit by 3° - and are called the NEB (North Equatorial Belt) and the SEB. Large glasses may also show the most famous feature of the planet, the *Great Red Spot*, a giant, swirling superhurricane three times the size of our entire world, whose colour actually varies from complete invisibility through dull orange-yellow to brick red. The spot, even when it is not visible, sometimes betrays itself by an indentation in the South Equatorial Belt, which it lies directly North of. Owners of large, good-quality binoculars may be able to see this feature, which is permanent enough to have been given its own bit of Jovian terminology; the GRSH (Great Red Spot Hollow). To some extent the Great Red Spot is a one-off; it is not the only spot on Jupiter, but simply the most permanent, though it is not one single entity. Spacecraft flybys have revealed complex, chaotic swirls around its edges where the turbulence is strongest, and in 1839 the observer South witnessed in real time the breaking up of a spot into smaller ones, so we know this happens. Spots are giant storms caused by rising gases from hotter regions of the lower atmosphere. When they reach the top of the cloud layer, these vast bubbles of gas are trapped between the swirling belts and roll about in the wind currents until they are torn apart by turbulence, for Jupiter sports the mother of all winds; speeds of 500 km per hour are not uncommon, so a feature has to be very robust and very large to persist for any length of time. The GRS is different in that it now appears to be self-sustaining. Certainly we know it to have existed since it was first recorded in the 17th century. Perhaps it was originally a conventional spot which was large enough to have survived anything the planet could throw at it. Bear in mind that 400 years is as absolutely nothing in the life of a planet, and that it could be gone in another hundred years' time. These things only appear permanent to us because our own life-rhythms are far less stately.

Sometimes, however, we are surprised by sudden events. In 1994 it was found that a comet called Shoemaker-Levy 9 was on a collision course with the giant planet. Astronomers, both professional and amateur the world over, were pleasantly aghast that such a thing could happen in their lifetimes. And this would not be just one big bang either; the gravity of Jupiter had pulled the comet apart, with the result that the planet would be struck not once, but over twenty times! Amateur astronomers too played their part, drawing and photographing the results of the collisions. Many of these occurred on the far side of Jupiter, so it was the results of the explosions that were seen, for the most part, rather than the collisions themselves.

On July 16th, the first stricken rock hit Jupiter at a speed of 200,000 kilometres per hour. The explosion ejected a giant expanding swathe of debris three thousand kilometres above the top of the Jovian cloud layer. Falling back into the atmosphere, this then raised its normally-chill temperatures of -100°C up to +700 degrees, hot enough to melt lead. But gradually, the debris clouds cooled and settled, destined to be torn apart by the relentless planetary winds over the course of the following weeks and months. A week later, the comet was no more, and Jupiter was a few billion tons heavier. Looking at it today, you would never know that a few years ago destruction of the most apocalyptic variety had rained down upon the giant planet.

In fact, we may owe our very existence to Jupiter and its fellow gas giants, since their enormous gravity 'shepherds' many comets away from our part of the Solar System. It has long been known that Jupiter has its own family of comets, for instance; and over the past few years, Jupiter-like planets have been discovered orbiting other stars as well. Maybe such worlds are an inevitable part of the planet-formation process itself, which bodes well for the existence of intelligent life having evolved around other stars. In many ways in fact, Jupiter is a mini-Solar System in miniature, and chief among its retinue are its many satellites.

Jove has a large family of Moons, though only four need concern us here. They are *Io, Europa, Ganymede* and *Callisto*. The others (new ones are continually being identified) are small and are named, like the main moons above, after the numerous amorous liaisons of the mythological Jupiter! Known collectively as the Galilean satellites, the four large moons are so called as they were first seen by Galileo with his little telescope though there is some evidence that the ancient Babylonians may have noticed them with the naked eye; and in the clear Mediterranean skies of a cleaner age that may well come as little surprise. Even in these times of industrial- and light-pollution, some claim to have seen them thus, though binoculars are definitely recommended for seeing them well. The most enjoyable pursuit as regards the Galileans is to watch their ever-changing waltzes around Jupiter, as they disappear behind the disc or into and out of the long shadow that the giant planet casts behind it. Sometimes, a satellite will transit the disc of Jupiter, though unfortunately your average binoculars will not show this. Just simply drawing their changing positions from one night to the next is an interesting and fun exercise. Magazines such as *Astronomy Now* will tell you not only which is which, but will also give details of the above phenomena too.

With the visit of numerous space-probes to the region of Jupiter over the past twenty or so years, the importance and interest of the Galilean satellites has threatened to overtake that shown in Jupiter itself! The innermost galileans Io and Europa are without a doubt the most interesting, though they are slightly smaller than crater-marked Ganymede (the largest satellite in the Solar System) and Callisto. It is their very closeness to Jupiter that has made them such interesting places. Most satellites in our Solar System are, like the Moon, dead and uneventful places, their hot cores having long since cooled down. But the immense tidal force exerted by the gravity of Jupiter has ensured that its inner moons' interiors have never had a chance to settle down, and are churning reservoirs of hot magma. This is especially true of Io, the most volcanically active place in the whole Solar System. For example, in November 1999 the camera of the amazing little Galileo probe was overloaded with a two-kilometre high fire fountain from the thought-to-be-dormant formation called Tvashtar Catena. The lava erupted at a speed of 350 kilometres per hour along the entire length of a fissure almost 40 kilometres long, and the eruption was visible from the Earth. So high is this, and other material flung that much of it breaks free of the satellite altogether, and either encircles Jupiter in the form of magnetic particles or is deposited on the tiny innermost moon *Amalthea*, coating it in an orange-red dusting of Sulphur and making it the reddest body in the Solar System.

If you thought Io was something of a happening place, you've seen nothing yet! The first spacecraft to encounter the Galileans was *Voyager* in 1979. It returned a view of Europa that was markedly different to the others. We have a thick surface of water ice, wracked by a complicated and extensive system of cracks, some of them thousands of kilometres long. Clearly Europa was a dynamic world too, since any cratering, which there must have been at some point long ago, has since been obliterated by the constantly-changing surface - the same reason, if not quite courtesy of quite the same causes, as to why our own world is so minimally-scarred by impact craters.

Of course, the deep cold of Jovian space means that the surface of Europa, though composed of water ice, is far too cold to be 'interesting'. However, below the surface it is a different story. There, due to the tidal heat generated at the moon's core, we have, in all probability, a satellite-wide ocean of liquid water! And that is not all; from what we know of the history of the early Solar System, this ocean could well include not just water but also the carbonaceous compounds necessary (we believe) for the existence of life. Even the lack of sunlight reaching this water is not a problem - witness the discovery a few years ago of whole ecosystems - of real, full-sized moving creatures, not just lowly lumps of jelly -

on the deep ocean bed, reliant not on sunlight but on the chemicals emerging from volcanic fissures. And Europa, as we have seen, is a tectonically-active world. Who would have suspected even a few years ago of looking for life in such a strange place? But already, plans are being made to explore the oceans of Europa. Think of all this when you next look at the satellites of Jupiter with your humble binoculars.

SATURN

This, the sixth planet of the Sun's retinue, is famous for its system of rings. While all the giant planets are now known to have rings, those of Saturn are the only ones readily visible from the Earth. The strange shape of Saturn was first noted by (who else?) Galileo, though his small telescope was not powerful enough to show the rings, and his drawings show one large globe flanked by a smaller one on either side. The first astronomer to actually see the rings was Huyghens in 1659, using a type of telescope popular at the time; one with a very long focus. Though the objective's diameter was binocular-sized at 50mm, its *focal length* (the distance between the objective and the eyepiece) was seven metres! Huyghens used a magnification of 100 with this instrument, and the rings duly became visible. Can they be seen with binoculars? I have to say that they can - but not necessarily as rings! Most glasses will show that there is something 'not quite right' about the dull yellow starlike dot that is Saturn, and larger ones may, in fact, show a bright yellow ring around the planet on occasion. I have seen this myself with 20x70's; in fact, the ring you will see is merely the combined image of the two brightest of Saturn's three main rings visible with amateur instruments. The Rev. Webb, writing in the early years of the last century, saw it using a power of 20, so using this magnification or higher will probably show you the ring proper, and because of the generally higher quality of modern optics you may even catch the rings with a slightly smaller magnification than this. Between the two bright outer rings is a gap, discovered in the 1675 by Cassini, and named after him. I have not heard of this being seen with binoculars, but I do not necessarily consider it impossible, at least with large glasses.

The visibility aspect of the rings is constantly changing, whether from the position of Saturn relative to the Sun, Earth's relative to Saturn, and the various permutation of all three. Since the axes of both our own planet and Saturn are tilted to the plane of their respective orbits, and the rings are in the plane of Saturn's equator, there will be occasions when the rings become either edge on to the Sun (so they reflect no light back) or Earth (so they appear too thin) and become invisible, since although the diameter of the rings is over 270,000 kilometres, their width is far less and can be considered for our purposes as simply two-dimensional. Space missions have revealed what many of the eagle-eyed observers of the past two centuries suspected; that in fact Saturn has many rings. The concentric appearance of one of the rings was likened to the steps of an amphitheatre by the French astronomer Lacaille, and other observers of the period traced fine black lines in the area. Also noted were radial spokes running across the rings' surface, another phenomenon confirmed in our own times. Of course, these will not be visible with binoculars, but the point is that you will never know what either you or your optics are capable of unless, like these observers, you occasionally push them to the limit. Since Saturn takes nearly thirty years to orbit the Sun, each round of aspect changes for the rings will cycle once for each face of the ring-system, and obviously the binocular owner will find it easiest to see the rings when they are farthest open.

The rings themselves may look like flat, solid objects, but they are not. In reality they are made up of countless pieces of rock, ranging in size from mere grains up to lumps maybe weighing several tons. This idea was suggested by Cassini over 300 years ago. Various theories have been put forward to explain their origin, though it is fair to say that we cannot yet be completely sure; though it is surely no coincidence that rings are, at least to our present knowledge, found only around large, gaseous planets with an extensive existing family of satellites. They could either be coeval with the gas-giant formation process or be the result of the destruction of a former satellite which never formed properly or was broken up because of the tidal effects of the giant planets' gravitational fields.

Physically, Saturn is very similar to Jupiter with a yellowish disc traversed by dull brownish belts, though these are *not* ever to be considered as binocular objects. The surface is also home to swirls and spots, though again these are not nearly as well-developed as those on Jupiter, due probably to the smaller temperature differences at the distance of Saturn, and these phenomena tend to show a closer relationship to the 29½-year 'year' of the planet. Similarities between the two giants of the Solar System can also be found in the high speed of rotation (just over ten hours in the case of Saturn) and the consequent flattening at the poles. The atmospheres are broadly similar also, though Saturn appears to possess more of the light gases like Hydrogen and Helium, and in fact these light gases make Saturn the least-dense of all the planets.

Like its big brother Jupiter, Saturn has an extensive family of satellites, the largest of which is *Titan,* largest satellite in the Solar System after Ganymede, and only slightly smaller than the fully-fledged planet Mars. Titan has a thick, poisonous atmosphere; in fact, it has more atmosphere than the Earth! From Titan's surface, the sky (including Saturn of course) is never visible. So interesting is Titan that the Cassini probe to Saturn carries a separate craft, Huyghens, to specifically investigate it. The main topic of interest is the high proportion of hydrocarbons present, the key ingredients for carbon-based biochemical processes. Titan is easily seen in most binoculars as an eighth-magnitude attendant of Saturn, and diagrams showing its position throughout the month are given in the main astronomical magazines.

The tradition for naming satellite families from their connection with the mythological parent is continued here; all the chief moons of the planet are named after the offspring of Saturn himself, whom, in his role as the God of time, he ate! The second-brightest moon *Rhea* may be seen with larger instruments, reaching the tenth magnitude when well-placed. The other satellites are interesting in their own ways, but are never visible with binoculars.

URANUS

Butt (sorry!) of many risqué puns, the world discovered by Sir William Herschel in 1781 provides some unexpected scope for binocular observers. This may appear strange, seeing that a telescope of some size is needed to show the tiny greenish disc of Uranus at all. However, this means that we can regard it observationally as a star - albeit one that moves - and estimate its brightness by comparing it with nearby real stars whose magnitudes we already know. Uranus is normally of the fifth magnitude, but it is known to change in brightness as its peculiar axial tilt of 98° shows us alternately the bright polar areas and then the darker regions. You only need to look at it once a month, as it takes over eighty years to encircle the sky once. The method of estimating the brightness of a star is found later under *variable stars.*

Uranus has a large family of small moons named after characters from Shakespeare, none of which are ever visible in binoculars. However, they do demonstrate that the far-flung reaches of the Solar System were once dangerous places! The innermost large moon Miranda, over 500 km in diameter, would appear spherical but for the presence of a vast slice off one side as though it had been gouged with a chisel the size of a planet; and this is not the only major accident to have befallen this area. A gigantic collision may also be the only way to explain the crazed tilt of the axis of the main planet itself. Add to that the fact that Triton, the large moon of Neptune, rotates backward around its own mother planet, and you can see that some pretty catastrophic events have been going on in the distant, cold depths of the Solar System.

NEPTUNE

This planet is in many ways a twin to Uranus, possessing faint rings, a gaseous atmosphere tending to form bands, and a greenish-blue tinge. Its size is also comparable; both worlds are about 60 times the size of the Earth, with Uranus slightly larger than Neptune, In 1989 Voyager 2 showed us the Great Dark Spot in the atmosphere, though this proved to be nowhere near as enduring as the Great Red Spot of Jupiter. The rotation periods also are comparable, faster than our own but not as rapid as those of Jupiter and Saturn.

Neptune is three magnitudes fainter than Uranus, but can still be seen in most binoculars. In the early years of this century, both Uranus and Neptune lie in the constellation of Capricornus, and will remain in that area for some years yet. Neptune has fewer satellites than Uranus, but its largest moon, *Triton,* is another planet-sized satellite, though too faint to be seen with binoculars, and has shown displays of volcanic activity. It has a thin atmosphere of Nitrogen, though the temperature (at -235°C Triton is the coldest world in the Solar System) ensures that this is in the form of frost. Positions of Uranus and Neptune for the coming years are given in the Appendices.

PLUTO

Though never visible with binoculars, it seems a shame not to mention Pluto at all, though many doubt its meriting the very name *of planet,* and following recent deliberations it has now lost that label. Pluto is in fact more of a double planet than a single one, for it has an attendant moon called Charon, which is half the size of Pluto - by proportion, the largest satellite in the whole Solar System. Another peculiarity is that both these worlds have captured rotations, always keeping the same sides facing each other, and revolving around their common centres of gravity as opposed to one object circling the other.

It is quite likely that the Pluto-Charon system are objects from the *Kuiper Belt,*(pronounced 'cowper') a large wide region of small rocky bodies at the rim of the Solar System. The recent discovery in the Kuiper Belt of the largest presently-known asteroid makes this even more likely, as does the peculiar orbit of Pluto. Its orbit is not only highly eccentric - on occasion it can be closer to the Sun than Neptune - but is also inclined some 17 degrees to the main plane of the Solar System which means that it can be found in several non-zodiacal constellations. What effects this has on the tabloid astrology industry I shudder to think. Just imagine your horoscope for 2070: "Pluto is in Cetus; expect to have a whale of a time..."

LOCATING THE PLANETS

Mercury, as has been said, is always near the Sun and is best seen in the East just before Sunrise or in the West after Sunset. It appears as a pinkish star which may look quite faint due to the inescapable bright sky background. Venus will leap out at you, so presents no problems. Mars, on the other hand, can become relatively faint to the extent that it may be confused with some similarly-hued stars in the zodiacal constellations such as Aldebaran or Antares. The name of the latter star in fact means "rival of Mars". Jupiter, like Venus, is always too bright to be confused with any star, though Saturn and Mars can easily be confused with a star (as well as with each other) so check the positions of these planets first. Many magazines, even regular good newspapers, will carry details of any visibility of bright planets, and there are easily-obtainable computer applications for viewing the virtual sky, some shareware, some free, and some worth the money.

At certain times, two or more planets will be in the same area of sky, an effect called a *conjunction.* The best candidates for an impressive view are Venus and Jupiter as they are so bright. I remember an occasion from my childhood in 1964 when these planets appeared as one brilliant, gleaming mass. The fact that it happened near Christmas was an added bonus, and claims that the second coming was imminent were rife. More frequently, a planet will disappear behind the Moon as it moves across the sky. These *occultations* are fascinating to watch through binoculars, and astronomical magazines will give details of any coming occultations. Rarer still are transits of inferior planets across the Sun. Transits of Mercury are fairly common, though those of Venus far less so; there is a Venusian transit coming up in 2012, though none for more than a hundred years after that!

ASTEROIDS, METEORS AND COMETS

These are essentially the leftovers of the Sun's family, though this fact should not be seen as diminishing their interest for us. They are all the small pieces of rock which never came together to form planets when the Solar System first began to take shape, and today we know there is much less difference between the three classes of object than used to be thought, although the origin of comets is rather separate.

The asteroids orbit mainly in a doughnut-shaped zone between the orbits of Mars and Jupiter, and Jupiter is the key to their existence. The planets were formed by the coming together of the countless specks of dust in the vast sheet of material encircling the newborn Sun. As this accretion process went on, the dust grains became larger, to the point where at diameters of between 0.1 and 10 kilometres, they were sufficiently massive to attract smaller particles by gravitation. This process could only go one way of course, and these *planetesimals* grew and grew until they became the planets we see today. However, the gravity of Jupiter was so dominant that it interfered with this accretion process. Each would-be planet now tended to disperse its potential planetesimals rather than accrete them when collisions took place. These disrupted chunks of rock wandered either inward or outward; the outward-bound ones crashed into either Jupiter or its large satellites, producing cratering on their surfaces. Some of these rocks settled into orbit around the giant planets to become satellites themselves, where they remain to this day. The inward-headed chunks helped to crater the inner Solar System; Mars is heavily cratered and its two tiny Moons must surely be captured asteroids. Those that largely stayed put remain the asteroids we see today, many of which still bear those early scars of collision in their own cratered surfaces. In spite of their often coming last in terms of description, they afford some useful and interesting binocular study. The larger asteroids are about 1,000 km across but there are only two or three of these; the vast majority are much smaller. The larger ones are vaguely spherical, but as we head down the size ladder, the shapes become ever more irregular. Eros, for instance, measures 7 by 25 km and is sausage-shaped, causing its brightness to change as it tumbles along its orbit and in this regard, the orbits of several asteroids are peculiar. Most stay in the

Fig. 13: Motion of Vesta, 25 Oct to 9 Nov 2001 on the Orion - Taurus border

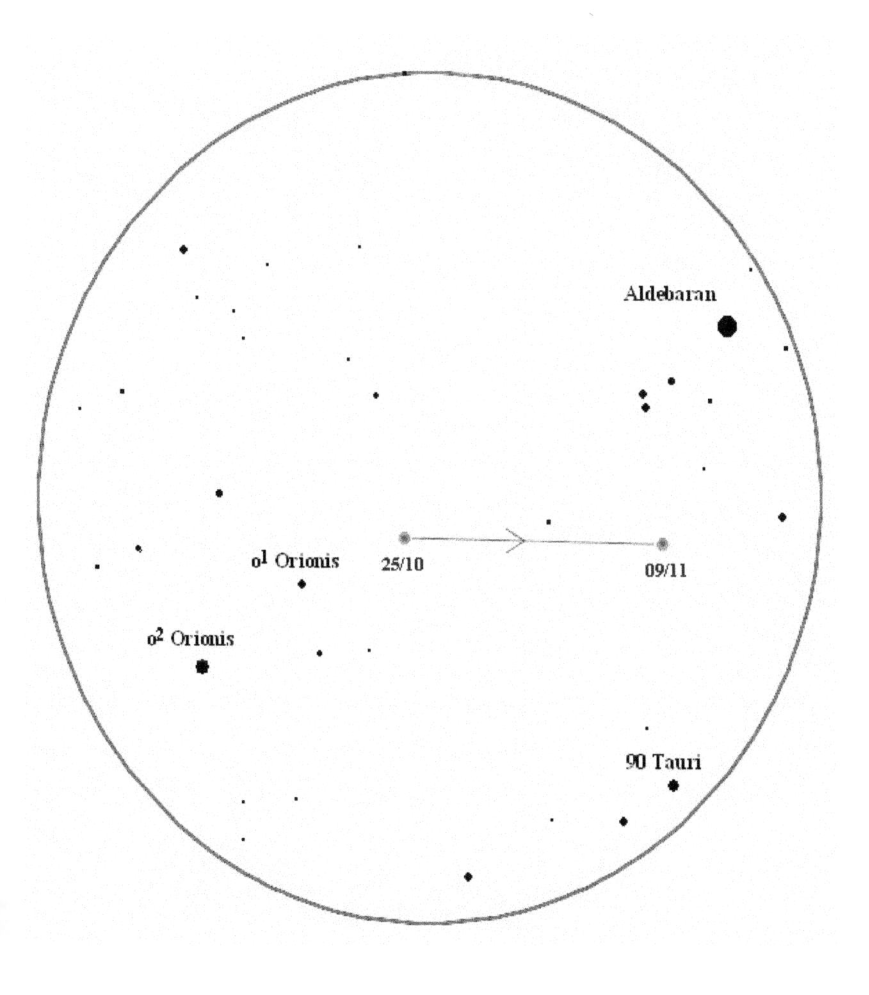

well-defined area between the paths of Mars and Jupiter, though several head in towards the Sun. Icarus swings within the orbit of Mercury, while Apollo, Adonis and Hermes can approach the Earth's orbit. The latter of these can come to within less than a million kilometres of us, and this leads us into the realm of the NEA's, or Near-Earth Asteroids. These have become quite the flavour of the month of late with the realisation some years back that one of these was almost certainly the cause of the climatic conditions which exterminated the dinosaurs 60 million years ago. "Watch the skies" has become a buzzword again, though now the invaders take the form not of little green men in their flying saucers threatening Civilization As We Know It, but big lumps of rock instead.

First find your asteroid. Many computer charting programs include the brightest ones, and astronomical magazines and handbooks also give details of their appearance. The word 'asteroid' means *starlike* - and they do indeed seem indistinguishable from stars. Therefore, the thing to do is to make a sketch of the area of sky centred on the position at which the asteroid is predicted to be. Return to the same field the following clear night and draw it again. On comparing it with your first drawing you will hopefully find some sort of change in the field's appearance. Maybe one of the 'stars' has moved or disappeared altogether, and this will be the asteroid. Fig.13 shows the motion of Vesta between 25 October and 9 November 2001. The size of the field is about 7° and the limiting magnitude is about 7.5; this part of the sky as you would expect to see it with 7x50's. The brightness of Vesta around this time was around magnitude 7. The four large asteroids Ceres, Pallas, Vesta and Juno are the brightest, though in 2001 a new asteroid was found which is larger than any of these, but it lies in the far depths of the Solar System and is too faint for our purposes here. Vesta, in fact, can reach naked-eye visibility though the others are slightly fainter; I remember following another one, Eunomia, in 1975 as it passed by the Square of Pegasus. It was fascinating to watch as it slowly tracked against the background of distant stars in the Autumnal sky. Most good sky-charting software such as *Cartes du Ciel* (downloadable free!) will plot all the major asteroids for you.

METEORS

Most of us have, at one time or another, seen a shooting star - maybe you've even wished on one. It is a harmless enough activity. Meteors are not of course stars, but tiny lumps of rock whose brief career of starriness is a result of friction with the air, as our gravity pulls them in and overwhelms them. They are, for the most part, tiny bits of rock left behind by the passages of comets through our corner of the Solar System, or sometimes merely smaller fragments of previous astronomical collisions or accidents.

If you spend long enough taking note of meteors during the course of a night, you may notice that many of them seem to emanate From a definite region of the sky. This area we call the *radiant*, and when a well defined radiant exists, we have a shower of meteors. One very well-known meteor shower radiates from a point in the constellation Perseus, so is called the *Perseids*. Radiants themselves are purely an effect of perspective caused by our heading in that part of our orbit around which the meteoritic material has been deposited by its parent comet.

Observing meteors was long considered a purely naked-eye activity, but for about the past thirty years several observers have used binoculars to great advantage, and much that is new has been so learned, such as the existence of whole new meteor radiants. Before you decide to take up this pursuit, however, be prepared for a long night coupled with rather more tedium than is usually associated with observing: meteors are best seen after midnight, for then we are turning against, rather than with, their paths and their speeds and brightnesses thus have a tendency to be greater. It is probably a truism that most binocular meteors are actually happy accidents seen when looking at something else, but if you do decide to pursue this field of study, you will need to note the following: (1) The magnitude of the meteor: (2) Approximate position - use a star map to determine the coordinates: (3) Direction and aspect - did it come from outside the field of view and disappear inside it. or vice versa? (4) Speed, colour and other similar details.

Some meteors, especially the brighter ones, can leave a transient, wispy train behind them. If you see a trained meteor (so much better, I think, that one with no idea of what it's doing!), note its shape and how long it remained visible. In the late 90's during an especially good display of the Leonid shower, I saw a train persist for several minutes, though not the meteor that produced it, which must have been extremely bright!

We assess the richness of showers by determining their ZHR or *Zenithal Hourly Rate*, in other words the number per hour we would see if they appeared at the overhead point or Zenith. These corrections are necessary since a radiant near the horizon will *appear* to produce fewer meteors, as many will ignite below the horizon or be lost in mist, haze and so on.

COMETS

It is in the study of comets, however, that the binocular observer can really (if you can forgive the pun) shine. Comets are among the flimsiest members of the Sun's family but at the same time some of the most spectacular, and have long aroused awe among mankind as being seen to be responsible for plague, pestilence, tragedy, doom and just good old bad luck, though this last clearly depends on who you are! Halley's comet appeared around 1066 and was shown in the Bayeux Tapestry as a "wonderful star" - though King Harold may well have called it something rather more Anglo-Saxon and monosyllabic.

A comet is made up of two basic parts, the head and the tail streaming behind it. The tail material is extremely tenuous in comparison with the partly rocky head. In the nineteenth century - a wonderful one for bright comets - the star Arcturus could be seen glowing its usual merry orange self through the tail of Donati's comet. Here it needs to be pointed out that few comets are the great curved scimitars of light shown in some illustrations, as shown by the rather feeble displays put on by comets Kohoutek and Halley in the 1980's, though the last few years of the twentieth century gave us two fine comets in Hyakutake and especially Hale-Bopp. The early years of the new millennium saw some good ones too, and the heavens were considerate enough to organise one for the southern hemisphere (comet McNaught, a fabulous object) with comet Holmes for us northerners. At one stage this comet was the largest object in the Solar System - *including the Sun!*

With binoculars, you may notice that the simple division into head and tail may not prove to be quite so simple. The light of the head may be concentrated in one place, the *nucleus*, or then again it may simply take the form of a dim smudge of mist. There may be no tail, or more than one; occasionally a Sunward-pointing 'spike' may develop, though these are not in reality true tails, but merely sunlight reflected off cometary dust, and are only seen when the Earth-Comet-Sun alignment is suitable. Tails may be of varying lengths, from zero to many degrees, and easily visible with the naked eye. The great comet of 1843, which was seen in broad daylight, was calculated to have a tail some 1,000,000,000 kilometres in length! The head of a comet is not one solid lump of rock, but is made up of numerous chunks held together by their mutual gravitational attraction. As the comet approaches the Sun, its material is heated and gaseous matter is released as the chemical ices on the surface evaporate, a process known as 'outgassing'. Water ice is quite a common component in comets, so much so that many see them as having helped to inaugurate the life processes on the early Earth from the frequent cometary collisions that must have taken place in those remote eras. Naturally, the more times outgassing occurs, the smaller the reservoir of gas becomes. That is why these cyclical, *periodic* comets like Halley's tend to get worse, in terms of a spectacle, with each return. The particles produced by the heating and outgassing are extremely light; so light in fact, that the force exerted by sunlight itself is enough to affect them. A comet's tail, consequently, always points away from the Sun.

The periodic comets are called that because we have seen them keep coming back for more. Less common but more exciting, since they have not been subject to such frequent outgassing, are the long-period comets. Their origins lie in deep space either in the Kuiper Belt beyond the paths of the planets, or the even more distant *Oort Cloud* on the boundaries of interstellar space, and their periods are suitably horrendous. The famous Comet Kohoutek which disappointed so many in the 1980s is one of these, and we can expect to be disappointed with it again in another 70,000 years from now. The much more impressive comets Hyakutake and Hale-Bopp have periods of the same general order. Interestingly, in his *Celestial Objects for Common Telescopes*, the Rev. Webb gives a figure for what we would call Solar System or Kuiper Belt comets of over 17 million. Multiplying this figure by a factor of one thousand would start to approach the true number of comets we suspect to exist if we include also the longer-term Oort Cloud comets as well!

On those occasions when a comet does show itself, binoculars are probably the best optical aid to use. Telescopes have a small field of view, and may not show as much detail in the comet as will binoculars. Having hopefully found your comet, draw not only it but also the surrounding star field from night to night - comets move in a stately fashion from night to night; they do not streak across the sky as some people think. Your first comet may actually take a bit of finding, as the positions predicted for them may be slightly out, due to perturbations by interfering planets. The standard method of locating an intractable comet is that of 'sweeping'. Looking in the general area of the object, move your bins slowly in a horizontal plane, then raise them slightly and head back in the opposite direction. By using this method, you can cover a fair area of the sky and hopefully spot your quarry. It is helpful to use averted vision here, since motion of a faint object - as it may well prove to be - is easier to detect by this means. The same technique is also used for the discovery of comets, and it is here that the amateur armed with binoculars (the larger the better) is at an advantage, since professional astronomers spend no time whatever trying to discover comets.

Whether you are trying to find a comet, or even to discover one, what is essential is a good knowledge of the night sky, especially of those many sky objects which look like comets but aren't! It was these objects that led the comet-hunter Messier to draw up his famous catalogue of objects-to-avoid, and while his comets have retreated into obscurity as well as into the depths of space, Messier's list of fuzzy night-sky objects is probably the most famous astronomical catalogue of all time. Most of the Messier or "M" objects are listed under their respective constellations in the following chapter, and you can get valuable experience by identifying and drawing them. Not only will you end up as a more skilful comet-hunter, but you will also gain a deeper knowledge and love of the stars as well. Oh yes - you will also save yourself the embarrassment of contacting a major observatory or the local paper to announce the discovery of a new 'comet' just below the Belt of Orion or the top half of Hercules. Bear in mind, though, that Messier's list is not exhaustive - there are many more nebulous gleams in the sky that can fool you. A good star atlas such as *Norton's* is needed for your cometary pursuits. For example, I once found a comet in the little Summer constellation of Delphinus, the dolphin. I knew for a fact there was no fuzzy deep-sky object there, neither was one marked in my trusty old *Nortons Star Atlas*, so I realised it was a comet. Sadly, comet Poxon was not to be - it was discovered by someone else the day before. All in all, comets and comet-hunting represent a rich field for the binocular observer, but it is helpful to join your local astronomical society if there is one, for they will probably have others who are in the same boat as you, and will keep you informed of any discoveries. It can be a cold, lengthy and often unproductive pastime loooking for comets, though you will gain a great knowledge of the changing starry sky - and that can be no bad thing.

CHAPTER 4. THE STARS AND BEYOND

Whatever astronomical object you are interested in, there is one place you can be sure of finding it - in the sky! If you are a novice astronomer, trying to find your way around even the naked-eye night sky without some sort of guide can be like trying to find a certain single tree in a forest, so here we kick off with a quick tour of the night sky as seen from latitude 45° North.

To begin, we need two things to help us: first is a sense of scale. It is amazing how hard many beginners find the task of relating what they see in a book to what is in the night sky as regards size. Secondly, we need a sense of direction. Both of these are furnished by first finding the one constellation that nearly everyone knows - the *Plough* or Big Dipper. We will start our sky tour in the Spring, when the Plough is more or less overhead during the hours of darkness; so bend your neck back, lay on the ground, or perform some other similar contortion so that you are looking straight up and can see the pattern in *Figure 14.* (If you think of the starry sky as a giant inverted bowl, then the horizon line in the maps which follow represents the rim of the bowl). Note that, strictly, the Plough is actually part of a much larger constellation, *Ursa Major* (the Great Bear) but we will stay with the familiar seven stars for the moment. Extend your arm and spread your fingers over the Plough, as though you were making a "No! Get Back!" sort of gesture. You should just about cover it, since a hand-span spreads about 25 degrees in the sky on average. Orienting yourself so that the pattern looks the same as in the map, use the two stars, which are about 5° apart, at the Western (right-hand) end as pointers to the Pole Star. As we have seen, this star indicates North and is always in the same place in the sky. As the Earth

Fig. 14

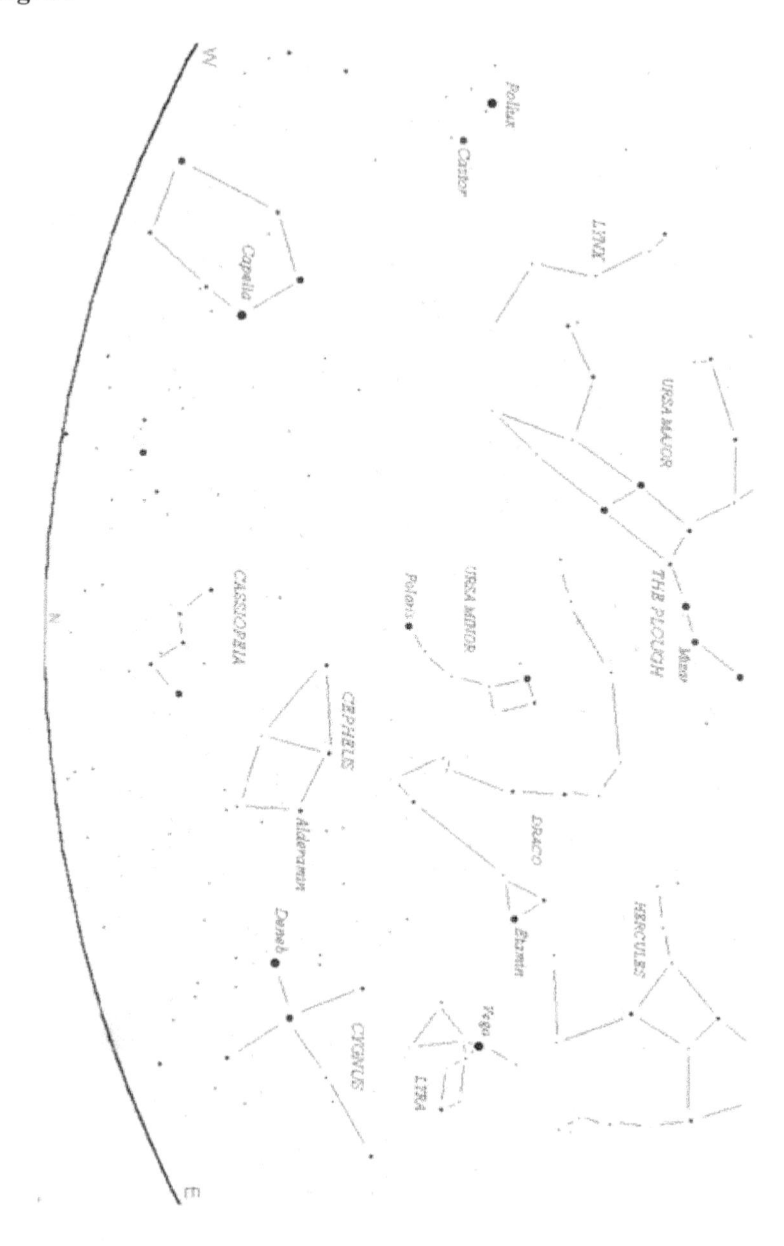
W
Pollux
Castor
Capella
LYNX
URSA MAJOR
THE PLOUGH
Mizar
URSA MINOR
Polaris
CASSIOPEIA
N
CEPHEUS
Alderamin
DRACO
Eltanin
HERCULES
Deneb
Vega
CYGNUS
LYRA
E

Fig. 15

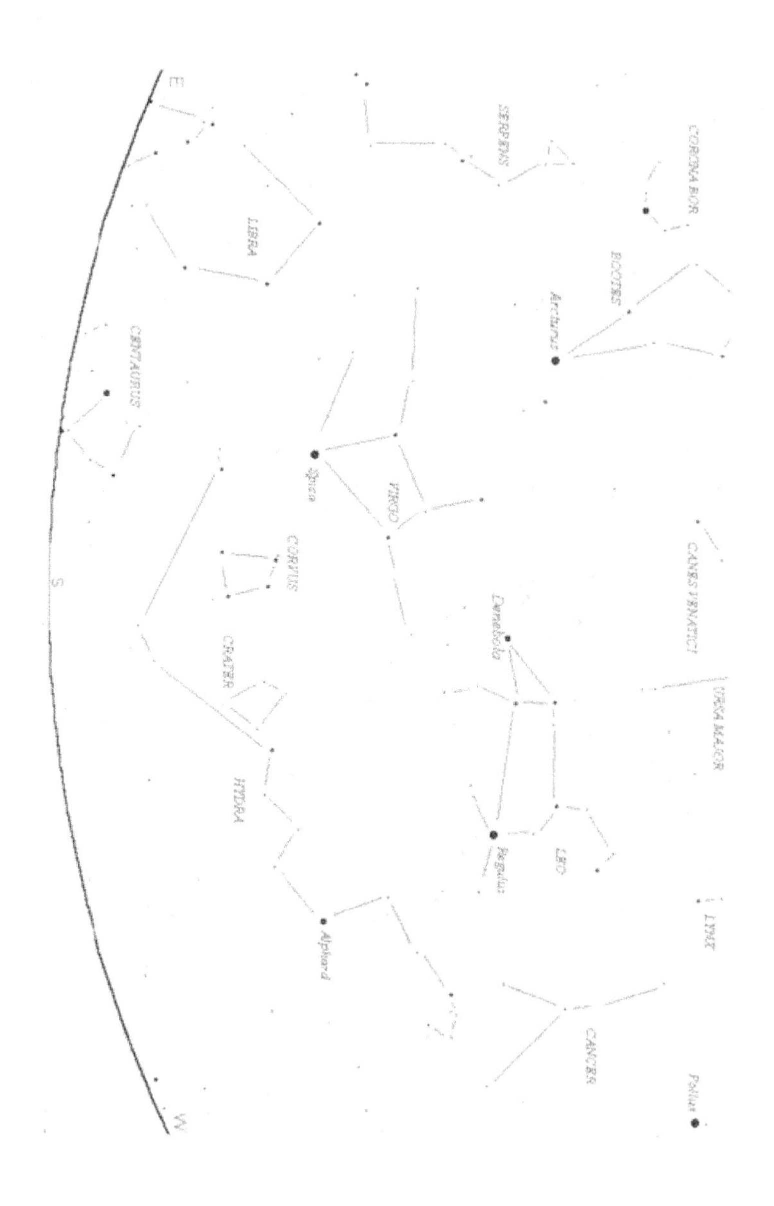
E
S
W
CORONA BOR
SERPENS
BOOTES
Arcturus
LIBRA
CENTAURUS
Spica
VIRGO
CORVUS
CRATER
HYDRA
Alphard
Denebola
CANES VENATICI
URSA MAJOR
LEO
Regulus
LYNX
CANCER
Pollux

turns, all the other stars appear to wheel around the Pole Star. Found the Pole Star? Good - we now have some idea of scale and a sense of direction, since if you are facing North, West is on your left and East on your right. Now we can start to find our way around the sky properly. The technique is to recognise a few distinctive shapes first - and speaking of distinctive shapes, return to the Plough. Find *Mizar*, the middle star of the handle, and look at it closely. Anything strange? Can you see a much fainter star almost touching it? If you can, congratulate yourself! You have found your first naked-eye *double star* - two stars that look so close together in the sky that they seem to form a single object. If you look at this pair of stars with binoculars, they will be easy to see separately.

Now draw a line from here through the Pole Star. The same distance beyond it, you will see a W-shaped group of stars called *Cassiopeia*, and since it is more or less opposite the Plough, it is always at its lowest when the Plough is overhead, and vice versa. If you were writing the letter W that the stars of Cassiopeia form, the two stars making up the last stroke point straight to another star of about the same brightness. This is Alderamin, the main star of *Cepheus*, in mythology the husband of Cassiopeia. The stars of Cepheus form a distinctive shape, rather like a steep-roofed house. Note the little triangle of stars between Alderamin and the W. Farther round to the East, we have two bright stars rising. The higher and much brighter one is *Vega*, a brilliant blue star, while between it and the horizon is *Deneb*, the leading star of Cygnus (the Swan) sometimes called the Northern Cross. In spring evenings, the cross is lying somewhat on its side. Vega and Deneb form a large right-angled triangle with a fainter star called Etamin, leader of the long constellation of

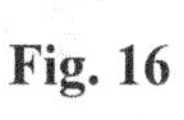

Fig. 16

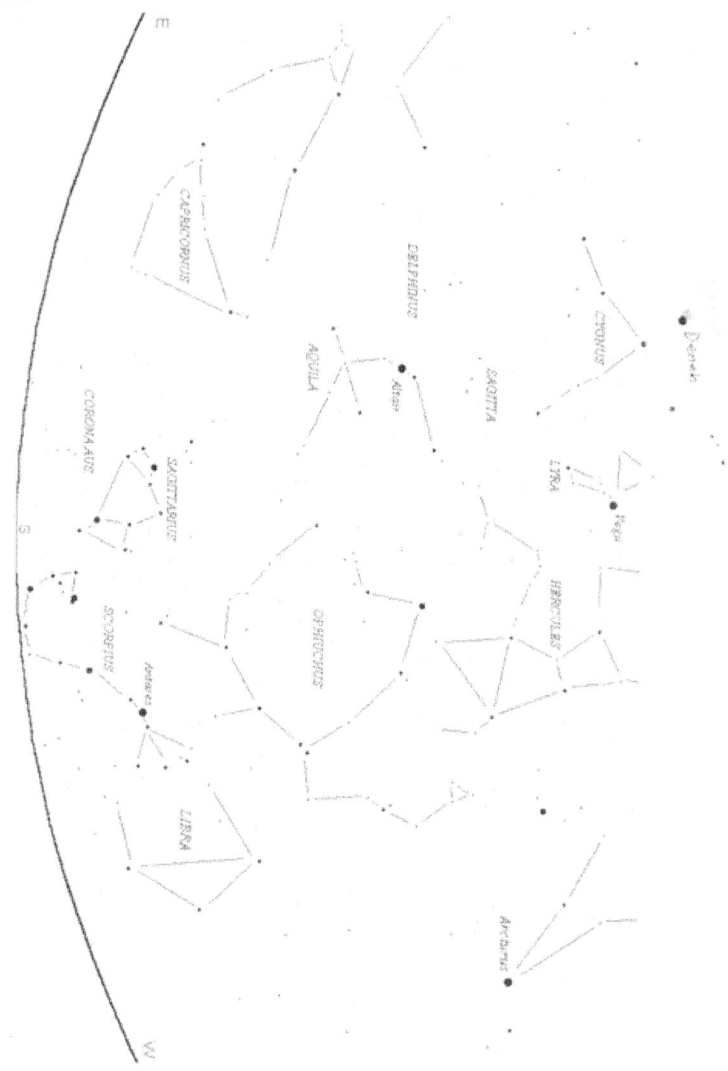
E
CAPRICORNUS
DELPHINUS
CYGNUS
Deneb
AQUILA
Altair
SAGITTA
CORONA AUS
SAGITTARIUS
LYRA
Vega
S
HERCULES
OPHIUCHUS
SCORPIUS
Antares
LIBRA
Arcturus
W

Draco (the Dragon) which winds from here all the way around the Pole Star until it terminates just North of the Plough.

Face the Pole Star and turn through a half-circle. You are now facing South (Fig.15) and you should notice, slightly to your right maybe, a distinctive group of stars in the form of a reversed question-mark, with a bright star for its dot. This star Regulus is the brightest in the constellation of *Leo* (the Lion). Look a bit to the left (East) and you will find that Leo's hind quarters are marked by a fairly bright triangle, the main star of which is called Denebola. 'Deneb' is quite a common part of star-names, which were bestowed on them by the Arabic-speaking peoples. Deneb simply means 'tail'. If you now draw a line from Denebola through Regulus and carry it on, you come to a little group of fainter stars forming the head of *Hydra* (the Water Snake). Hydra is actually the largest constellation in the sky, though apart from Alphard its leader, the stars of Hydra are not terribly bright or noticeable. Perched on the snake's head, look for the rather dim group of *Cancer* (the Crab) to the West of which you will find the bright twins, Castor and Pollux. In the middle of the faint stars of Cancer you may see a beehive-shaped patch of mist. Not unreasonably, this object is called The Beehive and is actually a large clustering of stars which, though not individually visible to the naked eye, together become visible as this glowing patch of light. Binoculars will reveal many stars here.

The South-Eastern part of the Spring sky is dominated by the brilliant orange star *Arcturus*. You can also find it by following the curve of the Plough's handle, and continuing the curve on past Arcturus will bring you to another bright star, Spica. This blue star is the brightest in *Virgo*, though generally the stars in this part of the sky are rather few and far between.

Fig. 17

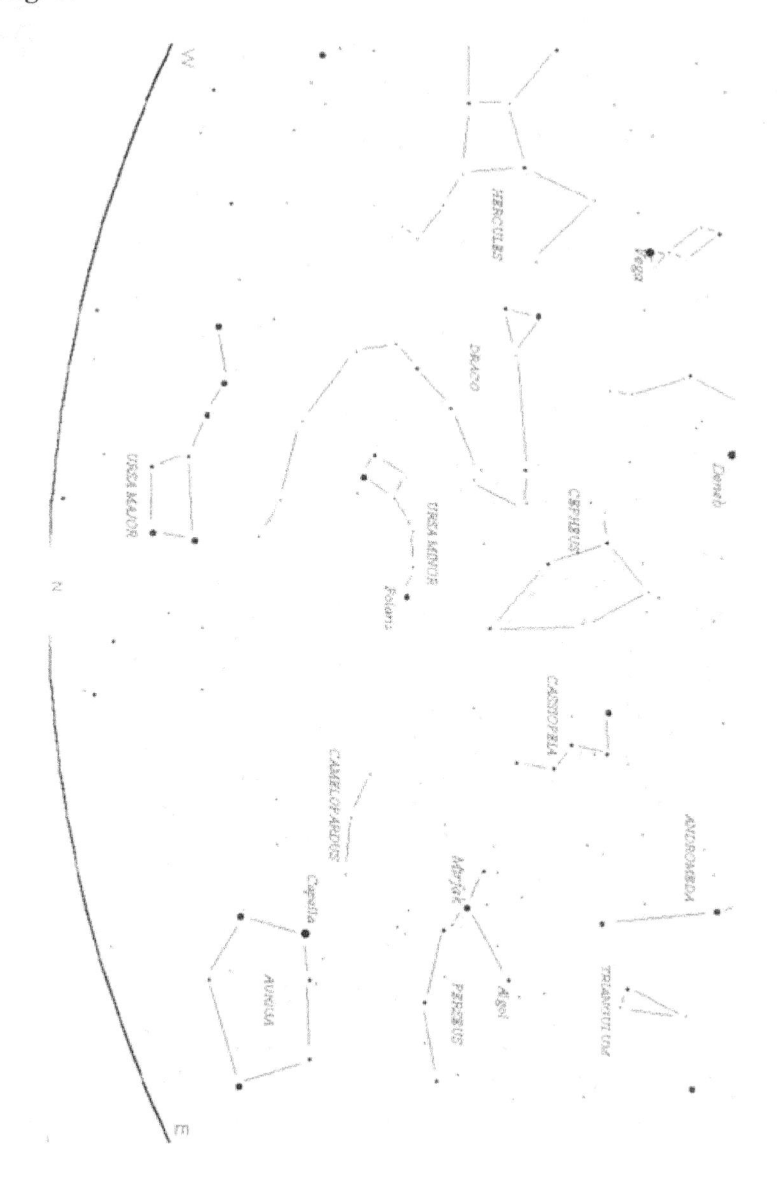
W
HERCULES
Vega
DRACO
Deneb
URSA MAJOR
URSA MINOR
CEPHEUS
N
Polaris
CASSIOPEIA
CAMELOPARDUS
ANDROMEDA
Mirfak
Capella
Algol
AURIGA
PERSEUS
TRIANGULUM
E

Three months on, the world has gone a quarter of the way around the Sun, and on Earth, we are now looking out towards different stars, those best seen in the short darkness of Summer. However, we are also looking in the direction of the well-populated centre of our own Galaxy, and there are far more stars to see. Looking North we can see that the Plough has now swung rather nearer the horizon, and the head of the Dragon is now overhead. Turning to face South (Fig.16) high in the sky are Vega and Deneb; lower down is a third bright star, *Altair* in *Aquila* (the Eagle). These three stars are commonly known as the Summer Triangle, and can be used to find other objects. Following the axis of the cross of Cygnus down almost to the horizon will reveal the baleful red light of Antares in *Scorpius*. Note that both Altair and Antares have a faint star to either side, but you can easily tell the difference, since Altair is usually higher in the sky, and is white rather than red. Between Vega and Arcturus is the very large constellation of *Hercules*. Its stars are noticeable though not outstandingly bright but do form a distinctive shape, and between Hercules and Antares is another large group, *Ophiuchus* (the Serpent Bearer). It is rather a 'hollow' group of stars for its size. At the same altitude as Antares but farther to the East lie the bright, though not brilliant, stars of *Sagittarius* (the Archer) in a shape which has been called 'the Teapot' for obvious reasons. The stars of the Archer lie in the direction of the centre of the Galaxy, and stretching from the horizon right up to the zenith you may be able to see the beautiful silver path of the Milky Way, which is merely the combined light of the countless millions of stars which form the central plane of our Galaxy, too faint to be seen individually with the eye alone.

Another three months on, the leaves are falling from the trees, and the nights are darkening earlier; in the North (Fig.17) the Plough is

Fig. 18

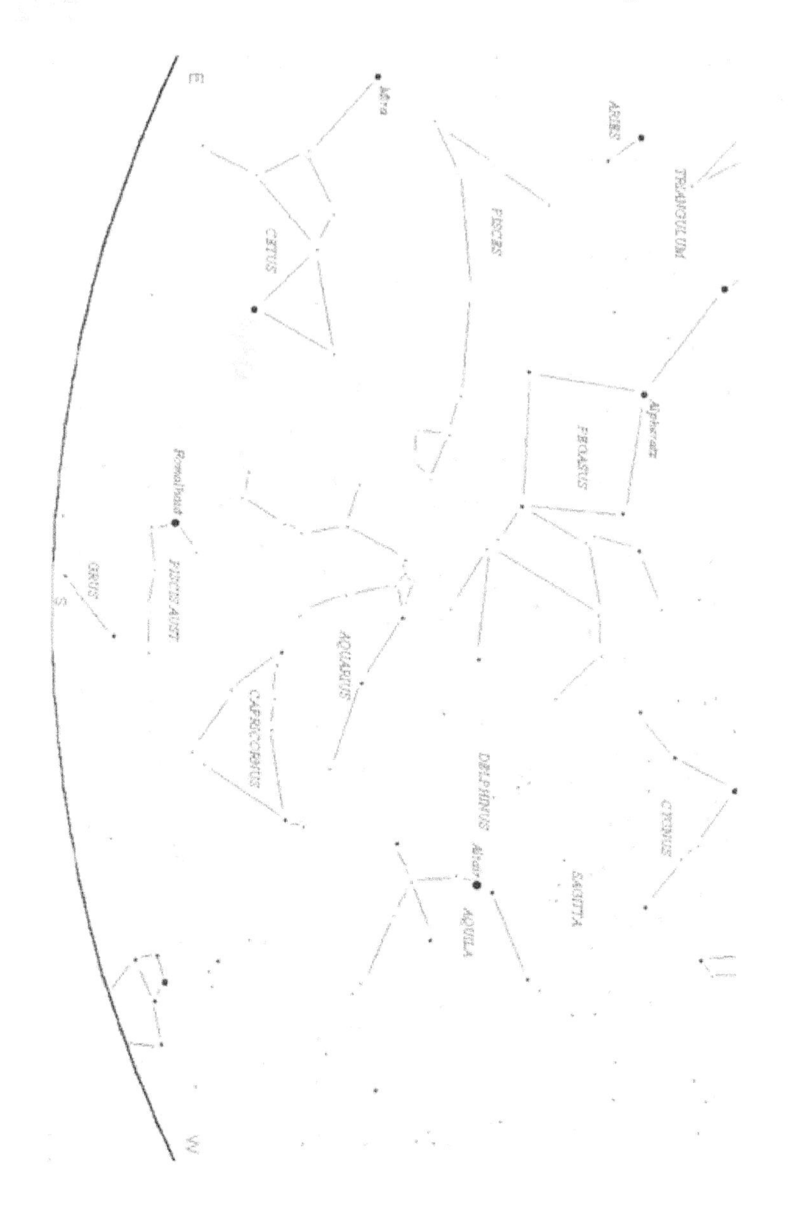
E
Mira
ARIES
TRIANGULUM
CETUS
PISCES
Alpheratz
PEGASUS
Fomalhaut
GRUS
S
PISCIS AUST
AQUARIUS
CAPRICORNUS
DELPHINUS
CYGNUS
Altair
SAGITTA
AQUILA
W

as low as it can get, skimming the horizon below the Pole Star. That means Cassiopeia is now high up instead, with Cepheus and Deneb almost overhead. And as Vega sinks from its high place, so now the equally bright star *Capella* starts to climb up the Eastern sky. The milky way still crosses the zenith, through Cygnus, Cepheus and Cassiopeia thence to *Perseus*, where it is especially bright. Perseus lies more or less between Capella and the W of Cassiopeia, and one of its main stars is Algol, a star which changes regularly in brightness over the course of a few days. Forming an isosceles triangle with the two faintest stars of the W (i.e., if you were writing the letter W, the stars that make the first stroke), you may see a glowing patch with the eye. Binoculars reveal this to be, in fact, two clusters of stars close together; the appropriately-named Double Cluster in Perseus. To the South (Fig.18) the sky is graced by very few brilliant stars. The most notable constellation is *Pegasus,* in the form of a large square, with sides about ten degrees (half a handbreadth) long. From the top left-hand corner trails a regular line of brightish stars, delineating *Andromeda.* This line can be extended to Mirfak, the brightest star in Perseus, and bent slightly to end with Capella. Mirfak is, by the way, a star worth turning your binoculars on if you have them to hand, since it is surrounded by a circlet of fainter stars. Around Pegasus are several large, dim constellations - Pisces, Aquarius and Cetus, and low down near the horizon (use the right-hand stars in the square as pointers) is the bright star *Fomalhaut*. Not to be confused with this is Diphda, the brightest star in *Cetus* (the Whale). From time to time, you may catch sight of Mira, also in Cetus. Like Algol, this is a 'Variable star' whose brightness changes periodically, but Mira's light-changes are much slower and less predictable than those of Algol; Mira can become as

Fig. 19

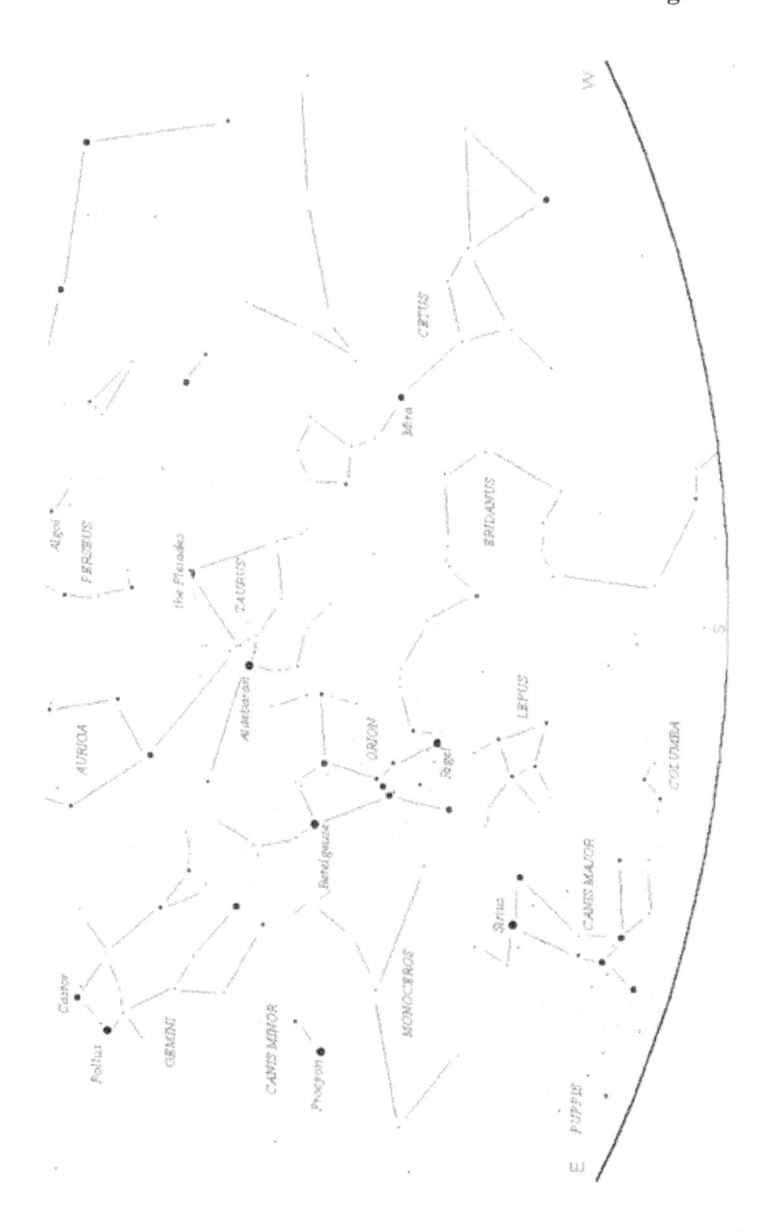
W
CETUS
Mira
ERIDANUS
S
LEPUS
COLUMBA
CANIS MAJOR
Sirius
PUPPIS
E
Algol
PERSEUS
the Pleiades
TAURUS
Aldebaran
AURIGA
ORION
Rigel
Betelgeuse
MONOCEROS
Castor
Pollux
GEMINI
CANIS MINOR
Procyon

bright as Diphda or become so faint that the naked eye cannot see it at all.

Carol-singers and snow, robins and corny old films on the TV. Winter is with us now, with its long nights eminently suited to stargazing. Capella is now overhead, occupying the position that Vega was in during late Spring to early Summer. In the North, Vega has gone, just below the horizon, with the Plough now pointing straight upward at Capella. The Southern view (Fig. 19) is dominated by the wonderful constellation of *Orion*. A careful look at Orion is enough to show that there is definitely something a bit different between its two chief stars, *Betelgeuse* and *Rigel*; whereas the former is an orange-red colour, the latter is a definite bluish-white. These colours are a guide to the temperatures of the stars. Red stars like Betelgeuse are cool; orange ones such as Arcturus are slightly hotter, while yellow stars such as our Sun are in the middle of the heat league, with white stars such as Sirius hotter still. Even hotter than the white stars are the blue ones like Rigel.

Binoculars are especially useful for revealing star colours. Orion's famous belt of three bright stars is another good direction-finder. Pointing them upwards brings us to the orange star *Aldebaran* that is seemingly in a V-shaped group of stars called the *Hyades*. This is a cluster of the same type as the Beehive, but closer to us and thus more spread out to our eyes. Prolonging this line still farther alights on the gleaming mass of stars that make up the Pleiades, another cluster, that looks like a tiny Plough. Six or seven stars should be visible here, though keen-eyed people have seen double this amount under good conditions. In the opposite direction the belt stars point to the brilliant, twinkling *Sirius*, which is the brightest star in the sky. The twinkling has little to do with the star itself but happens because Sirius is fairly low down in the sky from these latitudes, and we look at it through the unsteady air of Winter.

Forming a large and nearly equilateral triangle with Betelgeuse and Sirius is another bright star called *Procyon.* Just as Vega, Deneb and Altair make up the Summer Triangle, so these three stars form the Winter Triangle. Most of the star-names, as we saw, are Arabic in origin, but Sirius and Procyon come from Greek. Sirius means "the Scorcher" while Procyon means "before the Dog". Sirius is in the constellation of *Canis Major* (the Great Dog) and is thus called the Dog Star. Procyon, which is in *Canis Minor* is normally seen before Sirius so acts as a kind of herald to its big stage entrance. Directly above Procyon are a pair of bright stars, the twins Castor and Pollux in *Gemini.* You may remember them from the Spring skies, for our year is just about to come full circle. As the night wears on and our world turns beneath the stars, it is they that appear to move across the sky. If you wait long enough, the Sickle of Leo will be seen climbing grandly over the Eastern horizon and the Plough will swing ever higher about the Pole Star as the glowing starlets of the Pleiades begin to sink into the West.

ABOUT THE STARS

Water droplets in the air refract Sunlight and show us its constituent colours as a rainbow; a device called a spectroscope does the same thing by using artificial means such as a clear glass prism rather than a drop of water, and it turns out that by looking at the 'rainbow' which we call a spectrum produced by a light source such as a star, we can find out much more about the stars than we ever could by observing them with telescopes, however powerful. The spectrum of a star can not only tell us its temperature, but also its chemical composition, its size, whether it is single or not, whether it is approaching us or receding, and many other things besides. Astronomers classify stars according to the type of spectra they produce, and label the different types of spectrum with capital

Type	Colour	Temp (°C)	Example
B	bluish	25,000	Rigel
A	white	12,000	Sirius
F	yellowish	7,500	Deneb
G	yellow	6,000	Sun, Capella
K	orange	5,000	Arcturus
M	red	3,500	Betelgeuse

letters. This system was originally alphabetic, but over the years the order has managed to get itself jumbled up, so that it now reads as in the sequence above.

In addition to these types, which make up the vast majority of the stellar population, there are a few others outside the range above; at the hot end are the very powerful, hot blue stars of spectral types W (very rare) and O with temperatures above 35,000°, while at the cool end the deep red stars of types R, N and S are especially good to view with binoculars because of their colours. At the moment that is as far as we need to go into matters spectral; all you need to know is that when a star is mentioned as being of type M, for example, then you will know it is both (relatively) cool and of a red cast.

DOUBLE STARS

We have just used Orion to search out coloured stars, and it is time to use the other constellation that everyone knows to look at double stars, the Great Bear (Ursa Major) - or more specifically, the seven bright stars within it known as the Plough or Big Dipper. We saw earlier that the middle star in the 'handle', when looked at with the eye, may be seen to have a fainter one very close to it - almost touching, in fact. The brighter of the two is called Mizar and the fainter Alcor. Together they form a naked-eye double star that used to be known as the "Horse and Rider". This is actually not a terribly good example of a double star since Mizar and Alcor merely happen to lie in the same part of the sky; think of them as a tiny, two-star constellation. Real double stars are actually physically associated with each other, revolving around their common centre of gravity as do the Earth and Moon - though, since the centre of gravity in this case lies inside the Earth itself, we say that the Moon revolves around *us*. Mizar-Alcor type doubles, called 'optical doubles' are actually very much the exception, as most double stars are 'real' doubles, or *binaries*, where the stars are physically associated with each other.

Double stars afford an absorbing and almost inexhaustible field for the binocular owner. Though nothing 'useful' can be done observationally, looking at them and noting their colours or general appearance all helps to increase your enjoyment and experience of stargazing. There is something for everyone to see, and you will find hundreds of examples in the next chapter, from wide pairs of the Mizar and Alcor type to real tests for your optics, and even some that may be just that little bit too much for them!

Double stars of whatever sort are defined by three parameters: firstly, their magnitudes. Secondly, their distance apart in angular measure - nearly always given in seconds (") of arc, and thirdly, the bearing of the faint star with respect to the bright one. For this last parameter, telescopic observers use bearings of degrees from the vertical called *position angle*, but the binocular owner will not need anything so precise, so I shall be using simpler directions in this book.

One of the common peculiarities of doubles is their tendency to accentuate the colour of one component at the expense of the other, an effect known to artists as 'complementary contrast' - thus a faintish white star close to a bright yellow one will, in effect, receive the bright star's complementary or 'opposite' colour (purple) and take on this hue for itself. Yellow-purple and red-greenish pairs are quite common in the sky. There was even a "Star Colours Section" of the British Astronomical Association in the nineteenth century. You will meet many examples in this book, but sometimes I will encourage you to make your own estimates of colour.

A smaller proportion of the stars are made up of more than two members, and triple stars are by no means as rare as you might think. As with humans, the standard-sized family is more common than those with many members, and in fact the norm is the double, rather than the single, Sun-type star. So common are they that special catalogues were prepared for them, a field cornered by the two families of the Herschels and Struves in the eighteenth and nineteenth centuries. Sir William Herschel was probably the greatest observer of all time, and his son John followed in his footsteps - largely South of the equator. In Europe the Struves, Friedrich and Otto, established a similar tradition with their numerous exhaustive catalogues of doubles, still in use today even though the best-known standard modem catalogue is not of their making. In this book I have stayed with the older tradition for two reasons: one, I have to say, is pure sentimentality - but the other is practical. This is a book about binocular astronomy, and the modem catalogues leave out many of the wider doubles suitable for binoculars, since they have, without exception, concentrated on the telescopic objects. The catalogues I have used are known by the following abbreviations:-

Σ	Friedrich Struve's catalogue (mostly telescopic, though quite a few that are suitable for binoculars).
OΣ	Otto Struve's catalogue (fewer binocular pairs than S).
OΣΣ	The Pulkova catalogue (nearly all binocular doubles).
Hh	Sir William Herschel's double star discoveries, pub lished by his son John.
Ho	Catalogue of G. Hough.
S	Catalogue of J. South
β	Catalogue of S. W. Burnham (too close for binocular users as a rule).

Most of these catalogues list double stars in the Northern skies; the far-Southern constellations include the catalogues of:-

h	Catalogue of Sir John Herschel (several good binocular stars).
I	Catalogue of R. T. Innes.
Jc	Catalogue of W. S. Jacob.
L	Catalogue of N. de Lacaille
Δ	Catalogue of J. Dunlop (some good binocular doubles).

By the way, don't forget that sometimes, behind each of these impersonal-sounding catalogue symbols there are interesting human stories; Burnam for example worked in the Chicago legal system as a court reporter, but spent his nights discovering, observing and (of course) cataloguing some of the most difficult double stars in the sky - and all that with by no means a giant telescope.

You will find one more 'catalogue' in this book that you will find nowhere else, however! Whilst researching it, I found many binocular doubles which had not been listed elsewhere, so have tentatively given them a number in my own 'catalogue' preceded by P.

The observation of a close double star is a critical test for both eye and binocular. Begin your double-star observing career by learning to walk before you can run, start therefore with wide, easy doubles - I have reserved a special section in each constellation for these. Then move on to some objects listed under 'Close Doubles', some of which may be quite easy by now for the more experienced you! Gradually home in on the more testing of these, bearing in mind the often-forgotten and not always controllable variables such as difference in magnitude, observing conditions, altitude and so on.

NEBULAE, CLUSTERS AND GALAXIES

Along the borders of the curdling, pearly band of light that is the Milky Way there lie many objects whose real nature could only ever be guessed at with the eye alone. Their forms are many; we meet with misty glows, tantalising speckles of suggested starlight, or from time to time, bright splashes of stars against the black sky. These groups of stars are the *Open Clusters* of which the most famous example is the Seven Sisters or Pleiades. Many stories and legends surround this beautiful group of stars, and Alfred Lord Tennyson wrote of them:

Many a night I saw the Pleiads
rising through the mellow shade
Glitter like a swarm of fireflies
tangled in a silver braid

Tennyson, *Locksley Hall*

These lines describe wonderfully the appearance of these stars to the eye, though the fireflies will fairly swarm the moment you point your binoculars their way. Most people see six or seven Pleiads, though the record is said to be nineteen. The open clusters are probably some of the most beautiful sights you will ever see with your binoculars, and it is worth noting some of the best now, like the Double Cluster (Perseus), M.35 (Gemini), M6 and M7 (Scorpius), while of the many wonderful far Southern examples, NGC 3114 (Carina) deserves mention. The *M* numbers show these objects to come from the comet observer Messier's 18th-century catalogue of "Objects-that-are-not-to-be-confused-with-comets" while *NGC* is the standard abbreviation for the *New General Catalogue* of nebulous objects drawn up at the end of the nineteenth century by J.L.E. Dreyer (known presumably as *Herr Dreyer* to his associates), who brought together all the existing catalogues, such as those of Messier and the Herschels, into one volume. There are some 7000 NGC objects, and the catalogue has since been added to by additions, known as the IC or 'Index Catalogue'. Some of these IC clusters are binocular objects, since their large size rendered them unnoticed to telescopic observers. In this regard, the apparent sizes of clusters varies as much as do their individual appearances; sometimes it is hard to tell just what is a cluster and what is simply a chance rich field of stars. Astrophysically, a cluster means that the constituent stars must be gravitationally bound together as a unit. To observe them to advantage, especially the fainter ones, use averted vision to catch every last faint gleam of light. Many of the coarser clusters such as the Perseus double cluster contain several red stars, so use direct vision here. Averted vision, you may remember, gains us light but loses us colour.

There is another type of cluster, though these are definitely more suited to larger telescopes. The members of *Globular Clusters* are arranged, not in 'sweet disorder' but, as the name suggests, in great balls of stars. Many are visible in binoculars, though the magnifications necessary to show the thousands of faint stars of which they are composed are available to telescopes only. Some, like 47 Tucanae and omega Centauri in the far South, seem to blaze; while others such as M13 in Hercules resemble balls of cotton wool. The globulars surround the Galaxy at some distance from it, though they are gravitationally bound to it, and as such are considered as part of it.

Just below the famous three-stars-in-a-line that is Orion's belt can be seen two spots of light. The lower of these is the star called iota Orionis, but just above it is a rather less starry gleam. This is the famous, and photogenic, Orion *Nebula*, number 42 in the catalogue of Messier. A nebula is a vast cloud of gas, the birthplace of Suns. Even as you read these words, new stars are being born inside the Orion Nebula and other places like it.

In those situations where we have a gas cloud but no bright stars to illuminate it, we have a *dark nebula*, the most famous of which is the Coal Sack in the Southern Cross. Most dark nebulae are much smaller than this naked-eye example, however. The particles which make up nebulae, whether light or dark, are tiny and thinly-spread - but there are so many of them that nebulae, at the distance from which we view them, are very good at masking off the light of anything on their far sides. Thus it is that dark nebulae, especially, look more like holes in the fabric of space than anything else, as one was in fact called by Sir William Herschel - though of course they are not.

All the nebulous objects we have encountered so far have in common the fact that they are members of our own vast star city, the Milky Way Galaxy (with a capital G). However, many of the faint gleams of light we see in the sky are themselves cities of stars, Milky Ways in their own right. These *galaxies* (with a small g) have been called 'island universes'. In the previous two centuries, it was not known whether these gleams of light were within our own star system or without. Several observers claimed to have resolved them into stars, but we realise now that this could not have been, and these people were merely seeing what they expected to see. A salutary lesson: bear in mind that even the best observers can be misled! No, the galaxies are well and truly 'without' - the famous Andromeda galaxy, so famous in fact that it is sometimes called just 'Andromeda', and which is the nearest normal galaxy to us - lies some two million light years away, so when you look at that smudge of light in the Autumn sky, bear in mind that you are seeing it as it was at the dawn of Mankind itself!

Many galaxies show structure. A lot of them resemble our own in having a bright, starry centre from which dusty arms spiral outward rather like the swirling milk in a cup of coffee. Some are elliptical, composed almost entirely of stars but with very little free gaseous material from which new stars are made. Some are irregular, like the interesting

M82 in Ursa Major. Binoculars will show several galaxies, though unfortunately very little in the way of detail can be seen. With only a few exceptions, the external galaxies are to be found well away from the plane of the Milky Way in constellations like Ursa Major, Virgo and Sculptor. This has nothing to do with the actual distribution of the galaxies in space, however, but is solely due to the fact that their relatively feeble light is dimmed by the background dust, gas, and general star-stuff which predominates around the main plane of the Galaxy.

We have already seen how stars have a tendency to cluster together, and so indeed do galaxies. Our Milky Way is one of a large, so-called 'Local Group' that includes the Andromeda galaxy, and there are others far out in deep space. In fact, we now have catalogues not only of galaxies themselves, but catalogues of clusters of galaxies! Observations of galaxies are sometimes carried out by amateurs with large telescopes or CCD cameras attached, in the hope of finding spectacular supernova explosions, though it is probably fair to say that the binocular observer can forget this field. However, there is one area of "serious" astronomy that is well and truly one of his domains - that of Variable Stars.

VARIABLE STARS

Fortunately, our own Sun is an average, well-behaved, run-of-the-mill (oh, very well! Even maybe slightly boring) star. Its moods are usually as predictable as the stereotypical bowler-hatted accountant on the suburban train, but there are some stars which are more anarchic in their behaviour. Their output of light can change over the course of hours, weeks or days (sometimes even minutes) - and not always in a regular manner either. These errant objects we call Variable Stars and I have left them until the end, as they are not only interesting objects for owners of binoculars, but also to professional astronomers with their giant telescopes and orbiting astrophysical satellites.

In recent years, amateur astronomical bodies such as the AAVSO (American Association of Variable Star Observers) have been given funding by groups such as NASA to help with observational programmes. This should give you some idea of how useful amateur observers can be - NASA does not give its funds away lightly! If you want to make some sort of contribution to science, and of course want to have an interesting time too, then variable stars are what you should observe.

We do not know who discovered the first variable star; it was probably the old-time Arabic stargazers, who named a bright variable star *Algol*, the Demon, though whether they called it that because they knew it to be somehow strange or not, we will never know. Ironically, we now know that Algol is not a 'real' variable star, but only *appears* to vary because it is a double star in which the darker member lies between ourselves and the bright member of the system. As it passes in front of the bright star, the dark one cuts off its light just like the Moon does when passing in front of the Sun during an eclipse. The Algol-type stars, of which there are many, we therefore call *Eclipsing Binaries*, and will return to them later.

The first useful observation of modern times was of a strange kind of variable star indeed. The Danish astronomer Tycho Brahe observed in 1572 a *Supernova* - a large star that has come to the end of its cosmic sell-by date and completely destroys itself in a vast explosion. Up until this point the heavens were seen as 'perfect' and unchanging, and Tycho's discovery merely served to hasten a change in cultural belief, long prepared by the Renaissance, away from abstract perfectionism towards a more scientific view of the world. Sixty years later came the discovery of the first, what would be now thought of as a 'true' variable star. Fabricius shew in 1638 that the star in the appropriately large constellation of Cetus the Whale, named omicron (o) Ceti, was reappearing out of invisibility, reaching its greatest brightness in a cycle of about 11 months. This star he named *Mira* 'the wonderful' and even today we still call such objects Long-Period Variables or 'Mira stars'.

With the use of optical aid, more variables were found, and the sky began to look like a much more dynamic place. It soon became evident that these special stars needed a special type of name. The great 19th-century astronomer Argelander devised the capital letter system whereby the first variable to be discovered in (say) Cygnus would be called R Cygni (it happens to be a well-known Mira star visible on occasion with binoculars). The next Cygnus variable would be called S Cygni (another Mira star, but this time never visible in bins), and so on. The single letters A to Q were not used for this purpose, as they had already been employed in naming ordinary, non-variable stars in many of the far Southern constellations. When the ninth variable in a given constellation was reached, he went back to R but now used double letters - RR, RS through to RZ; then SS to SZ and so on. This system gives you 54 variables per constellation, but even that proved to be insufficient for star-rich areas like Cygnus, so we now return to the beginning of the alphabet with AA through to AZ, then BB through to BZ, always omitting J so as not to confuse it with I. You would have thought that this scheme, which provides for 334 variables by the time QZ is reached, would be enough, but no! However, the system we use after QZ is simpler - the next variable to be discovered in that constellation was called V335 Cygni, and so on. In 2001, a nova appeared in Cygnus and it was given the designation of V2275 Cygni!

Once a variable has been followed for a period of time, we may be able to plot its light-changes on a graph. Astronomers call these graphs *light-curves*, and its form can often tell us which of the many types of variable star we are dealing with. All curves, though, have certain features in common; maximum is that part of a star's variations when it is brightest, and minimum obviously the opposite; amplitude represents the difference between the two. The period is the time between two successive maxima or minima. Sometimes this period is constant, though most of the stars followed by amateur astronomers are the more interesting ones, in which the period is subject to varying degrees of irregularity - as may be the other parameters mentioned above, too.

The accuracy of a light-curve improves as more observers' results are used to compile it, but it is great fun to make your own. Let's do it now. The star we are going to use is a very interesting binocular variable called CH Cygni, whose light curve is shown in Fig.20. It needs observing on every possible occasion, since it is not only irregular in its variations, but is also subject to shorter-term fluctuations as well. Over the past 25 years in particular it has shown both maxima and especially minima way outside its official range. I shall be using my own observations, made with 7 x 50 and 10 x 50 instruments. But before we can use the observations, we should have them to hand, of course! Record your findings at the eyepiece and then write them up, in either soft or hard form, later. Include the date, and, for eruptive stars like CH Cygni, the time to the nearest minute. The type of curve you draw will depend on what you want to show. Observations by one single observer will not be frequent or consistent enough to show the short-term variations, so instead we shall just look at the behaviour of the star over four years, from

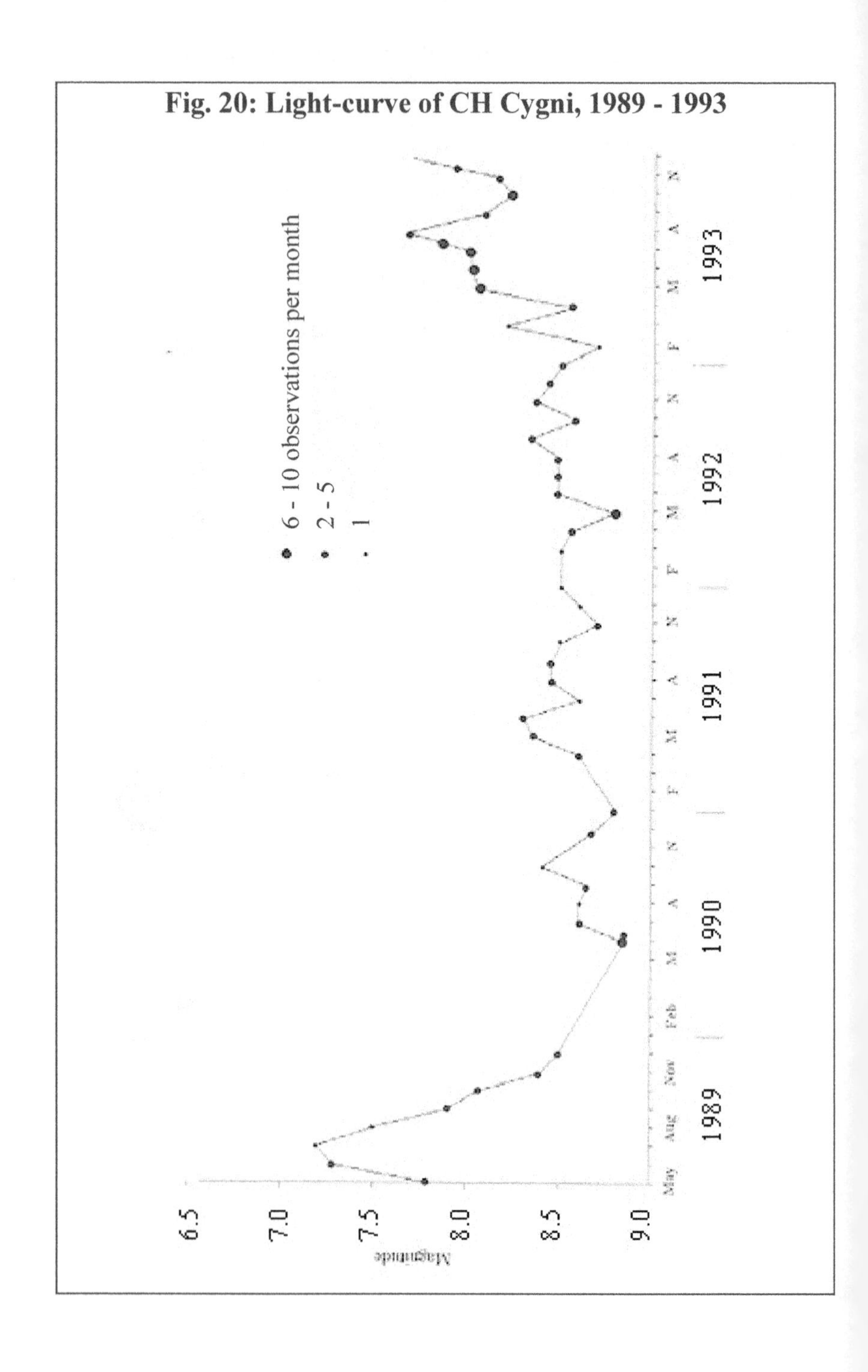

Fig. 20: Light-curve of CH Cygni, 1989 - 1993

Month	No. Obs	Mean Mag.	Month	No. Obs	Mean Mag.
May **89**	5	7.82	Oct	1	8.50
Jun	4	7.28	Nov	2	8.70
Jul	1	7.20	Dec	1	8.60
Aug	1	7.50	Jan **92**	1	8.50
Sep	3	7.90	Feb	0	
Oct	3	8.07	Mar	1	8.50
Nov	3	8.40	Apr	4	8.55
Dec	1	8.50	May	8	8.79
Jan **90**	2	8.55	Jun	4	8.48
Feb-Apr	0		Jul	6	8.48
May	8	8.79	Aug	4	8.48
Jun	2	8.85	Sep	3	8.33
Jul	2	8.60	Oct	4	8.58
Aug	1	8.60	Nov	5	8.36
Sep	2	8.65	Dec	3	8.43
Oct	1	8.40	Jan **93**	1	8.50
Nov	0		Feb	1	8.70
Dec	3	8.70	Mar	1	8.20
Jan **91**	3	8.80	Apr	4	8.55
Feb-Mar	0		May	10	8.04
Apr	2	8.60	Jun	8	8.01
May	4	8.35	Jul	6	8.00
Jun	3	8.30	Aug	4	7.65
Jul	1	8.60	Sep	3	8.07
Aug	4	8.45	Oct	6	8.22
Sep	2	8.45	Nov	2	8.15
			Dec	5	7.68

1989 to the end of 1993. Daily observations would be swamped in this case, so I shall plot only the average value for each particular month.

We now need three sets of data:- the date, the mean monthly magnitude, and the number of observations made in each month. Having listed these in the following table, we can then draw a simple graph, with magnitude as the vertical (y) axis and time as the x-axis. Much can be gained by looking at some of the figures opposite, even before we plot them; note for instance how few observations there are around February, when the star is at its lowest point in the sky. However, the value of just one observation at such times is borne out by the result for February 1993, since the star then appeared to be at a minimum - though this was based on only one observation so may not be as reliable as most of the other results. Also, since the object was observed on every possible clear night (I happen to know this as a 'fact of life') we may be able to get some sort of historical information from the figures; for instance, it appears as though the time around May tends to produce more clear nights - at least based on this brief sample. The final result for December 1993 is italicised as it does not fit on the graph, even though it has been included to show a possible brightening of the star. In this regard, it is best to wait until you have a good string of results before you draw your graph, otherwise you may become biased by what you think the star should be doing next!

TYPES OF VARIABLE

There are many different types of variable star, but for the moment we can divide them into three broad groups - *eclipsers, pulsators, surprisers*. I need hardly add that these are merely sweeping temporary labels of mine, and are of course not the official names! We will come to those soon.

The eclipsers are of two main sorts - the Algol variables, where one star is rather cooler and so appears 'darker' than the other, and so causes fairly deep eclipses when it passes in front of its brighter fellow, but where the eclipse is much shallower when the positions are reversed; and the sort typified by beta Lyrae. Here the stars are equally as luminous, and we get constant, but less extreme, changes in light-output from the system as a whole. Beta Lyrae itself is easily followable with the naked eye.

Eclipsers are quite common, though there are some highly peculiar stars to be found among their ranks; XY Persei, for instance, is an eclipsing system made up of two stars, each of which is believed to be independently and irregularly variable! This interesting star varies from magnitudes 9.2 to 10.6, rather too faint for average glasses, unfortunately. Other strange systems include ζ and ε Aurigae, binaries made up of a highly luminous but small star, and a vast companion many thousands of times larger, so that we have not so much an eclipse as a transit by the bright member. Amateur observation of eclipsing variables is fairly recent, but has proved useful by revealing hitherto unsuspected changes in their periods, for instance.

Among the largest group of variables, the pulsating stars, is the important subgroup named after delta Cephei called, not unreasonably, the *Cepheids*. Though amateur observations of these stars is not terribly relevant, they are good objects on which to practise, since many of them are quite bright. δ Cephei itself is completely a naked-eye variable and also a rather attractive binocular double star too. One interesting facet of these stars is that the longer the period, the more powerful the star, though the Cepheids are all reasonably luminous objects anyway, with periods that start at about a week and run up to stars like DN Arae with a period of 82 days. As the periods increase, so the sizes of the stars do also, along with increasing redness. DN Arae is a huge orange-yellow powerhouse of a star, though at magnitude 13 much too faint to concern us any further. At the other end of the spectrum (literally!) are the very short-period Cepheids called the *RR Lyrae* group, white stars whose periods are measured in hours for the most part rather than days. On the whole, light-curves of Cepheids have a characteristic shape, that of a comparatively steeper rise to maximum than the ensuing fall to minimum.

After the Cepheids come the *Mira Stars*, or LPVs (Long Period Variables). We saw how the colours of the Cepheids became redder with increasing periods, and here the stars with longer periods than any Cepheid are also much redder than they, though the Miras are a far more diverse class of stars, as well as being not nearly as predictable as the highly-regular Cepheids. Nearly all Mira stars have ranges in excess of 2.5 magnitudes and are often recognisable by their red colour. Mira itself is a good example of the group; at maximum it has been known to touch the first magnitude, whereas at others it has struggled to reach the fourth, although minima as a rule are not so dissimilar, being between magnitudes 9 and 10. Like many of its class, Mira will sometimes show humps on the rise to maximum, as the variations subside for a few days. Its period is about 330 days, but again, this is never followed with absolute regularity. Periods quoted for these stars are statistical averages rather than exact values. Mira and, in the South, R Carinae, are good examples for the binocular owner to follow.

Even less regular than the LPVs, and of great interest to us here, are the *Semi-Regular variables*, since many of them are binocular objects. Their colours can be even redder than those of the Mira stars, and their amplitudes invariably smaller, rarely exceeding two magnitudes. Period-wise they are even more irregular than the previous class, with normal service sometimes being interrupted by periods of constant light, or reduced ranges of variation. A good binocular example is AF Cygni.

Related to these objects are the *Red Irregular* variables, with usually even smaller amplitudes than the SR stars, sometimes less than a whole magnitude and consequently not quite as interesting to follow in my opinion as the Semi-Regulars. Many go through long periods of complete inactivity, though all the red stars - Miras, SR's and red irregulars - should be observed only once a month; their variations are quite stately, and more frequent observations are usually unnecessary.

The last group of pulsating stars are those typified by *RV Tauri*, itself not a binocular object, though a few of them are. This class of star lies astrophysically somewhere between the Cepheids and the Mira stars. Their colours and periods are more Cepheid-like, though they are not so regular. They make very rewarding objects to follow, and their light-curves are typified by double minima: a shallow, brighter minimum divides the maxima pair, after which there is a fall to the main, deep, minimum. This is very much an idealised scheme, however, with sometimes minima being missed, interchange of maxima, and all sorts of chaotic fun. An excellent star of this type for binoculars is AC Herculis. Observe these interesting stars once a week if you can.

Coming to the last large group, the stars which surprise us, we begin with what used to be called the *Nebular Variables*, but for which I prefer the term YSO (Young Stellar Objects) as while all are newly-formed stars, not all of them are necessarily associated with nebulae, even though many are found in regions where nebulae, whether bright or dark, predominate. They are all unstable stars not long born from the nebular material itself, and their variations can be both irregular and sudden. Some spend most of their time at maximum with unpredictable falls and fluctuations, and are white in colour like the binocular variable AB Aurigae. Others show constant, often extreme, variations, while others, the T Tauri stars, show smaller and less drastic changes but are among the youngest of all stars, and very interesting as regards the early stages of stellar evolution. AB Aurigae above actually has a companion, SU Aur, that is one of these stars, though at ninth magnitude it is rather faint for most binoculars and also too close to the brighter AB Aur for binocular study, as well as having a rather small range of variation.

Markedly different are the *R Coronae Borealis* objects. Again, they are completely irregular and unpredictable, with most of the time being spent at maximum light. Every now and then, however, parts of their carbon-rich atmospheres will condense into giant clouds of what can only be described as soot, and they may plummet to depths of faintness that can only be rendered visible with large telescopes. Binocular observers can play a useful role here in announcing news of these falls. R Coronae and its Southern counterparts RY Sagittarii and V854 Centauri are easily visible with binoculars at maximum.

The *Nova-like* stars include our recent friend CH Cygni, and are all doubles composed of a hot blue-white star and a cooler but larger orange-red one. Their changes in light probably arise from effects resulting from the exchange of gaseous material between them. Z Andromedae is a similar star which can sometimes erupt from its 11th-magnitude resting place up to binocular visibility, as it did in the early part of 2001 and 2008-9.

Coming to the *Novae* proper, these are not, as was once thought, new stars, but more like new bright stars. They are all doubles of the same type - a small 'white dwarf' smaller than the Sun, and a larger yellow-orange star, which because of evolutionary pressures it is undergoing, unloads large quantities of its material in the direction of its smaller, hotter neighbour, where it swirls in a thin disc around it. This material eventually builds up on and around the small star to such an extent that it detonates in a mighty nuclear explosion, though the stars themselves do not explode. Many such systems must exist, and where the explosions are smaller, we can actually observe them happening with some regularity. We call such stars *Dwarf Novae*, and though only three (SS Cygni, U Geminorum and VW Hydri) are ever visible with binoculars - and only then at a bright maximum - amateur observations of them are in demand from the professional community. The only reason we do not see the regular Novae explode more frequently is that their eruption cycles are far longer (the three dwarf novae above erupt on average once every month or two) with the result that Novae only explode once over the course of many human lifetimes. But remember that the Sun can never become a Nova, since it isn't a double star of course!

Supernovae are a different proposition altogether. All they have in common with Novae is the fact that a bright star becomes visible where none could be seen before. However the cause of this explosion is entirely different. Here the explosion is not merely skin-deep, and most of the star is destroyed. For involved astrophysical reasons, a large and massive star at the end of its life becomes unstable and can no longer radiate energy properly. It contracts in on itself rapidly, generating vast internal pressures and temperatures which then cause a titanic cosmic explosion, blasting the star's atoms out into space in all directions. These atoms, fortunately for us, include those of the chemical elements from which living things are made such as Carbon; so if it were not for the Supernovae, we would not be here to thank them for our existence. We are truly children of the stars.

OBSERVING VARIABLE STARS

In the field of VS observing practice makes, if not perfect, at least better; so let's practise now. First we need to obtain a chart of the area of our star, with non-variable stars, whose magnitudes we already know, included. We will use these as *comparison stars* with which to estimate the brightness of our variable. Having acquired the chart, we must then actually find the real version in the sky! Often, especially with the brighter stars visible in binoculars, there may be a recognisable naked-eye star already present for us to locate the field by. If not, you can nearly always "star-hop" by beginning at a well-known star and gradually homing in on the field's area. You will soon find yourself recognising distinctive patterns of stars such as lines, triangles, right-angles and so on. Start with some of the charts in the Appendices, then try some of the written descriptions in the next chapter. Appreciate what a certain magnitude 'looks like' and get to recognise differences in the magnitudes of stars, whether variable or not.

Having found the field (bear in mind that the variable itself may be too faint to see!) identify the patterns of the brighter comparison stars. Can you see the variable? Which comparison is it obviously brighter than? Which one is it fainter than? Let's use the make-believe Fig.21 as an example. In view 1, you have mentally placed the variable, shown as

Fig. 21: Estimating the brightness of a Variable Star

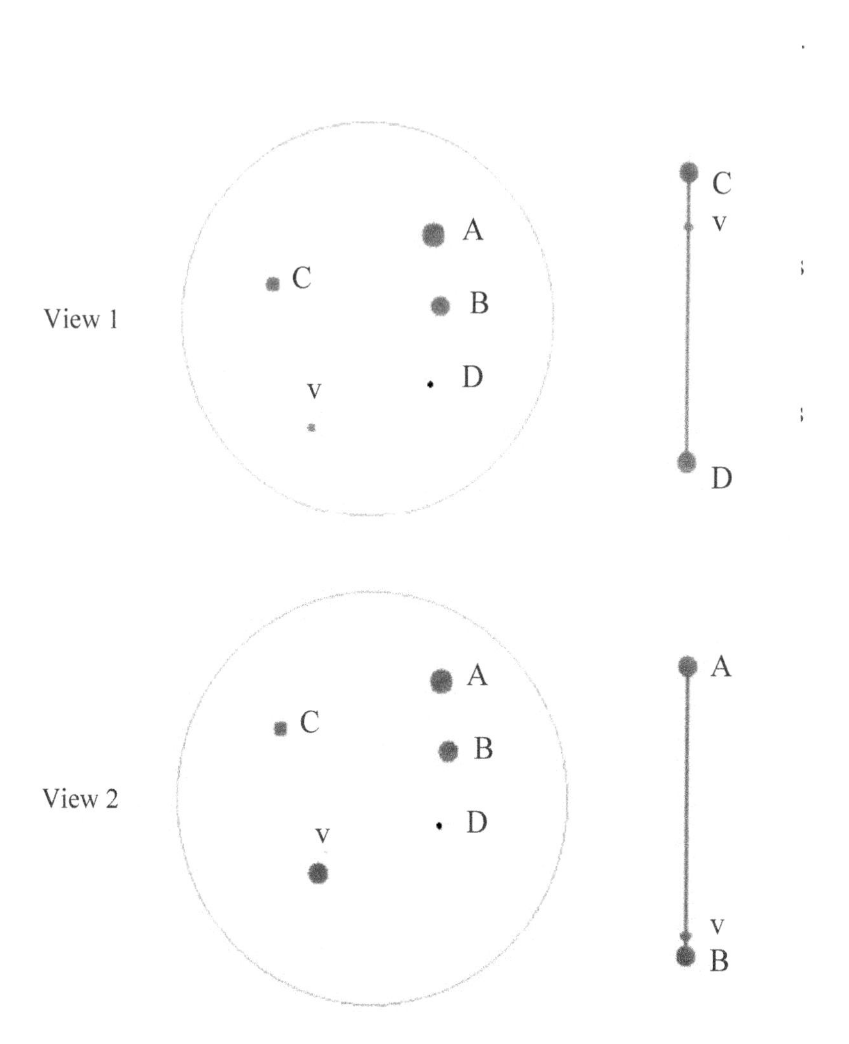

'V' here, somewhere between stars C and D in brightness. Divide the C-to-D interval into fractions if you can; (personally I find it hard to do this, so there's no need for you to despair if you can't either, since we will meet another way of doing it shortly). Decide where V comes in this interval. Perhaps you divided the C-D gap into fifths and think that V is one fifth fainter than C and four fifths brighter than D. Record this as C (l) v (4) D. Sometimes, I even record my estimate graphically, as in the representation at the side of the page. Congratulations ...you have just made a variable star estimate using me Fractional method.

The other version, the Step method invented by Pogson (no relation to me as far as I know) seems easier, at least to me, but just takes a bit of training first. You need to recognise actual qualitative magnitude differences between stars, which you will be able to do with practice possibly down to 0.1 magnitudes. This is why I stressed earlier the point about practice being so useful. The step method requires that you make direct estimates between two stars at a time, in what you hope are 0.1m steps. Recording is much easier; you would write C-1 V+4, for instance. Deriving the magnitude can be done later and needs only arithmetic. Suppose the gap between C and D is 0.8 magnitudes, and that star C is 7.2m with D being 8.0;

Fractional method: each fraction is thus 0.8 / 5 (you divided the interval into fifths, don't forget) or 0.16m, so "C minus one fraction" gives us 7.36. "D plus 4 fractions" gives us the same figure, i.e., (0.16 x 4) = 0.64m brighter than D. (Don't forget that the magnitude system works 'backwards' so you have to subtract 0.64 from the magnitude of D to get a brighter result, and vice versa with C).

Step method: A step of 0.1m from 7.2 gives us 7.3; four steps from 8.0 gives 7.6. Take the mean of these two values to give a result of 7.45. In both cases, the derived magnitude of the variable would be rounded to the same value, 7.4 - since visual estimates by one person are not sufficiently accurate to warrant greater accuracy. However, you have recorded the estimate (haven't you?) so you can analyse it later if needed.

Most of the time your estimates with two comparison stars will not, in fact, come out quite as clean and tidy as in the Fractional example above, and you will need to take the average value produced. In view 2, perhaps a couple of months later, the field looks a bit different, since our variable is now much brighter, so we now need to use different comparison stars, say A and B. The variable star may continue to brighten even more so that you can only use A alone, and in such cases you have to make do with just me one star to use.

There are two ways of showing the magnitudes of the comparison stars; either they are indicated as in Fig.24 by letters, with the actual values shown elsewhere (i.e., C = 7.2 D = 8.0, etc.) or the magnitudes are shown as labels themselves, omitting the decimal point so as not to confuse it with a faint star. Our C and D would then be labelled on the chart as 72 and 80 respectively. I use both methods in this book.

DIFFICULTIES

Sometimes the variable is too faint to see at all. In this case record the faintest comparison star that you *can* see. These negative results can be useful; if you are watching an eruptive star, your "too faint to see" observation may be the only one to show that the star was not erupting on that date, which could be significant. On those occasions when the star is too bright, I simply tend not to observe it at all, especially when it is a slowly-varying red star. Red stars actually look much brighter, the brighter they become! This is known as the *Purkinje* Effect (pronounced poor-kin-yay) after the Czech scientist who drew attention to it. One way to overcome the effect is to use defocussed binoculars on the offenders, as this has the effect of spreading the light over an area, thus diluting its colour. Be sure to always use this method with that star, however, or you will get discordant results when changing from one method to the other.

Another source of error is misidentification, especially when the pattern of some of the field tends to the fractal, i.e., maybe there is a triangle of stars, one of which is also a corresponding member of a smaller, but similar triangle. This has caused even highly experienced observers to make mistakes.

Bias is yet another problem. If you have been watching a Mira star slowly brighten for a couple of weeks, you will (consciously or otherwise) expect it to continue doing so. The star, on the other hand, may have ideas of its own and decide to undergo one of its occasional slowdowns, or something else unexpected. One way to overcome bias is to have as many stars on your observing programme as you can follow. This way, you will not only get to know the sky better, but will also find it easier to forget what any given star was previously doing. Some people estimate red variables once a week; this is much too frequent in my view. Once every three weeks to a month is quite sufficient for these (as a rule) slowly-varying stars. Rapidly-varying or eruptive stars need the time noting to the nearest minute, especially if they are in an especially active phase, though just the date is actually sufficient for all the red variables.

To close, some refinements: the constant wish of stellar astronomers everywhere is *More Light Please!* However, it might be better to rephrase this as "less of the wrong sort of light". Sir William Herschel, no less, used a photographer's black hood to cut down stray light and generally keep things as dark as possible. Us moderns could improvise using a black bin liner (unused, of course!) which can be easily shaped. The discomforts of wind, and they are many, can be eased by the use of something such as a beach windbreak, though foolish Spartan folk like me just grin and bear it. If you do use a windbreak, make very sure that it is well-secured - otherwise it could end up savaging you, your binoculars, charts, and everything else!

Very well: we have spent no little time describing the wide variety of objects in the sky, how to observe them and what we need. Now it is time to look at them in detail. In the next chapter you will find detailed charts of every constellation, and descriptions of all binocular objects in each.

CHAPTER 3. THE CONSTELLATIONS

ANDROMEDA

This fine constellation contains interesting objects of all types, the best-known probably being the Andromeda Galaxy (colloquially known these days as just "Andromeda"). Binoculars give a view of this object as a fuzzy oval patch, but do not let that distract you from the many other sights here.

Groups of Stars

1. The line of 3, 5, 7, 8 and 11, the last of which is a wide double. This group is visible with the slightest optical aid, and there are other stars near 7.

2. 63, 64, 65 and 66. These form a much smaller, and rather fainter group. 64 is noticeably orange in colour.

3. 00h, +42° Small collection of 6th and 7th magnitude stars in the form of a rough V, pointing in the general direction of ι, κ, λ and ψ. This little group used to be known as a separate constellation called *Gloria Frederici* or "Frederic's Glory". It has since been omitted from the official constellations but is still of some interest to astronomical historians.

Wide doubles

o & 2. These form a bright wide pair of magnitudes 3 and 5.

62. Magnitudes 5.1 and 6.1, both white.

Close doubles

ΣI 1. A pair of 7th-magnitude stars separated by 46 seconds of arc.

56. This is a brighter and wider object. Mags are both 6, and the separation here is 182". Close to the fine star cluster NGC 752.

Variable stars

R (6.9-13.3) This red Long-Period Variable is easy with binoculars at maximum, which usually takes place around magnitude 6, though occasionally much fainter. Coming maxima of this star are listed in the appendices. R Andromedae itself has a neighbour of 6.9m which is perfect for estimating when the variable is around maximum.

TZ (7.6-9.0) TZ is situated in a small equilateral triangle near group 3 above, and the other stars in this triangle are 7.9, 8.9 and 9.3, so they are good comparison stars.

VX (7.8-9.3) A deep red star of spectral type N, with some good comparison stars in the area.

AQ (8.0-8.9) Easily found near a prominent Y-shaped group made up of ρ, σ and θ, this variable is part of a neat group of seventh and eighth magnitude stars. Like the preceding variable, it is quite red.

BZ (7.5-8.4) A brighter variable in the North of the constellation. It lies between π Cas and an impressive line of 6th and 7th magnitude stars.

Clusters and Nebulae

M.31 (NGC224). This is the famous Andromeda Galaxy, easily visible to the smallest binoculars. Indeed, I can see it easily with the naked eye whenever the Moon is out of the way, and it was known to the ancients, though of course its true nature eluded them. Its elliptical outline extends over two degrees of sky, and binoculars give a better view of its size than do most telescopes. Try various observing techniques on M.31; move the binoculars around slowly, use averted vision, combine these two techniques, and so on.

NGC 752. A cluster described by the Hungarian observer Bela Szentmartoni (alas now deceased) as "a great star-cloud, containing faint stars". Well worth the finding.

NGC 7662. A planetary nebula which looks like a faint bluish star. Owners of small binoculars will need to use the chart here to find it.

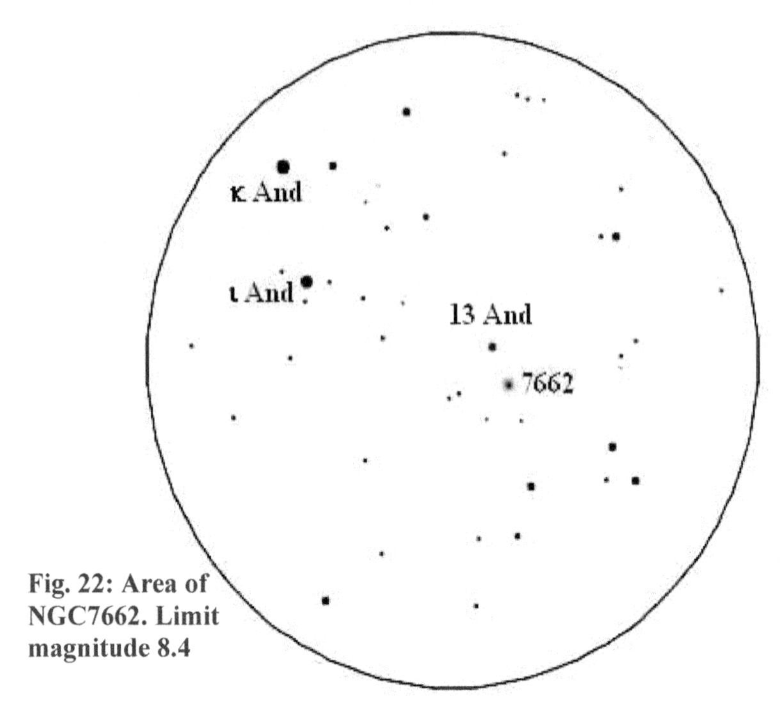

Fig. 22: Area of NGC7662. Limit magnitude 8.4

Planetary Nebulae are remnants of once-giant red stars which, in the course of their evolution and growth, threw off most of their outer atmospheres in a giant shell of gas, which we see as the nebula. All that is left of the giant is a small star at the centre, usually very faint and invisible with binoculars.

AQUARIUS

A large group with many fine fields, double stars and some interesting nebular objects. Unfortunately most of them are beyond the range of normal binoculars!

Groups of stars

1. Sparse group of 15, 16 and 21 plus 20, a further star of 6.4m. This group is located just North of β Aquarii.

2. 22h, -5°. Larger collection of 6th-magnitude stars in addition to many fainter ones. In a wide field, the area is very attractive.

3. $\psi^{1,2,3}$. These form another fine group with some fainter outliers.

4. The stars 86, 88 and 89 form an imposing group along with numerous fainter stars. Not far away is a slightly smaller collection centred around 99 Aquarii.

5. Beautiful group of 103, 104, 106, 107 and 108. Fewer fainter stars here, though.

6. On the border with Cetus is an attractive curved line of 6th and 7th magnitude stars. To find it, locate ι Ceti (3.7) and sweep Westward along the horizon by about 5 degrees.

Wide doubles

4 and 5. Magnitudes 6.0 and 5.5, in an attractive area near 3 (4.6). See if you can spot which are the orange-coloured stars.

β . This star has a 5m companion directly South.

σ and 58. Two stars differing by a magnitude in an interesting area.

δ and 77. This is a slightly fainter edition of σ above.

Close doubles

Σ2809. A good object for large binoculars, the stars of magnitudes 6 and 9 being separated by 31 seconds. Near the cluster M2.

Σ2993. This is associated with group (3) and is even harder than the previous object. Distance only 26".

Variable stars

R (6.8-10). This interesting star can be followed for much of its cycle with binoculars. R Aqr is actually a star in two separate classes of variable! It is embedded in faint nebulosity and is a "symbiotic" variable like CH Cygni, made up of two stars which influence each other's evolution and behaviour. Most of the time, however, it behaves as a fairly normal Mira-type star. Even so, you could do worse than observe it once every three weeks. Predicted times of maxima are given in the appendices.

Z (7.2-9.8) A star with quite a large amplitude, though you will need dark skies and large binoculars to catch it near minimum. It lies between the 6m star 1 Ceti and the bright group around R above. It is just to the East of a 6.4m star and a pair of stars half a degree South of 7.6 and 9.2 will prove useful.

Clusters and Nebulae

M.2 (NGC 7089) A globular cluster, appearing in 6x30's as a starlike nebulous point. Improves with altitude.

NGC 7009. A planetary nebula near ν , visible as a small spot with 8x30's.

NGC 7293. A large planetary, which class of object Aquarius seems to be well-provided for. 7009 is called the Saturn nebula, this one is the Helix. These evocative names will not unfortunately convey very much to the binocular observer, as high magnifications are necessary to show these objects to any advantage. But you can at least say that you've found them!

AQUILA

A constellation with a great store of interesting objects for the binocular observer. Its leader, Altair, is quite close to the Sun, at a distance of 16 light-years. It is in a field of bright stars and dark nebulae are nearby too, but you will need a transparent night and no light pollution to see them.

Groups of stars

1. η Scuti (5.0), the orange star 12 Aql, λ (3.6) and 14 & 15 (both 5.5) form a brilliant group.

2. The region SW and E of δ (3.4) is strewn with bright lines and small groupings of stars.

3. 20. A beautiful fan between this star and 12, though rather nearer the former.

Wide doubles

10 and 11. This 6m pair forms a triangle with ε and ζ. 10 is a small-amplitude variable with an official designation of V1286 Aquilae.

χ and 46. Close to Tarazed (γ Aql) in a fine region.

56 and 57. An interesting wide pair, since 57 is itself a binocular double. 56 is orange and 57 is blue. Together they point South to a wide triple including the 5.6m star 51.

θ and 66. Another coloured pair of white and red. Have a look and see if you can see which is which.

Close doubles

Σ2425. A rather close double (32") of magnitudes 7 and 8.

15. Wider and brighter; the companion is orange.

OΣΣ 178. Magnitudes 5 and 7, separated by 90 seconds of arc.

28. This is one of a bright triangle. It has a faint companion 60" away.

Σ2497. A difficult object of mags 7 and 8, separated by 30 seconds.

57. Magnitudes 5.9 and 6.5, distance 36". Webb *et al* have remarked on the colours; those of the primary have been seen as pale yellow or white, whilst the fainter star seems a bit wilder - pale lilac, bluish, greenish or azure white (!) After all this, you really will have to have a look for yourself!

OΣΣ202. An easy pair separated by 43".

S749. A fainter, wide pair in a rich field. Distance 60".

Variable stars

R (6.1-11) This interesting star can be followed for much of its range with binoculars, and times of maximum are given in this book. It is peculiar in that its period has undergone definite changes over time.

V (6.7-8.2) A beautiful deep red star which makes a triangle with 6.9 and 8.2 objects. Easily found when bright because of its colour.

TT (7.0-8.9) A Cepheid type star with a period of 13.75 days. There are several 8th-magnitude comparisons around, but it has to be said that Cepheids are not always the most exciting stars for amateurs to follow for long because of their predictability, though they are good practice objects.

UV (8.3-9.3) Better suited to powerful glasses, this forms an equilateral triangle with 10 Aquilae and a star of 8.9m to the SE which is useful as a comparison. With large instruments, you may detect this star's deep red colour.

V450 (6.3-6.9) A red star at the right-angle of a triangle (var, 6.4, 7.0) not too far from Altair, although the 6.4 star is slightly variable, so probably best not to use it. Suitable for the smallest instruments.

Clusters and Nebulae

NGC 6709. A cluster which is best seen in large glasses. Imre Toth sees it as (my translation) "stars around two diffuse parts.... about 4 stars can be distinguished". This observation was made with 10x80s.

ARIES

A rather dull group marked by the stars Hamal, Sheratan and Mesartim. A neat little triangle in the East of Aries used to be called Musca Borealis (the Northern Fly).

Groups of stars

1. Three doubles in the same field; 10, 11 and 14.

2. A very attractive area is enclosed by the stars μ, ν, 26 and 27.

3. ξ. Sweep from here to another xi, ξ^2 Ceti (4.3)

4. The large group of ο, σ, π ,40 and RZ is worth looking at, and there are fainter stars visible in the area. RZ is of course variable, but its amplitude is too small to concern us here.

5. Another fine group, including τ, 63 and 65.

Close doubles

λ. A close pair near Hamal. Mags 5 and 7, distance 38". It makes a wide pair with 7 (RR Ari).

14. A triple star of magnitudes 5, 8 and 8, distances 93 and 103 seconds.

30. A fine pair of 6.6 and 7.4, 39" apart. Possibly variable.

AURIGA

A marvellous Milky Way constellation with endless groups of faint stars. The Eastern reaches are quite barren in comparison with the area bounded by the bright stars. Capella, its leader, was known to the Arabian sky-watchers as the "Guardian of the Pleiades". Sir John Herschel thought it had brightened during his lifetime, but there is no evidence for this. Near Capella is the little triangle of ε, η and ζ known as the Kids, held in no mean dread by classical writers. An old couplet runs:

Tempt not the winds, forewarned of dangers nigh,
When the Kids glitter in the Western sky

Groups of stars

1. Beautiful collection of hot-looking bright stars around 16 Aur.

2. 5h 12m, +31°. Radiant curving line of 7th- and 8th-magnitude stars. A degree South is a similar, though less striking, line.

3. The area round the bright star β Tauri is worth sweeping. Though not technically in Auriga, this star is clearly part of the pattern, and was actually known at one time as gamma Aurigae.

4. Between μ and σ there are sprinklings of stars of assorted brightnesses. An impressive area.

5. λ . Another star in a fine region, with lines, diamonds and circlets of small stars.

6. The area around χ and ϕ abounds likewise in fans and lines of faint stars.

7. Bright reversed Y between 40 (5.3) and theta.

8. Prominent group of four stars all bearing the Greek letter ψ and distinguished by their superscript numbers of 2, 4, 5 (56 Aur) and 7. Try and arrange these in order of brightness - there is not much to choose between them! Also, the area around the second of these stars is rich in faint stars.

Wide doubles

5. Makes a wide coloured pair with 6 nearby, yellow and red.

ψ^8 , 59, 60. Wide triple. All these three are roughly equal in magnitude, though 59 is slightly variable and is also known as OX Aur.

Close doubles

Σ698. In a fine area, this is a close pair (31") of magnitudes 6 and 8.

56 (ψ^5). This coloured pair has a separation of 48" of arc.

Σ994. A faint, close double. Mags 7.3 and 8.0, distant by 26 seconds.

Variable stars

TU (8.0-9.1) TU Aurigae lies between ψ^4 and 47 (6.0). Few bright comparisons here, but I have supplied a chart in the appendices.

UU (5.1-6.8) A very red star, suitable for the smallest glasses.

UV (7.4-10) An object for large glasses, this is a symbiotic variable, or possibly a Mira star, near group (2) above.

TW (7.8-9.1) Very easy to find near β and π. Two stars between the latter object and the variable are of 6.8 and 8.0 magnitudes, and just East of the 6.8 star is a fainter comparison of 8.9. TW Aur illustrates the necessity of actually looking at the sky when doing research of any kind. I had originally planned to select this star for telescopic observation, but on looking at it, found it too bright, and better suited to binoculars!

WW (5.7-6.4) A bright eclipser well-removed from the main group. The wide triangle of 49 (5.1) 53 (5.5) and 54 (4.8) can be used here.

AB (7.3-8.5) A nebular variable - not terribly active unfortunately - which underwent a decline in early 1976 among others. It lies between two stars of 6.8 and 7.4m, and has two companions; one of 7.5 directly S, and another which is in fact also a nebular star of the T Tauri variety to the N. This is SU (9.0-9.6) which owners of large glasses might like to look at. I have just glimpsed it with 10x50's.

AE (5.4-6.1) A white variable, in this sense not unlike AB above. AE Aur is an example of what is known rather spectacularly as a runaway star. The theory is that it, along with two other similar stars, 53 Arietis and μ Columbae, were hurled out from the part of the sky now marked by the Orion Nebula by a supernova explosion.

NO (5.8-6.3) This lies close to 26 (5.5) but is rather unsuitable for visual observers because of its small range of variation.

Clusters and Nebulae

M.38 (NGC 1912). Even in small glasses this shows up as a bright oval spot in a rich field.

M.36 (NGC 1960). You may be able to see some individual stars in this cluster. Also in a fine area.

M.37 (NGC 2099). The largest of the Messier objects in Auriga, this appears as a large oval nebulosity. Very worth while looking at.

NGC 1857. Visible as a haze beyond an unequal double, one component of which is reddish in hue. Again, a cluster in a very rich area.

BOOTES

A large group marked by the brilliant Arcturus, mentioned in the Bible. A constellation containing several good double stars, but no nebulae - at least for the binocular observer.

Groups of stars

1. θ, ι, κ. Sweep this area, noting a small, faint triangle directly South of iota.
2. 15h 10m, +39°. A region of several double stars.
3. ζ. Sweep the area a few degrees west of this star.
4. Bright quadrilateral of ω, ψ, 45 and 46.
5. 14h 21m, +8°. Line of three bright stars, including a red one.
6. 14h 08m, +15°. Delicate line of faint stars.

Wide doubles

ι. Mags 4.8 and 6.1. The bright star is also a close double of 4.8 and 7.7 magnitude and 38 seconds separation.

6. The companion to this star is ΟΣΣ126, a beautiful pair 86" apart. On the other side of this double is another (7th mag.) star.

ζ. A wide double of 3.9 and 6.0, both white.

ν. This makes a superb wide pair with its neighbour 53. Both equal in brightness but not in colour. Have a look and note your colour estimates.

Close doubles

Σ1850. This 6m star has a fainter *comes*, 26" away.

Σ1921. Two seventh-magnitude stars separated by thirty seconds of arc.

δ. A wide pair (105") of contrasting colours. In the same field is 50 (5.4) with two 8th-mag neighbours, one SE and the other NE.

GC 20252. For all its designation, this is an ordinary 6th-magnitude star, but it does have a faint companion to the NE.

$\mu^{1,2}$ Makes an imposing pair 88" apart. Mu has a proper name, *Alkalurops*. Most of the old star-names come down to us from the Arabs, but this one is a mixture, part Arabic and part Greek. Al- is simply the Arabic word for "the", found in many star-names, whereas the rest of the name comes from the Greek *Kalaurops*, meaning a shepherd's crook or staff.

Variable stars

RV (7.5-8.8) This is near the similar star, RW (7.5-9.2) Both of these are shown on the chart in the appendix.

RX (6.9-9.1) An object of similar type to the above, close to a bright comparison of 6.2m. A wide pair of 7.0 and 7.4 are not far away. Note the red colour of RX.

UV (8.0-8.7) An eruptive variable forming a little triangle with RX and its bright comparison. Unfortunately it is not terribly active, and its range is rather on the small side.

ZZ (6.8-7.6) This eclipsing star forms a right-angle with 9 and 11 Bootis. Some useful stars lie to the SE.

CAMELOPARDUS

This group, faint to the eye, contains a wealth of interesting objects. Clear nights will reveal its chief stars, only of the fourth magnitude; but this is a constellation with a great number of stars just beyond the reach of the naked eye.

Groups of stars

1. 03h 40m, +63°. Bright 5th-magnitude line. Two little lines of faint stars radiate from its easternmost member. Known sometimes as *Kemble's cascade*.

2. 03h 34m, +59°. Beautiful semi-circle of 7th- and 8th-mag stars.

3. GC 6288. This 5.4m star is the centre of a good area for sweeping.

Wide doubles

BK. Also called A Cam., this forms a fine pair with OΣ52, a telescopic double star.

11. This 5.3m star forms with its neighbour 12 a pretty object of contrasting colours, yellow and blue.

Close doubles

OΣ54. A difficult pair of 7.4 and 9.1 with a brighter double directly S.

Σ396. A slightly brighter, but very close double at only 20 seconds distance.

OΣΣ36. This double, separated by 41", is the apex of a triangle that points to group (1). A similar South-pointing triangle is close by.

OΣΣ39. Composed of two stars 59" apart whose magnitudes are both 6. Can you see any colour here?

β. An easy object for average-to-large glasses. Magnitudes 4 and 8, distance 80 arcseconds. A blue and yellow pair.

K. Closer at 34". Note a strange little group a degree south.

OΣΣ90. A triple star of magnitudes 5, 8 and 8. Rather isolated.

OΣΣ117. Again an isolated object. The mags are 6 and 8, the distance 65".

Σ1694. A close pair at only 22" but bright and roughly equal. A fainter but wider double lies closely NW.

Fig. 23: Red stars in Camelopardus. Field size is about 5° and the magnitude limit is 9.

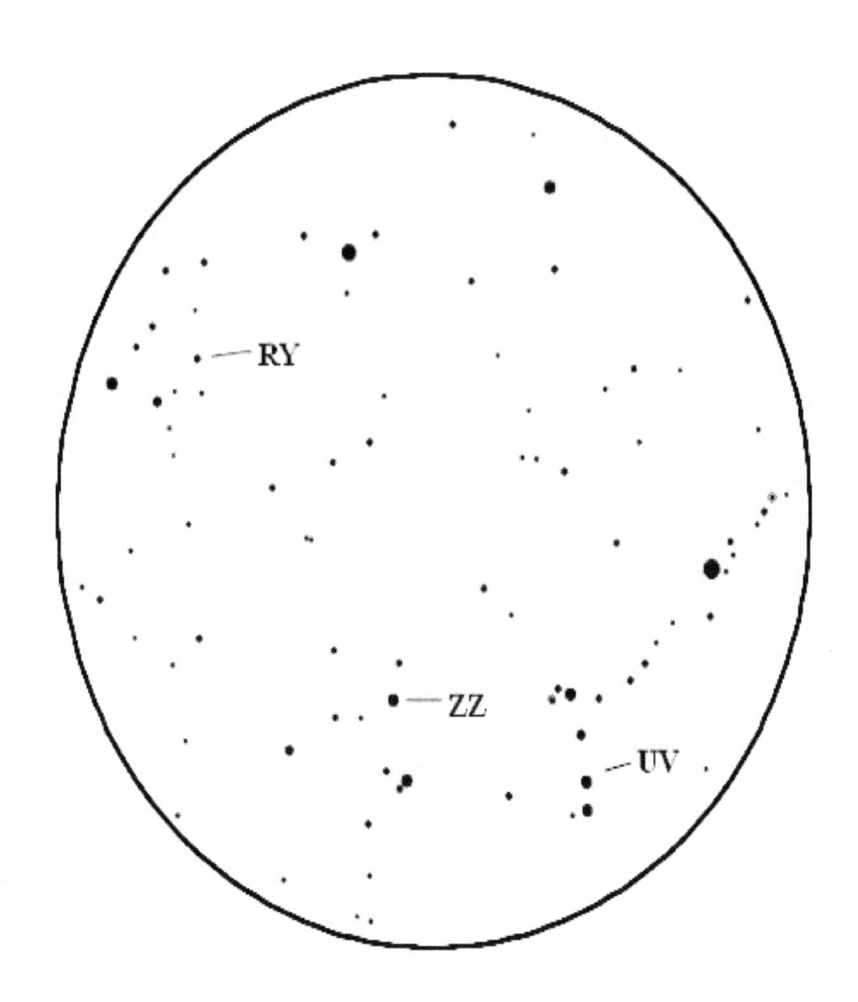

Variable stars

U (7.7-8.7) A beautiful red star with a rough period of 400 days, so you only need to make one observation per month. It is near group (1), as are UV and ZZ following.

RY (7.3-9.4) This lies near the Eastern member of a line of three bright stars. The figure on the previous page shows the immediate area.

ST (6.0-8.0) A very easy star to observe, as it lies in a small pentagon near α, whose other members are of 7.0, 7.1, 7.3, 7.8 and 7.6m.

UV (7.8-8.4) This is one of a Y-shaped cluster near group (1) although its small amplitude makes observation less easy than some.

UX (7.8-8.8) This star is 3° North of α, not far from ST Cam, in a little, faint inverted Y. It has a *comes* of 8.8m which hampers observation with many binoculars, and an 8.6m star is closely South.

VZ (4.7-5.2) You might like to observe this star with defocussed binoculars because of its brightness and red colour. A star for the smallest of glasses, or even the naked eye. Too small a range for my liking. Look at it once a month.

ZZ (7.1-7.9) A red star lying in a neat little cross-shaped group. It is the most Northerly star in the cross. At the centre is a star of 7.7m which is useful for comparison purposes, with the star at the Eastern end (the foot of the cross, as it is lying on its side) being of 7.0m.

Clusters and Nebulae

NGC 2403. A spiral galaxy visible in 7x50s as a brightish oval nebulosity.

CANCER

A nebulous group to the eye, but containing many fine fields in addition to a notable cluster. In some 17th-century star maps, poor old Cancer is depicted not as a crab, but as a lobster - with a small counterpart, indeed a shrimp, *Cancer Minor*, between Cancer (Major) and Gemini!

Groups of stars

1. β. 1° W of this orange star, and closely S. of a 6m, there is a small circlet of the 7th magnitude downwards.

2. 07h 58m, +10°. Beautiful little group.

3. 07h 58m, +17°. Large collection of bright stars, including 3 (5.8) and 5 (5.9).

4. 07h 54m, +12°. Symmetrical arc of five faint stars.

5. $\sigma^{1,2,3}$. Included with these are many fainter stars. There are other attractive groups near the borders with Gemini and Lynx.

Wide doubles

ω. A wide pair in a fine field.

β. This 3.8m star has a bright neighbour to the SE. Directly East is a delicate little triple.

ρ^1 . Of 6.1 magnitude, this makes a pair with BO Cnc, a red variable. Well worth looking at with small glasses.

ξ. This forms a much closer pair with 79 (6.1).

Close doubles

ι. A beautiful object of yellow and blue. Distance 31 seconds.

P.224. A good, easy object - magnitudes 6.6 blue and 6.7 yellow. The distance is 229".

P.226. Much fainter and closer, this is composed of 8m stars 144" apart.

P.228. Slightly more testing at mags. 6.8 and 8.0. The main star is orange, and the separation is 194".

P.230. A difficult triple star. The faint members, both of 9th mag., are 132 and 179 seconds from the 7.3m main star. There is another faint triple a degree Southwest.

Variable stars

X (5.9-7.3) A deep red star, lying near d between two comparisons of 6.3 and 6.8, with a 7.1m above the latter. An excellent star for the beginner.

RS (5.5-7.0) Another bright red variable well-provided with useful comparisons.

RT (6.9-8.0) This semi-regular variable is indicated by two stars in a line to the South which point straight at it; their mags are 8.6 and 8.0.

RX (7.5-8.9) A similar object in a crowded field. A chart is provided for this star and also the nearby, and brighter, BL Cancri.

VZ (7.2-7.9) Easy to find, this is an RR Lyrae star with a regular period of 0.178 days, or only 4 hours 15 minutes! This means you could follow a whole cycle during one evening's observing. It lies exactly halfway between 36 and 49 Cnc, the latter of which is slightly variable and is also called BI Cnc. It is a shame that VZ has few useful stars nearby.

Clusters and Nebulae

M.44 (NGC 2632) Known as the Beehive, this is the only cluster in the sky to have a hairdo named after it. A perfect object for small glasses, which will reveal many triples, doubles and streams of stars. Looking at this cluster with the naked eye certainly shows you why it is called the Beehive - but you need binoculars to see the bees!

M.67 (NGC 2682) A small, but prominent, cluster between 50 and 60 Cnc. Appears as a misty patch.

CANES VENATICI

An apparently dull group to the eye, but holding some fine doubles and bright variables. The hunting dogs themselves were called Asterion and Chara. β is still occasionally given this name in some astronomy books.

Groups of stars

1. The region around 9 and 10 is quite rich in stars.

2. 12h 32m, +37°. Line of three faint stars below a 7m.

3. 2, 4 or AI, 6 and β form a large quadrilateral with many fainter stars within its perimeter.

4. 20 (4.7). Other bright stars around this object.

5. The area bounded by Cor Caroli (α), 20 and 14 is quite rich - at least for this part of the sky!

Wide doubles

7. Magnitudes 6 and 8. One of a right-angle.

15 and 17. A wide, equal pair in a fine field.

Close doubles

Σ1607. Two 8m stars separated by 33 seconds of arc.

α. 2.9 and 5.4. A closer double of similar hues, i.e. yellow / reddish.

OΣΣ125. An unequal pair of magnitudes 5 and 8. Distance 71".

Variable stars

V (6.8-8.8) An interesting red variable lying between two 6m stars. Observe it twice a month.

Y (5.2-6.6) With a rough period of 158 days, this star is of a beautiful red colour, which led Secchi, in the last century, to call it *La Superba*. Y itself lies a degree North of a comparison star of 6.3m.

TU (5.8-6.3) This lies in the same field as Y, and is likewise a degree North of a useful star, this time of 6.2m.

Clusters and Nebulae

M.3 (NGC 5272). A globular cluster which appears in small glasses as a nebulous star.

M.51 (NGC 5294). This is a distant galaxy, to be found as a misty patch of light near Alkaid, the end star in the bear's tail.

M.94 (NGC 4736) Another galaxy, appearing in 6x30s as a faint starlike object.

CANIS MAJOR

A rich constellation pointed out by the brilliant white *Sirius*, brightest of all the stars. Sirius was sacred to the Egyptians among others, and gives us the "dog days" when its rising just before the Sun presaged stifling weather, presumably because then Sirius would be aiding and abetting the Sun! As Smyth pointed out in 1881, however, the British dogdays "often commenced a fortnight after the veritable dog-days were ended". Somehow this does not surprise me. Peculiarly, several classical writers asserted Sirius to be red in colour, but there are no good astrophysical reasons for such a change. One theory supposes that Sirius could have been veiled by a cloud of interstellar dust, which does have a reddening effect upon stars behind it.

Groups of stars

1. 06h 16m, -20°. Large triangle of bright stars.

2. ζ. Beautiful areas near this star

3. 15 and π. Wonderful sweeping around these stars, especially with powerful glasses.

4. η. This is one member of a brilliant curving ray that extends to ω.

5. β238. Many bright stars around this telescopic double.

6. 07h 12m, -31°. Fine, singular collection of 6m stars.

7. 07h 25m, -32°. Smaller, brighter group that includes some fainter stars.

Wide doubles

Σ3116. This forms a wide pair with a red star, and there are others nearby.

10. The companion to this star is a faint double.

o^1 . A red star with a 6th-magnitude companion directly S. Two more companions to the N and SW.

π and 17. These form a wide double, with 15 (4.7) nearby. There is another very wide, bright pair a degree or so to the North.

Close doubles

η . Magnitudes 2.4 and 6.9. Distance 180".

ν^1 . A very difficult object. Distance only 15", so only to be tried for with large glasses, and even then not for the faint-hearted!

h3945. A beautiful coloured pair, red and blue; but still close at 27".

P.231. A faint (8th-mag) pair separated by 151 seconds.

P.233. Very difficult, this is a triple star whose primary of 5.8m overshadows the 9th-magnitude attendants, which are 131 and 216" distant.

P.237. Rather easier this time - magnitudes 6.7 and 8.3; distance 99".

P.239. A close pair in a fine area. Mags. 5.7 and 7.7 but only 26" apart.

P.241. Just North of κ , this is an unequal pair of mags 5 and 8 separated by 42 arc-seconds.

P.242. Wider but fainter, these stars are of 7.9 and 8.9. Distance is 132".

P.243. A fine object lying in a beautiful little group of bright stars. The mags are 5.4 and 8.0, and the distance is 99".

Variable stars

R (6.2-6.8) An eclipsing binary with a 1.1-day period. A small line of three lies closely SW. Their brightnesses are (N to S) 7.4, 6.8 and 6.6.

W (6.9-7.5) Easy to find from its deep red colour, this lies inside a fine triangle formed by β328 (see above under 'wide doubles') and two other stars of 6.4 and 7.0. A fainter star of 7.7m lies directly North of it.

VY (6.5-9.6) A peculiar "slow" variable lying between two bright stars. It lies just NE of one of these, from which a small line of 7.0, 8.8 and 9.2 runs Eastward below VY. This star is what is known as a 'hypergiant' and is in fact the largest star we know of. If it were centred in the Solar System, all the planets out as far as Saturn would be orbiting inside it!

FS (7.6-8.6) Another out-of-the-ordinary star, similar to the "shell stars" typified by P Cygni. These are all very hot, luminous objects. It has a companion of 7.5 closely East, and another of 9.0 directly N, while closely West of these is an 8.3m star.

Clusters and Nebulae

M.41 (NGC 2287). This open cluster is a fine object, especially in large binoculars, which may resolve some of its members. In small glasses, it appears as a ragged, glowing patch.

CANIS MINOR

A small, unremarkable group containing few interesting objects for us. Procyon is, like Sirius, a near-at-hand, bright star with a dense white dwarf companion, though of not so extreme a type. Shown on the map with Monoceros.

Groups of stars

1. 6 (4.9) is in a fine field of bright stars.
2. 07h 37m, +08°. Three faint pairs in a curve.
3. 07h 55m, +10°. Some sprinklings of bright and faint stars.

Wide doubles

δ^2 . Makes a wide pair with δ^3 . There are two fainter stars SE of the latter.

Close doubles

14. A fine triple, suited to large glasses. The distances of the two faint *comites* are 86" and 117".

P.246. A difficult, unequal pair of 6.7 and 8.7m. The distance is 184", with a third star of 8.6m also nearby.

CAPRICORNUS

A large, triangular group, well supplied with binocular objects, though observers in the Northern USA and Europe may have some difficulty here with the low altitude.

Groups of stars

1. Very large, Y-shaped group that includes the 4.2m θ.

2. δ (3.0) is the leader of a fine parallelogram.

3. ζ (3.9) and 36 (4.6) are members of a large, wandering arc extending to η (4.9).

Wide doubles

α . A beautiful naked-eye pair. With binoculars, note also two fainter stars.

ρ and π . Make a wide triple with o.

46 and 47. The first of these is a close, difficult triple; the second is a small-amplitude variable, AG Capricorni.

Close doubles

o. Very difficult at only 22 seconds of arc separation.

Variable stars

RS (8.0-9.1) A red star, with a small triangle of 7.3, 7.5 and 9.3 just North. RS itself has a companion of 9.0m.

RT (6.4-8.1) Located conveniently closely to the 6.0m star 4 Cap., I have supplied a chart for this variable in the appendices.

Clusters and Nebulae

M.30 (NGC 7099). A bright, starlike spot, this is a globular cluster close to the 5th-magnitude star 41.

CASSIOPEIA

An obvious group to the eye, and full of memorable objects and marvellous fields. The shape of this constellation lent itself to all manner of mythological interpretations; to the Inuit it was a lamp of stone, to the Arabs an open hand; to the Greeks a lady in a chair, while in the cosmogony of J.R.R.Tolkien's Middle Earth it appears poetically as

Wilwarin, the butterfly in his Elvish language. While there are five main stars to summon up the five digits of a hand, it certainly looks more like a butterfly than a lady in a chair, so the prize for imagination goes to Tolkien.

Groups of stars

1. Line of 3 bright stars, τ , ρ and σ. The second of these is an interesting variable star. A fine region.
2. Sweep around kappa, where there is a fine angular Y.
3. 00h 28m, +55° 20'. Small faint circlet.
4. ζ (3.7). Sweep southwards from here.
5. κ (4.2). Interesting area N., that includes 16, an unequal pair.
6. Large triangle of RU (32 Cas) 35 and GC1426. All three are doubles of varying distances and brightnesses.
7. Fine field around 40, 42, 48 and 50, all of magnitudes 4 to 5.
8. Large, fine arc of 31, ψ , 43 and ω .
9. ε . This has three faintish companions directly E, plus a bright triangle to the NE. The region between this star and 35 is rich in doubles, lines, and groups of faint stars.
10. An area of brilliant groups between d and the previous star that includes M.103.

Wide doubles

1 and 2. A bright pair in the area of the well-known variable V Cas. The fainter of the two (i.e. 2 Cas) has an 8th-mag companion.

σ. Another wide, bright pair. The companion is of 5.7m.

ζ. This has a 5.1m star NW, which in turn has another again NW.

21 and 23. 21 is the variable YZ Cas. A poor field.

$\upsilon^{1,2}$. A faint star between the two is a good test for average-to-large binoculars.

Close doubles

35. A difficult pair (B rather faint in 20x70s) though 52" apart.

OΣ33. A more equal but closer pair at only 25"

Σ163. Quite a difficult object, but of contrasting colours. Note a bright little triangle to the East.

OΣΣ26. A much easier object, though harder to find. Mags are 6.9 and 7.4 and the separation is 64".

OΣ496. The primary is AR Cas, of small range, and actually a quintuple star. Only one of the four companions is seen with bins at 7.1m and 76" distance.

Variable stars

R (6-12) Even though this star reaches 6th magnitude normally, I have seen it as bright as magnitude 4! It lies about a degree North of a 6.8m star which is useful as a comparison, though there are few brighter stars in the area. R Cas is a very red star, and predictions are given in the appendix.

RZ (6.4-7.8) An eclipsing binary of period 1.2d. A star of 7.7m lying a degree to the W. can be used to estimate RZ when at minimum.

WZ (6.9-8.5) A beautiful deep red star lying in a triangle of 6.2, 6.4 and 6.6m stars. WZ has a close blue *comes* of 8.4m which can prove a nuisance with small glasses when the variable is near minimum.

IM (7.7-8.5) When I began drawing up a chart for this star, the figures in the variable-star catalogue led me to believe that this would be a telescopic variable. When I turned my telescope on it, however, it proved to be much brighter! I have supplied a chart for this red star.

V391 (7.6-8.4) This lies near group (7) as does the following variable. Two useful stars are close by; one between 42 and 48 of 7.7m, and another between this and the variable of 8.4. The three lie in a slightly irregular line.

V393 (6.8-7.9) A companion closely E is of magnitude 7.7.

ρ (4.1-6.2) This is a peculiar, immensely massive and powerful hypergiant star which tends to spend most of its time around magnitude 5. It is one of those objects which are too bright for most bins, but just too faint for the eye to estimate well. Sigma (4.9) and tau (5.1) serve as good comparison stars. Rho Cas is 10,000 light-years away, and some even believe that it may have already blown itself apart in a Supernova explosion, although the light has not yet reached us!

Clusters and Nebulae

NGC 225. A faint smudge in 10x50, but in 20x70, some stars can be seen, bordered by a 9th-magnitude curving line.

NGC 457. A beautiful oval gleam 'attached' to ϕ. A few faint stars can be seen in large glasses.

M.103 (NGC 581). In 6x30s this appears as a diffuse spot. Using 10x80 glasses the Hungarian observer György Zajacz sees a "very faint, hazy nebulosity in the centre of four faint stars".

NGC 663. A good binocular target. In 20x70s, I see numerous stars, including two pairs and a triple before a white gleam.

M.52 (NGC 7654). A rather faint cluster, but in a fine field.

NGC 559. Visible as a misty patch near a long triangle.

IC 1805. A fine field for most glasses and well worth locating.

CEPHEUS

Just as fine a group as Cassiopeia, though not so obvious to the naked eye. Many beautiful Milky Way fields. Probably one of my favourite parts of the sky, as it contains many interesting young, irregular variable stars; most of Cepheus is an area of intense and highly-dynamic star formation.

Groups of stars

1. 6 (5.2). Closely East is a small quadruple star.

2. A degree NE of 9 (4.9) is a 6m star with a tiny group to its North.

3. Large group including 11, 16 and 24. Note the small faint Y near 24.

4. Slowly sweep the triangle bordered by ξ , ι and ζ

5. 00h, +86°. Large bright group near the pole star, similar in shape to the Pleiades, but much larger.

6. The little naked-eye triangle of δ , ε and ζ is very beautiful with even the slightest optical aid.

Wide doubles

η . About a degree South is a wide double of 6.0 and 6.1.

Σ2970. This telescopic pair forms a wide double with a 6.4m star which is also a telescopic double.

7. There is a fine orange pair of the 7th magnitudes just SE.

Close doubles

OΣΣ1. A wide pair of 7.1 and 7.9. Colours are said to be red and yellow.

Σ2893. A close but easy pair near group (3). Distance 28".

δ . The primary is the typical Cepheid variable; but that aside, it is a lovely coloured pair whose yellow and blue stars are 41" apart.

Variable stars

T (6.0-10.3) You will need large bins to cover the whole of T's range, but any optical aid will show it at maximum. It lies just North of a straight line of 7.1, 6.7 and 7.7, and two little stars of 8.1 and 9.2 point right at it. Predictions for maxima of this star are given in the appendix.

W (6.9-8.6) A red star found between δ and two stars both of 6.4m. The variable itself is one of a little North-pointing isosceles whose other members are 7.0 and 8.7.

RU (8.2-9.4) A faint star in group (5), indeed rather too faint for most binoculars, I am inclined to think.

RW (6.2-7.6) This is one of a bright parallelogram whose other stars are 6.2, 6.4 and 6.6m. A good star for the smaller glasses and incidentally one of the most luminous stars known. Just to its West are two fainter stars of 7.3 and 8.0.

SS (6.7-7.8) A rather isolated far-north star which stands between a spoon-shaped line of 5-6m stars and a bright Y. Two useful stars of 7.2 and 7.6 lie to the NE.

VV (6.7-7.5) A remarkable eclipsing system with a period of twenty years! The large red component is more than 1000 times the diameter of the Sun, so that if it were placed in the centre of our system, all the planets out to Jupiter would actually orbit inside it.

AR (7.1-7.8) An easy star to find, in group (5). It is one of a line, lying between two stars of 5.9 and 7.1. A further 7.3m star makes an equilateral with these two. This star was thought to be an RV Tauri star at one time, though recent observations indicate a change of period from 324 to 364 days. More observations are needed to confirm this.

DM (7.0-8.2) Again easy to find, this red star has a distant companion of 8.1, with a wide pair of 7.4 and 8.8 just West of the nearby 24 Cephei.

EI (7.6-8.1) Though of small range, this bears watching as there is some evidence of variation outside its normal eclipsing behaviour. A 6th-mag star lies nearby, and between this and EI is a vertical line of 6.6, 6.9 and 7.2. The variable itself has a neighbour of 8.0 to the east.

FZ (7.0-7.6) Another small-amplitude star, this time a red one in a very dense field on the border with Cygnus. Worth finding for its colour.

GK (6.9-7.5) Near beta Cephei or Alfirk, this is a beta Lyrae star. A star of 7.2m lies on the opposite side of Alfirk, and two additional comparisons of 7.1 and 7.4 form a south-pointing isosceles with the 5th-mag 11 Cephei nearby.

Clusters and Nebulae

NGC 7160. A fine cluster, some stars being distinguished in binoculars. A wonderful region closely N and W.

CETUS

A appropriately large group for 'the whale' with some fine fields in its southern reaches, and of course the famous variable star Mira more or less at the centre of Cetus.

Groups of stars

1. Large group of seven, including AY or 39 Cet (5.5)
2. η (3.6). Note a group of four bright stars closely NW.
3. δ (4.0). A fine gathering just south of this star.
4. ν (5.0). An interesting region to the N.
5. The triangle of 2, 6 and 7 (AE Cet) outlines some attractive small groups of stars.
6. 00h 40m, -20°. A region of many bright stars, including a 5m triangle.
7. Striking trapezium led by π.

Wide doubles

77 and 80. The latter also has a faint star close by.

Close doubles

h323. A faint pair 65" apart. Note a faint diamond Northwards.

37. A fine green and blue double.

Σ150. Rather elusive at 36". A neat pair lies NE.

ΣI 6. A wide pair 81" distant. Easy with most glasses.

Variable stars

T (5.1-7.0) A chart for this bright, easy variable is supplied in the appendices.

o (1.7-9.6) These are the extreme values of Mira; the maxima in particular vary from one cycle to the next. All the well-known astronomical societies issue charts to follow it with; it is never out of binocular range for very long.

COMA BERENICES

A beautiful part of the sky to the eye, the lovely star-glow being caused by a particularly large cluster, around which is a much larger and far more diffuse sheen which is due to vast numbers of faint galaxies abounding in this area. Included in the map with Canes Venatici.

Groups of stars

1. Mel 111. This is the brilliant large cluster just south of γ(4.6), readily visible on a clear night. Use the smallest glasses possible on this group - a cluster in the last place you would expect to find one, as far removed from the plane of the Milky Way as possible!

2. 12h 50m, +19°. Small group of magnitudes 6 and 7. One of their number is triple (7.7, 8.2, 8.3).

3. α (4.2). Just to the NE of this star there lies a small collection which includes two wide pairs.

Wide doubles

3 (6.4). Closely S. is an unequal, wide double.

12 (4.7). This has two faint companions.

17. Otherwise called AI Com, this has a 7m attendant which is actually a telescopic double.

27. A 5.3m star with a close neighbour of 8.2. One of a bright triangle.

Close doubles

24. A difficult yellow and blue pair, only 20" apart.

Σ1678. Wider but fainter, this makes a triangle with 28 and 29. Near the latter is a slightly fainter close double.

Variable stars

FS (5.3-6.0) A rather underobserved star which would be good for owners of small glasses but for its poor light range. 35 Com (5.1) and 39 (6.0) make good comparison stars.

Clusters and Nebulae

There are innumerable distant galaxies in this part of the sky, where obscuration from the gas and dust in our own Galaxy is at a minimum, thus allowing us to see their faint light. Few, unfortunately, are good binocular objects. The most notable are M.64 (NGC4826), known as the "black eye" galaxy, and M.53 (NGC5024), a globular cluster. The two galaxies M.98 and M.99 are faintish, but could be worth hunting down.

CORONA BOREALIS

A group bearing a striking resemblance to its mythological model, though extending some way beyond the actual crown shape itself. Among its notable objects are two fascinating variable stars well-suited to binocular owners.

Groups of stars

1. Large, roughly triangular group of κ, λ and τ. An interesting area for sweeping around, especially near tau.

2. ε (4.2). Some faint groups to the SE of this star, near the strange variable T Coronae.

Wide doubles

ζ. This wide pair is made up of 5.0 and 6.0m stars.

η Two wide, but rather faint pairs a couple of degrees south.

ν A bright reddish pair with a third (8th-mag) star nearby.

Close doubles

Σ1964. A critical test-star for larger glasses, only 15" apart, and quite close to zeta. An eighth-magnitude companion lies NE. Both stars are of the same brightness, which makes separation easier than it would otherwise be.

Variable stars

R (6.0-14) The prototype of its class, affectionately known as "Sooty stars" because from time to time, and without warning, they eject clouds of Carbon which then condense into soot and cut off most of the star's light. Pollution on the grandest scale imaginable! This star thus needs constant watching - and amateurs with binoculars are the best people for the job. As I write these words, R Coronae is going through the deepest minimum in its history - magnitude 15!

T (2.0-10) In terms of its light-changes, this is a mirror image of the preceding star, and is known as a Recurrent Nova. It spends most of its time at its tenth-magnitude minimum (beyond the reach of most binoculars) but at long intervals undergoes a violent outburst. The last of these occurred in 1946, but the binocular watcher may well be the first to spot its next explosion. I have supplied a chart in the back of the book.

RR (7.1-8.6) Back to normalcy now, with a red variable which, together with its slightly less-variable neighbour SW CrB (7.6-8.3) can be compared against a nearby 8.1m comparison star.

CORVUS

A small though distinct group rather lacking in interesting objects. It is one of those constellations in which the Greek letter system seems to have gone somewhat askew, as its brightest star is not a, which is actually one of the lesser stars. Some nineteenth-century observers also had trouble with b, and thought it could be variable on a slow time-scale - a "secular variable". Have a look at Corvus and draw its shape, labelling the stars in order of brightness.

Groups of stars

1. Sweep around the area bordered by alpha, epsilon and zeta.

2. 12h 32m, -13°. Beautiful long inverted Y. The northern star is quadruple (6.8, 8.5, 8.7, 6.6) with another group of four 1° NE, whose members of 8.1, 8.3, 8.4 and 9.1 might be seen in large glasses.

3. 12h 36m, -19°. Bright 6th-magnitude trapezium, one of which is a wide double.

Wide doubles

ζ. This bright star has a 6th-mag companion.

δ and ε. Between these stars, in a barren area, is a 6m star that makes a fine pair with the Mira variable R Corvi (6.7-14) on those occasions when R is bright - a sort of "part-time double".

6. An attractive wide pair in a fine field.

β Note the delicate unequal pair to the SW.

Close doubles

P.249 This triple is wide but difficult because of the faint companions of 8.5 and 8.7, both about 160" from the 6.8m main star. Another 6th-magnitude star is close by.

Σ1669 A telescopic pair, this has a distant neighbour.

CRATER

Like Corona, this bears a good likeness to the object it is supposed to be; and like the preceding constellation, contains few interesting sights.

Groups of stars

1. 11h 00m, -12°. Large collection of fairly bright stars.

Wide doubles

ι This 5.6m star has a 6.8m orange attendant.

Close doubles

P.247 A wide, unequal pair (6.8,8.4; 205")

Variable stars

R (8.0-9.0) Easy with a telescope, but a close companion of 9.1, and the proximity of α, make this an object really suitable for powerful glasses with high magnifications. If so, you may be able to detect an even closer companion of 9.9m! Really a star for telescopes for these reasons.

S (8.2-9.2) A bit easier but still rather faint, and there are few useful stars in the area for comparison purposes.

CYGNUS

Certainly the best constellation in the whole Northern heavens for binoculars (or anything else for that matter) It contains many notable objects, including Cygnus X-1, prime candidate for black-holedom. Deneb, the leader of Cygnus, is a real celestial searchlight thousands of times more powerful than the Sun. It is also known as *Arided*, to my ears at least, a beautiful name for a star - but then Cygnus has it all!

Groups of stars

1. θ and ι . Brilliant fields around these stars.

2. 19h 50m, +47°. Rather singular bright Y of magnitudes 5 to 6.

3. Marvellous meandering line that includes ψ (4.9) and 20 (5.2).

4. Fine, large semi-circle ending at 33 (4.3).

5. NE of Deneb, near 55 and 57, are some of the most radiant areas of the Summer heavens. Dark skies and large glasses may pick out several nebulous gleams here.

6. π^1 (4.8). Note the beautiful line snaking away to the NW.

7. Large bright group, including δ and 14.

8. Long, sinuous trail, rambling through areas of breathtaking splendour, and stretching from 15 Cygni right through to 40.

9. γ . The central star of the cross marks probably the richest part of the constellation, and is a brilliant area to the naked eye. This whole region gives the impression of countless stars arranged layer upon layer, and the 19th-century Irish observer John Birmingham called it the "Red Region of Cygnus" because of the large number of red stars in the area. Excellent sweeping from here towards the fourth-magnitude η , which is in another magnificent region, star-clouds being visible with small glasses.

10. Large and bright irregular pentagon of 39, 41, 47, 52 and ε. Large bins, no light pollution, and dark skies will enable you to glimpse the Veil Nebula supernova remnant in the region of 52.

Wide doubles

θ . Directly E. is the well-known variable R Cygni which reaches mag. 6 at times. On the other side of theta is another star of 6.6m, and just West of this lies a large triple of 5.7, 7.8 and 8.3.

ω . A red star with a fainter associate. One of a bright group in a fine field.

γ . 1°N is a fine wide pair of 6.1 and 6.4, attached to the open cluster NGC 6910.

29. A star with a companion to the SW. Both these stars are also fairly wide doubles.

48. A beautiful equal pair, both white.

σ and τ . An unequal wide double forms a little triangle with these stars.

77. A 5.5m object with a seventh-magnitude attendant.

79. These stars are similar in brightness to the preceding objects.

μ . Directly North there lies a wide double of 7.3 and 7.4.

Close doubles

β . The famous Albireo is separable with binoculars. With colours of orange-yellow and blue-green, this is one of the sky's showpieces.

16. A fine, equal pair 37" apart. Beautiful in small glasses.

OΣΣ191. Distance 35", but the companion is faint.

26. Another unalike pair, though rather easier at 42" of arc.

31 (o^1) . A wonderful coloured triple. The faintest star is a beautiful clear blue, and the others are yellow. Amazing colours.

Ho 588. A faintish, closer pair near 39 (4.6). Separation 51".

OΣΣ207. A fine pair in a crowded field. Magnitudes are 6 and 8, 96" apart.

OΣ410. This is a rather dim, yellow double separated by 69 seconds.

61. A famous star as being the first to have its distance measured - and it was once thought to possess planets, so an interesting star for several

reasons. The binocular observer sees a fine, equal red pair only 25" apart, but even so, not too difficult.

Variable stars

χ (3 - 13). One of the first variable stars to be discovered - by Kirch in 1686. The range given here represents the extreme values we have observed. The maxima, especially, are not consistent, and usually chi will reach magnitude 4. At such times, it will probably be mentioned in astronomy magazines such as *Sky and Telescope* or *Astronomy Now*. The stars η (4.0) and 17 Cyg (5.0) will be useful around times of maximum.

W (5.0-7.6) A well-observed and interesting object easily found near ρ, with several good comparisons, notably two of 6.1 and 6.7 to the SE.

Fig. 24: Area of SS Cygni. Chart limit is about magnitude 9½ and diameter 3 degrees.

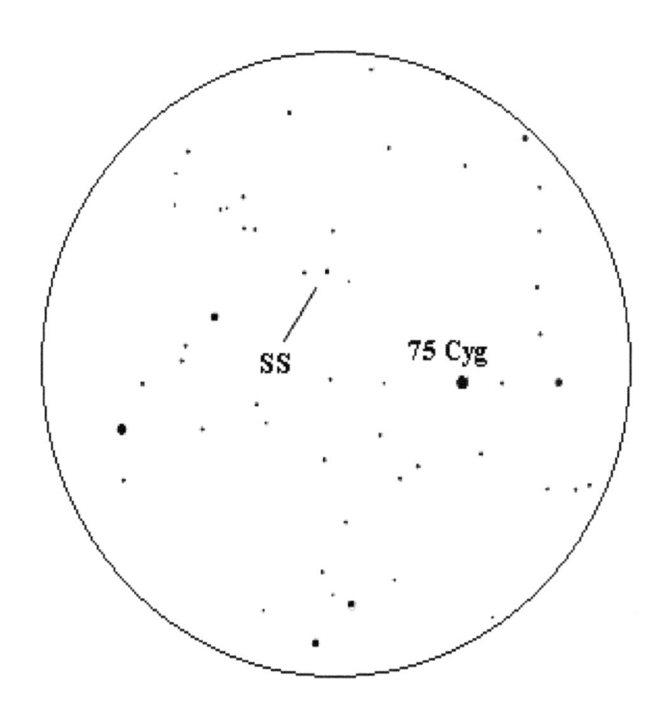

RS (6.5-9.3) A deep red variable in the "Red Region"; note the little triangle to the N. of 7.3, 7.5 and 9.0. Observation with small bins can be hard due to two rather close companions of 7.2 and 9.0. Observe RS once a month, appreciating its fiery appearance when bright.

RU (8.4-9.4) Probably too faint for average glasses, but in a fine area.

RV (7.1-9.3) In a small group around the wide pair 79 Cygni. The stars of 79 point to a star of 8.2, with another of 7.4 just to the South of RV. Another deep red star.

RW (8.0-9.4) A chart is supplied for this difficult object, yet again red, though not so obvious due to its faintness.

SS (8.0-12) This is the best-observed of the dwarf novae, and I have included it here since you might like to watch the area every night to see when SS pops up inside its little faint triangle. The diagram here will help you to find it; note the beautiful gold colour of 75 Cygni.

TT (7.4-8.7) Lying in a crowded area near the famous Mira star chi Cygni, this red star has a small line of three slightly SW whose two fainter members make good comparisons of 7.9 and 8.7.

AB (7.4-8.5) Rather isolated, this is the southern member of a vertical line whose other stars are of 7.9 and 8.2m. Closely North is a wide, equal double.

AF (7.4-8.7) Even though I use a 36cm Newtonian reflector to observe very faint variables, I still like to follow this star's variations, and certainly those of the next object. It makes a little isosceles triangle with stars of 7.1 and 7.6, and needs observing twice a month to give you an interesting light-curve.

CH (6.0-8.7) A very peculiar star, since a circular issued by the (alas now defunct) Binocular Sky Society called it an "Eclipsing Novalike Semiregular Variable"! I was doing a radio phone-in once when, as part of the intro, the presenter referred to "Celestial cannibalism". I was not too clear as to his exact meaning until he showed me the press cutting - CH Cyg had made the national newspapers! Once thought to be a run-of-the-mill red variable, we know now that this star is made up of two components which have evolved differently due to their different initial masses. Gaseous material passes between them which causes dramatic light changes from time to time. In 1968-69 this star brightened to magnitude 6, whereas in the late '80s and the early '90s it faded beyond its official limit to magnitude 9.2. An even more drastic fall took place two or three years after that, when it fell even farther to magnitude 10, way beyond its official range. A really fascinating and important star which you need to look at from one night to the next, recording times to the nearest minute.

V367 (7.1-7.7) An eclipsing binary, this is the W. member of a small right-angle whose other stars are of 7.5 and 7.9

V449 (6.3-7.1) Found near a 6.7m star, this is also one of a little right-angle, with the others being of 7.9 and 7.4.

V460 (6.1-7.0) A red variable, with a useful star of 6.6 to the NE.

V485 (7.2-9.0) Halfway between this star and eta Cyg you can find four objects of (N to S) 8.6, 8.4, 7.7 and 7.8. Another red variable.

V973 (6.2-7.0) An easy binocular variable, if of rather small range. It lies between two stars of 5.7 and 6.8.

V1070 (6.7-7.7) Another good star to observe. It makes an equilateral with stars of 6.1 and 6.4, with a fainter one of 7.5 the same distance to the East.

V1339 (5.9-6.5) This used to be used as a comparison for the nearby W Cygni until observers discovered its variability. Use the stars under W for this one.

V1500 (2.1-?) The strange Nova discovered by Honda in 1975, as well as by others (including the author). It was peculiar because of its very large range; most Novae have amplitudes of about 12 or 13 magnitudes but V1500 rose rapidly from below magnitude twenty right up to the second magnitude!

Clusters and Nebulae

NGC 6866. A cluster seen as a diffuse oblong patch.

NGC 6910. A misty spot near g in a superb field.

M.39 (NGC 7092) Visible with the eye as a brightening of the Milky Way, even the smallest bins reveal several stars. With 10x80, over twenty stars are visible. Telescopes are no good here as they have too small a field to give a good view.

NGC 7000. The North America Nebula, so called for its amazing resemblance to that landmass. Sandor Toth using 10x80s says "easy... hazy edges... brighter to the South and in the centre". A clear dark sky is an

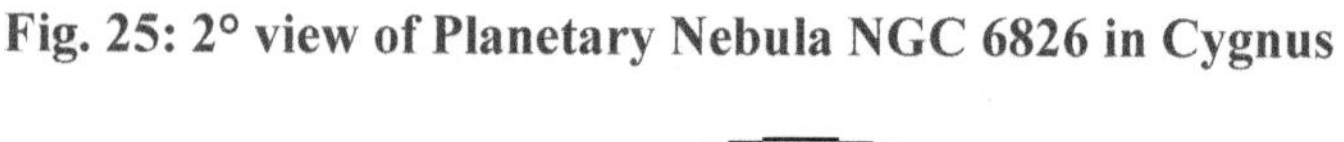

Fig. 25: 2° view of Planetary Nebula NGC 6826 in Cygnus

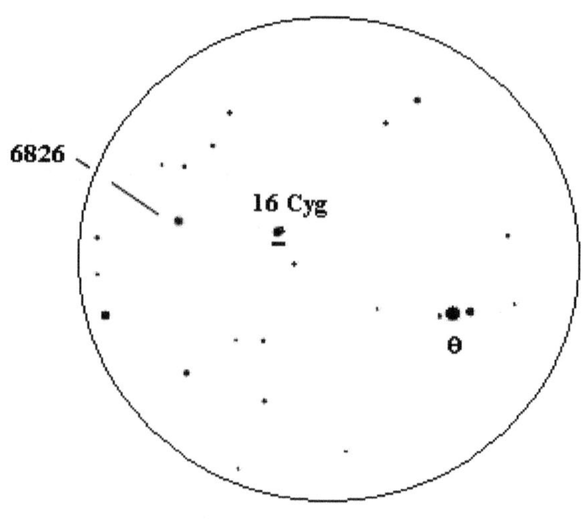

absolute essential for this nebula. The more elusive, and again well-named, Pelican Nebula close by has been seen with 10x80s.

NGC 6883. An open cluster, with three distinct stars seen with 8x30s.

NGC 6826. A planetary nebula in a fine area, for which you can use the chart here. Telescopes show a central star, about whose magnitude there has been some dispute, and the appearance of the star and nebula is such that when one looks at the star with a small telescope, the nebula becomes less distinct, and vice versa. For this reason it has become known as the "Blinking Nebula", although nebula and star will not be separately seen with ordinary binoculars.

DELPHINUS

A small but distinctive group near the Milky Way. Many fine fields and some good variable stars. However, one of the most peculiar things about Delphinus is that its leaders, α and β, have the peculiar names of *Sualocin* and *Rotanev* respectively. Antiquarians anguished over their etymology until wise old Rev. Webb pointed out that they spelled *Nicolaus Venator* backwards, the Latin version of the name of an assistant at Palermo observatory, Niccolo Cacciatore. (Or in English, 'Nicholas Hunter')

Groups of stars

1. Beautiful group, lying just west of 29 Vulpeculae (4.8).
2. 20h 40m, +19°. Small, square asterism. A pretty group with small glasses.
3. ε (4.0). Sweep from this star towards θ.
4. ρ Aquilae (5.0). A beautiful field of faintish stars just East of this object.
5. 13 and 14. Part of a large quadrilateral. 14 is a wide pair.

6. 20h 32m, +6°. Beautiful collection of 13 stars somewhat like the Greek letter Σ.

Wide doubles

Σ2665. A member of group (4). Mags are 6.9 and 7.2.

β. Note the tiny triple to the Northwest.

θ. Small wide triple E. A degree NE of this is a wide pair, while south of this is a small curve, of which the northern star is double. Directly East again is another close pair.

Variable stars

U (5.6-7.5) A well-loved red variable, with two useful stars of 6.3 and 6.8 between it and alpha Del.

CT (7.7-8.2) A red star with a rather small range for visual observers; but you can at least see when it is brighter or fainter than the 8.1m star to its Northeast.

CZ (7.8-9.0) In the same field, this has a companion of 8.9m, with a slightly brighter one of 8.5 closely Eastward.

EU (6.0-6.9) A well-followed red variable near U, whose comparisons can be used here as well.

DRACO

A long and rather dull group to the eye, but full of good fields and interesting objects for the binocular owner, especially in its Eastern borders.

Groups of stars

1. Bright triangle of 7 (5.7), 8 (5.3) and 9 (5.5) which also contains the red variable RY Draconis.

2. 15h 25m, +62°. A long, bright diamond.

3. Large trapezium that includes 15 Dra (5.0).

4. 27 and ω. A region of many bright stars to the binocular lenses.

5. χ. Fine sweeping in this area.

6. Brilliant groupings around 64 and 66 towards θ Cephei.

7. Fine, prominent group of delta, pi, epsilon and rho.

8. Large pentagon, including the bright stars χ, φ and tau.

9. 15h 44m, +55°. A fine region of several bright stars. Good for small glasses.

10. Sweep around the "head" of Draco; an area of many small stars.

11. γ. The brightest star in Draco is the guide to a fine sweeping area, especially to the East.

Wide doubles

η. This has a companion, itself a close double, just to the North.

19. With 20, forms a striking object for small bins.

16. Similarly forms a good object with 17, of the same brightness.

Close doubles

OΣΣ123. A lovely yellow and blue double.

ν. Said to be just splittable with the naked eye, binoculars show this to be a fine equal double.

ψ. Another good binocular pair. The colours are yellow and lilac and the separation is 31 arc-seconds.

Σ2273. A close (21") pair of the 7th magnitude. Use large glasses here.

Σ2278. Similar magnitudes, but wider at 39".

39. Rather difficult; mags 5 and 8, distance apart 89".

Σ2348. A hard pair of differing colours, and close at only 26 seconds.

46. A star with an attendant on each side.

o. This star has an eighth-magnitude comes thirty seconds away.

ΣI 44. A lovely gold and blue pair 77" apart. Note a faint foursome 1° to the west.

Variable stars

RY (6.7-8.0) Easy to find, this deep red star needs looking at once every three weeks. Use the stars in group (1) to estimate it when bright.

TX (6.8-8.3) Another easy star, one of a trapezoid near eta. There are two comparisons on either side, 7.2 and 7.9.

UW (7.0-8.0) A small right-angle just east of ξ (7.2, 7.5 and 7.8) is perfect for this star, which is orange instead of the usual red.

UX (6.2-6.9) One of the reddest stars in the sky, so beware how you estimate it. It lies near the star 59 Dra, on the other side of which is a good comparison of 6.5m.

VW (6.0-6.5) Quite difficult with small glasses because of a close companion of 6.7m. Like UW above, this is an orange variable.

AH (7.1-7.9) Another double, but rather wider. The attendant is South of AH, and is of magnitude 7.3.

AI (7.1-8.1) This eclipser makes a right-angle with two other stars of 7.1 and 7.7 that lie between it and μ.

AT (5.3-6.0) A bright star, well-suited to small glasses and lying halfway from η to θ.

Clusters and Nebulae

NGC 6543. A small but quite bright planetary nebula, with a wide pair 1°SE.

EQUULEUS

A small group with little to offer, figured in pictorial star-maps as not so much a Horse as a nag's head! Included on the chart with Delphinus.

Groups of stars

1. Large, fine group with attractive sweeping SE. It includes 1, 2, 3 and 4.
2. 21h 20m, +3°. Beautiful little group (6.6, 7.3, 7.4 ,7.6).
3. δ (4.6). Good sweeping to the N and NE.
4. 21h 00m, +13°. Small arc of eight faint stars.

Wide doubles

21h 07m, +07°30'. Small triple with an orange star of 6.4m to the S. A degree to the east is another wide pair of 7.2 and 7.4.

γ and 6. An easy wide pair of 4.8 and 6.0m.

ERIDANUS

A huge constellation with relatively few bright stars. Some of its length is too far South to be seen from most Northern countries, and is treated under the Southern groups.

Groups of stars

1. Bright triangle which includes the red variable CV Eri.
2. η, δ and 17. Sweep around these stars.
3. Large group, including 24 (5.1), 32 and 35.
4. o^2 and 39. Another area worth perusal.
5. Sweep the area bounded by Rigel, λ and β Eridani.
6. 60 (5.2). A region of numerous faint stars.

Wide doubles

ζ. This 4.9m star has a neighbour of 6.8. Forms a pair also with 14.

β. A star with two bright associates, 66 and 68. Attractive in small binoculars.

τ^6 . Between this star and γ is a fine wide pair of 6.4 and 6.6.

π Between this and 20 (5.3) is another pair, rather wider.

Close doubles

62. An easy object with a distance of 66". A is yellow.

P.255 Good for medium glasses, with mags of 7.2 and 8.4 and 145" apart.

P.257 This rather dim pair of 8.0 and 8.3 lies in a pleasant area.

P.258 Another faintish pair. Distance 122".

Variable stars

Z (5.6-7.2) This red star is one of a parallelogram whose remaining members are of 6.0, 7.1 and 7.9.

RR (7.0-8.0) Another red variable, two stars of 7.2 and 8.3 lying to the North-east, and making useful comparison stars.

CV (6.3-6.9) This is one of group (1) and is rather isolated.

GEMINI

A brilliant constellation with many striking Milky Way fields. Castor is a famous multiple star made up of no less than 6 components, though none of these are visible with bins. Gemini, as a constellation, has always been associated with a pair of something - whether Kids (Chinese) Boys (Greeks) Peacocks (Arabs) or Angels (some renaissance artists!)

Groups of stars

1. Long bright Y formed by ι, 59, 64 and 65.
2. Smaller, more regular Y between sigma and iota.
3. Large, bright arc near Castor, includes 70 and o.
4. Sweep from the red star 1 (4.3) to eta and mu, also red stars.
5. γ. Good sweeping from here towards Orion.
6. 41 (5.6), has beautiful clouds of faint stars to the S.
7. Tiny groups of faint stars also around 52 (6.0).

Wide doubles

39 and 40. A bright pair, with the latter a close double. There are two fainter and closer pairs 1° SE.

8 and 9. Two 6m stars with a bright triple (10, 11, 12 Gem) south.

ξ This star has two bright companions.

κ Directly West is a wide pair of 6.0 and 6.3.

Close doubles

OΣ 134. One of a small triangle. Distance 31 arcseconds.

ν An unequal but fairly easy pair (113") that makes a wide triple with 16 and 15. The latter is also a close double (6.5, 8.0; 30").

20. This equal pair is quite close but is in a superb field.

ζ. A fine binocular object, distance 94". The primary is a bright (naked eye) Cepheid variable.

Variable stars

SS (8.5-9.5) Owners of large glasses can follow this RV Tauri star by using the chart provided. Observe once a week.

TU (7.4-8.3) A red semi-regular, one of a trapezium.

TV (6.6-8.0). Near the star eta Geminorum in a field containing several red stars, which are shown on the finding chart on the next page.

WY (7.2-7.9). See under TV.

BN (6.0-6.6) A white variable not far from λ. The 6.3m 67 Gem is a good comparison.

BQ (5.1-5.5) The smallest of glasses will show this star, which like the previous object has rather too small a range for visual observation - though you can use 45 Gem (5.6) and 41 (5.8) nearby.

BU (6.1-7.5) 8 and 9 Gem (6.1 and 6.3 respectively) are useful when this star is near maximum.

Fig. 26: A field of red stars in Gemini, also showing the RV Tauri star SS Gem and the open cluster M35. Diameter 5°.

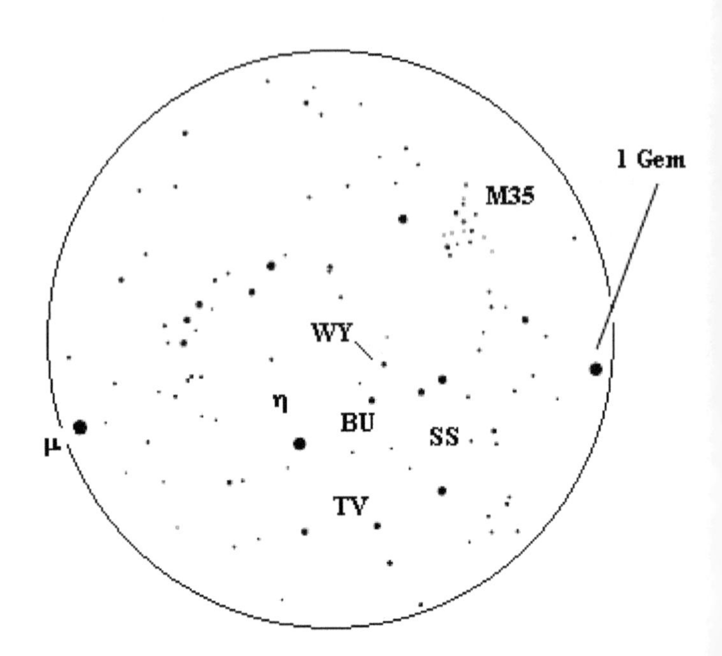

IS (5.3-6.0) Like BQ, suited to small glasses. SE of the nearby θ is a good comparison star of 6.0m.

Clusters and Nebulae

M.35 (NGC 2168). A magnificent sight in a telescope, and by no means disappointing in bins either. Anything larger than 8x30 will show some stars, superimposed upon a glowing gleam.

NGC 2129. Easily found near 1 Gem, large binoculars may reveal a few stars, though observers in this area tend to be more attracted by M.35!

HERCULES

A magnificent group brimming over with interesting objects including two notable globulars. It also contains the point in the sky towards which the Sun, along with the Solar system, is heading (the apex). Most old depictions of Hercules show him, for some reason, upside down; so it comes as no surprise to learn that the name of its leader, Rasalgethi, means "head of the kneeler" in Arabic, even though it lies at the foot of the constellation!

Groups of stars

1. The 4th-mag group of τ, φ, υ and χ. The latter has two neighbours, 2 and 4 Her, of contrasting red and blue colours.
2. 34 (6.2). Interesting sweeping to the S.
3. Range around ι (3.4) towards the head of Draco.
4. 16h 12m, +40°. Curved, inverted Y of 6th- and 7th-mag stars.
5. Large graceful curve of bright stars from ε to ρ.
6. 16h 50m, +43°. Beautiful collection of 6m stars.
7. Splendid group including xi and 99.
8. Beautiful group that includes 60 Her (4.9).
9. Another bright group, on the other side of Rasalgethi, that includes the small-range red variable V640.
10. Marvellous bright collection around 102 (4.3). This star used to be part of a now-redundant constellation called *Ramus Pomifer*, the apple branch, no doubt depicting the Apples of the Hesperides that our hero went in search of.
11. 18h 40m, +12°. Pretty line of four stars (7.0, 7.3, 7.9, 8.3).
12. Another magnificent group that includes 111 (4.4).

Wide doubles

42. A 5.1m star with a fainter associate. 2°S. is another wide pair.

77. A star with two fainter companions.

90. Another bright object, this time with 3 neighbours.

60. Note the wide double of 5.9 and 6.1 to the N.

8 and κ Magnitudes 5.3 and 6.1. The latter is a closer double.

83. A star with three bright companions.

17h 40m, +22°. Three wide pairs together.

Close doubles

κ A difficult pair; magnitudes 5 and 6, 31" apart.

γ. Slightly wider, but less equal in brightness.

36. This easy double has another pair to the SW.

Σ2277. A tricky pair in a fine field. Mags 6 and 8, distance only 28".

Variable stars

X (6.4-7.4) A well-known red variable, with a companion of 7.4. There is another good star of 6.6m a degree east.

ST (7.0-8.7) Also in the far North of Hercules, this is the eastern member of a small triangle whose other members are 8.6 and 8.8.

SX (8.0-9.2) A fainter, yellow variable, this has an equilateral triangle of 7.2, 8.4 and 9.0 to the East. Two other variables are nearby - RU, a Mira star reaching mag. 7 at maximum, and LQ, a small-amplitude red star, unsuitable for visual observing.

UW (7.5-8.6) Another yellow-orange variable, but of smaller range.

AC (7.0-9.0) This is one of the most rewarding binocular stars I know of, since it always seems to be doing something! Observe it once a week, as it is an RV Tauri star with quite a short period. A chart should be downloadable from the AAVSO.

IQ (7.3-8.2) There are several 7m comparisons close at hand for this red star, which is not too far from the preceding variable.

OP (6.0-6.6) One of a line of three (the others are 5.7 and 6.4m) which you will be able to pick out by its redness.

V566 (7.1-7.8) Again one of a line of three; the two stars to the South of it are 8.2 and 7.9m, this little line lying between two stars both of 6.3m. A good star to observe, though with a fairly small range.
V449 (8.0-9.0) This underobserved star has a northerly companion of 8.7m.
V640 (5.7-6.3) One of group (9). The other members are 5.2, 5.7, 5.9 and 6.2. See if you can work out which is which - a good exercise in estimating magnitudes.

Clusters and Nebulae

M.13 (NGC 6205). The famous Hercules globular cluster betrays a sizeable disc in binoculars, the bigger the better - though of course you won't be able to see its individual members, as high magnification is needed. Two 7m stars close by make recognition easy.
M.92 (NGC 6341). Though slightly dimmer and smaller than M.13, this one is still worth finding, even if it is in a more out-of-the-way area.

HYDRA

This largest of the constellations contains several interesting variable stars, though little else for its size. Of the many constellations which sit on its coils, two have since been dismissed from the lists. These are *Noctua* and *Felis,* otherwise known as the Owl and the Pussycat. I think their demise is rather a shame.

Groups of stars

1. Sweep the head of Hydra; some fine areas.
2. τ^2 . Interesting area to the SE.
3. Alphard. Beautiful sweeping around this fine orange star.
4. λ (3.8). Another region that rewards careful examining.
5. C (4.1). This forms a fine group with 1, 2, and other fainter stars.

Wide doubles

23. A star with two fainter attendants.

27. Note a 6m star close by. Can you see any colour here?

37. Two degrees West is a similar pair.

χ^1 This makes a fine wide pair with χ^2 .

Close doubles

Σ1255. A test object for binoculars; mags 7 and 8, distance 27".

P.260 A pretty pair of 5.9 and 7.1, both orange. Distance 67".

P.262 This is the western member of a little lozenge-shaped group of 5m and 6m stars. Separation is 150" and mags are 7.6 and 8.4.

P.265 A difficult pair in a fine area. Mags 6.9 and 8.4, 71" apart.

Variable stars

R (4.5-10) This was one of the first variables to be discovered - in 1704 by Maraldi. It is also interesting in that its period has decreased in these 300 years from 500 days then to about 400 days now. A very red star, for which predictions are provided.

U (4.8-5.8) An easy variable inside a wide trapezium of 5.5, 5.9, 5.9 and 6.3.

W (6.0-9.7) The best guides to this star are actually 1, 2, 3 and 4 Centauri! Come 4° North and you will find two stars of 6.1 and 6.3 with which W forms an isosceles triangle. When bright, this star is easy, but it has a close 9th-mag companion which needs a telescope for proper estimates to be made. Note a useful little line of 8.2, 8.4, 8.9 and 8.7 just S of the 6.1m star.

Y (6.9-7.9) This is one of a large Y (others are 6.3, 7.2 and 6.1). A red variable.

RV (7.5-8.7) Rather isolated, this forms the western corner of a diamond whose other stars are of 6.6, 6.9 and 7.8. RV itself has a neighbour of 8.7m.

RW (8-9) A reddish "symbiotic" variable, which should not be confused with an 8.4m star closely NE, though this is a good comparison star. The variations of RW Hya are usually small, however.

TT (7.4-9.2) An eclipsing binary making a triangle with two stars of 6.4 and 7.0m, with a pair of stars to the SE which are both 8.6. Its period is about one week.

FF (6.8-8.5) The two stars of 5.9m (see U Hya above) sandwich a 6.9m object. Closely N. is a 6.5m star near a wide pair of 8.2 and 8.8. A red semi-regular variable.

KN (7.0-9.5) A chart is given for this recently-discovered Mira variable.

Clusters and Nebulae

M.48(NGC 2548). A cluster marked by a few faint stars.

LACERTA

Although a small constellation, Lacerta is supplied with many beautiful fields, as it lies in a rich region of the Milky Way.

Groups of stars

1. Large bright group of α, β, 4 and 9. A fine sight, with a wide, 7m pair between the two first stars.
2. A beautiful line of bright stars linking 9 with EW (5.0-5.3).
3. 22h 51m, +40°. Brilliant group around a wide double.
4. 10 (4.9). Some magnificent groups to the S.
5. Sweep the triangle bordered by 11, 13 and 15.

Wide doubles

Bright wide triple (5.3, 6.2, 6.4) southwest of the variable AR Lac (5.9-6.7).

5 (4.6). Note the tiny quadruple closely SW of this orange star.

Ho.187. A faint telescopic pair. Note a wide triple just S, and a small faint arc W.

Close doubles

8. Good glasses show one, perhaps two, companions.

h1823. Easy; magnitudes 6 and 7, distance 82".

Variable stars

RX (7.5-9.0) A red star in a fine region, and with two useful stars to the NW of mags. 9.1 and 8.0.

SX (7.7-8.7) This lies in a crowded area between two stars of 7.0 and 7.2. It also forms a south-pointing equilateral with 8.1 and 9.1m objects.

AR (5.9-6.7) The triple mentioned under "wide doubles" is useful for this eclipsing binary.

Clusters and Nebulae

NGC 7209. A fine cluster, in which a few stars before nebulosity appear using 7x50 binoculars.

NGC 7243. The same calibre of glasses will show a fine, large star-group containing between ten and fifteen stars.

LEO

A magnificent group containing several good doubles and fine fields, considering its distance from the plane of the Galaxy.

Groups of Stars

1. Note faint collections of stars between ε and μ.
2. 10h 05m, +22°. A 6m star surrounded by many fainter ones.
3. o (3.8). Interesting sweeping, particularly W.
4. δ and β . Fine region between these bright stars.
5. Brilliant area which lies inside ϕ, υ, τ and 58.

Wide doubles

ζ . This has two fainter companions. Nearby gamma (Algieba) also has a fainter attendant in 40.

18. A 5.9m star that makes a pair with 19. When R Leonis nearby is at maximum, the three together make a good show in small glasses.

34. This has a fainter star close by.

44. A wide coloured pair of 5.9 (red) and 7.7 (yellow).

β 2°N. is a wide pair of 6.0 and 7.0.

τ and 83. Each member of this wide pair is actually a binocular double as well. A sort of "double double"!

Close doubles

α . Owners of good, large glasses might like to try Regulus, when feeling in a hopeful vein. Difficult (some would say impossible) due to the magnitude difference, but wide at 177". Mags 1 and 8.

83. A fine close pair of 30". The colours have been said to be yellow and lilac. Close by is τ, wider, though slightly fainter.

7. An easy to find, though rather faint pair. Separation is 42".

Variable stars

R (5-10) This popular variable, famous for its bright red colour, can be followed for most of its period with binoculars. Predictions are supplied for this variable.

Clusters and Nebulae

This is a fine group for the telescopic observer, as there are a great many distant galaxies here. Those which may be seen in bins are: Messier 96, 105, 65 and 66, but they will appear only as specks of light.

LEO MINOR

A small, insignificant group containing some binocular pairs, but little else. The only greek-letter stars it contains are, peculiarly, β and o! One other odd point of star nomenclature concerns the attractive little line of three bright stars in the East of the constellation. One member of the line is 46 Leonis Minoris, another is 46 Ursae Majoris. A strange coincidence! A little-known fact about Leo Minor is that it is a "real" constellation in that most of its main stars are physically associated with each other, and are also relatively close to us in space as well.

Groups of stars

1. Fine sweeping within the area of 8 (5.5), 10 (4.6) and 13 (6.0).
2. 10h 13m, +32°. Beautiful little symmetrical group.
3. 30 (4.8). A fine field that includes a wide red pair, P.268 (both 7.3m).

Wide doubles

10. Besides having 3 faint attendants, this has a brighter star NW.

Close doubles

P.266. Difficult, as rather faint at 7.4 and 8.5. Distance is 120".

P.268. Mentioned under group (3), the separation of these is 208".

P.269. Fainter and closer at 8.0 and 8.4, and 116 seconds apart.

P.270. Magnitudes 6.9 and 8.0, distance 132".

LEPUS

A beautiful little group both to the eye and binocular, though the Southern reaches can be rather difficult from parts of the Northern US and Europe. Some fine groupings.

Groups of stars

1. ι. Marks an area of fine sweeping, notably to the SW.
2. 06h 10m, -22°. Small group of bright stars, including a red one.

Wide doubles

β and μ . Lying between these is a wide pair, the S member of which is a close double of 38" separation, both blue.

Close doubles

h3780. Owners of good glasses should try this quadruple star, the *comites* being 76, 90 and 127 seconds distant from the primary. Note a bright vertical line to the E.

γ An easy pair of mags 4 and 6, 93" apart.

Fig. 27: Globular Cluster M79 in Lepus. Field size 8°

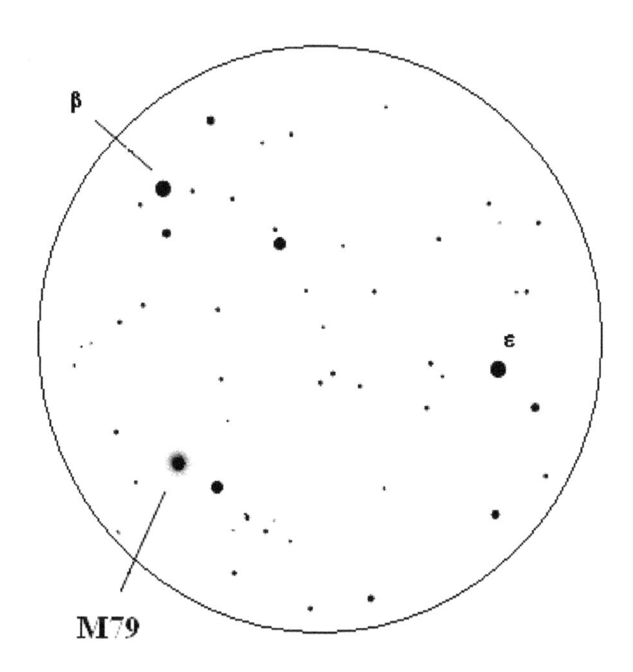

Variable stars

R (5.9-10.5) Called Hind's Crimson Star, this Mira variable is one of the front runners in the "reddest star in the sky" stakes. But you will need to catch it near maximum for the most impressive effects, so I have supplied predictions for it.

S (6.0-7.5) A good star for small glasses. A chart is provided.

RX (5.5-7.0) Another red star, forming a wide pair with the blue ι . A comparison of 7.0m lies close by.

Clusters and Nebulae

M.79 (NGC 1904). A globular cluster appearing as a faint burred star. Use the diagram on the previous page to find it.

NGC 2017. Four or five stars can be seen in this cluster, just E. of alpha.

LIBRA

A group containing little of interest to the binocular observer, though its leader, beta, is said to be one of the few stars which appear green. Try it with binoculars and see what you think. All the brighter stars of Libra have long names beginning *zuben-*, an arabic word meaning "claw" and harking back to the days when Libra was not seen as the scales, but the claws of the Scorpion just to the South.

Groups of stars

1. Rich area between 42 and κ.

2. 48 (4.7). An interesting region around this star.

Wide doubles

17 and 18. A fine sixth-magnitude pair.

α This has a 5m companion; good object for small glasses.

υ. A star with two wide companions.

Close doubles

Hh467. A difficult pair in a fine field. Mags 7 and 8, distance 47".

Variable stars

δ (4.8-6.2) An eclipsing binary, and one of those infuriating stars that are just too faint for the eye but too bright for bins! The stars 16 (4.6) and ξ^2 (5.6) are useful here.

ι^1 (4.3-6.0) An irregular variable. Its neighbour is of 6.0m, and a wide line of three (5.7, 6.1 and 5.7) lies to the South.

FY (7.1-7.9) This red star makes a triangle with ξ^1 and ξ^2 , and older editions of Norton's show it as *E-B 419*, so that it was in the catalogue of red stars compiled by two distinguished nineteenth-century observers, Espin and Birmingham. It is the northernmost of a little threesome whose other stars are 7.9 and 8.3m.

LYNX

An inconspicuous group to the eye, but possessing some fine fields in its circumpolar, North-West corner.

Groups of stars

1. 2 (4.4) is in a brilliant field, which includes a wide pair.
2. 06h 27m, +55°. Small group of 8 faint stars.
3. Bright parallelogram includes 22 Lyncis, N of which is a curved line.
4. 08h 02m, +36°. Large group of various magnitudes.
5. 08h 52m, +36°. A rich area resembling a large star cluster.
6. 42 (5.3) is in a fine region for sweeping.

Wide doubles

25 and 26. A fine pair of mags 5.7 and 6.4.

32 and 33. A similar wide pair.

Close doubles

OΣΣ93. A rather hard pair easily found from 31 Lyn. Magnitudes are 6 and 8 and the distance 77". What do the colours seem to be to you?

Variable stars

Y (6.9-7.4) A red star but of rather small range for us. Two stars just to the North are of 6.6 and 6.8m.

SU (8.0-8.9) Of similar type, this lies between two stars, one of which is 7 (6.4). This makes a triangle with two further stars of 7.7 and 9.2.

SV (6.6-7.5) A red irregular, with a line of three (6.2,6.3,6.6) to the S and a slightly fainter one of 7.1 to the East, with an 8.0 East of that.

LYRA

A small constellation but one with many objects of binocular interest - fine fields, doubles and variables. Some old star-charts also show this constellation as a vulture.

Groups of stars

1. ζ. Fine region all around this coloured double star.

2. γ. Another bright star in a rich area.

3. 18h 44m, +44°. Small group of 13 faint stars.

Wide doubles

ε. This is the famous double-double (though you won't see all four with binoculars). The naked eye can double this star, but bins give a much better view.

δ. A fine, contrasting pair, one of which is a small-amplitude red variable.

η. A faint, wide pair lies directly E.

θ. In the same field, this has a wide double to its SE.

Close doubles

Σ2380. A difficult pair in a fine area. Distance 26", magnitudes 7 and 8.

ζ. Yellow and blue with a distance of 44", this is a fine double.

β. The famous eclipsing binary has a 7m comes 47" away.

OΣ525. A faint pair near beta, with the same separation.

θ. A star with a faint companion 100" away.

Variable stars

R (4.0-4.7). This is really a naked-eye star, but bins will bring out its fine red colour.

XY (6.1-6.6) Another red star. Near Vega, with two comparisons of 5.8 and 6.3 just to the N. I remember one particularly clear night when, from my home in Norfolk, I observed this star with the
naked eye - the limiting magnitude was nearer 7 than 6!

HK (7.0-8.0) Difficult with small glasses due to a close-ish 8.4m companion, but in a fine area just south of the double-double, with a straight line of 7.4, 6.0 and 7.0 to its South.

Fig. 28: The Ring Nebula in Lyra. Field size is just 3°

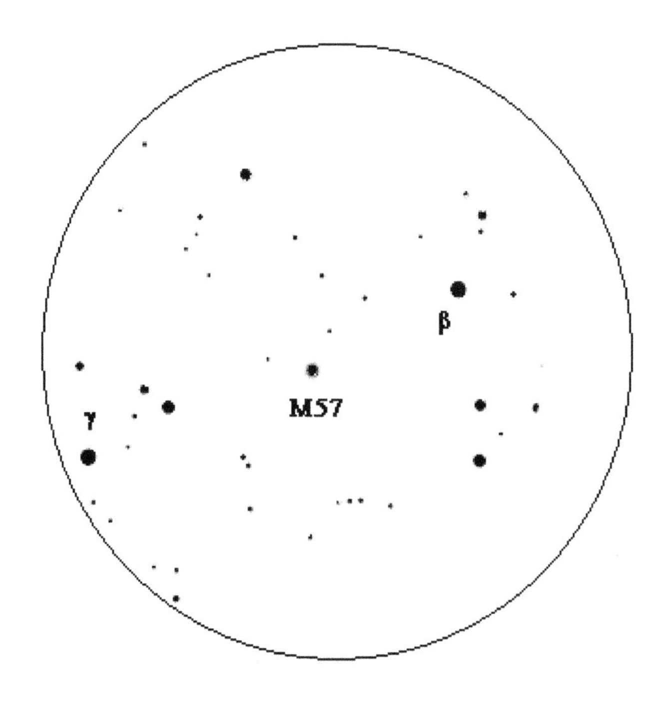

Clusters and Nebulae

M.56 (NGC 6779). A globular cluster, visible as a hazy spot.

M.57 (NGC 6720). The ring nebula. With large glasses, it can be seen as a faint star. Quite a difficult object, and in an area of several faint stars, so use the chart given here to find it.

MONOCEROS

A group replete with magnificent fields and clusters. It lies within the "Winter Triangle" of Sirius, Procyon and Betelgeuse.

Groups of stars

1. 12 (6.0), in the cluster NGC 2244, is surrounded by many brilliant formations.
2. 8 and 10. Sweep between these stars where you will find a bright group shaped rather like the constellation of Lyra.
3. 7(5.1). In a beautiful region.
4. Large triangle of 13, 15 and 17. Good sweeping here.
5. 06h 49m, -02°. Group of bright stars extending to 13 (4.5).

Wide doubles

07h 14m, -10°. Small triple, near the cluster NGC2353.

δ This 4.1m star makes a wide pair with 21.

Close doubles

Σ1183. A rather difficult, isolated pair. Mags 6 and 8, distance 31".

P.218. A faint pair in the central plane of the Galaxy. Mags 8.0 and 8.2.

P.219. A pleasant coloured pair of 6.5 and 8.3; 91" separates the faint star from its brighter blue primary.

P.223. A beautiful yellow and blue double of 6.8 and 7.0m. It can be found by drawing a line from ζ through 27 (5.1, orange) and prolonging it twice the distance.

Variable stars

U (5.6-7.3) An excellent star for the beginner, with good variations in a reasonable time. This member of the RV Tauri class is one of a little rectangle near α. The other members are 6.0, 6.6 and 7.0. Observe it once a week.

X (7.3-9.2) Another star giving good "value for money". A chart is supplied.

RV (6.8-8.3) A deep red star best found from the 6-7m arc a couple of degrees North. It lies between two stars of 6.5 and 7.1, and has a 7.7m neighbour.

RY (7.7-9.2) Another red variable, but with few good comparisons.

SU (7.7-9.0) A star of the rare spectral type S, 1° South of alpha. Take care to separate SU from its companion of 8.2 to the S. Just to the NW is a useful little triangle of 8.1, 9.3 and 9.6.

Clusters and Nebulae

NGC 2244. A brilliant cluster containing some bright coloured stars.

NGC 2343. Large instruments may show several little stars in this cluster.

NGC 2264. Appears as bright in the centre, with a prominent star to the South.

M.50 (NGC 2323). A large cluster, partially resolved in 20x70s.

NGC 2353. A really fine object, showing several stars in binoculars.

OPHIUCHUS

A really vast group, containing relatively few bright stars to the eye, but revealing many beautiful sights to the binocular lenses.

Groups of stars

1. The yeds, or δ and ε. Fine sweeping around these stars, whose full names are *Yed Prior* and *Yed Posterior*. Imaginative or what?

2. Beautiful sprinkle around 19 (6.0).

3. Another neat group is near the 4.7m υ.

4. The naked-eye T-shaped asterism of 66, 67, 68 and 70; this used to be a separate constellation called *Taurus Poniatowskii* back in the days when influential patrons of scientists were being elevated to realms celestial! A beautiful part of the sky.

5. θ. In a region of numerous bright stars, clusters and dark nebulae.

Wide doubles

72. Has two companions, one of which is orange.

14. This has an attendant to the NW.

θ. Note the orange star of 6.6m near this object.

Close doubles

Σ2166. A rather close pair (26") near Rasalhague.

53. An easier pair of magnitudes 6 and 7, 41" apart.

67. 55" apart, this is hard due to the 9m companion.

ρ. Binoculars reveal two companions. This star is in a large complex of dark nebulae which abound in this area. Best seen at high altitude and in dark skies.

ΟΣΣ164. A faint pair 50" apart, readily found between 71 (4.7) and a curving line of bright stars.

Variable stars

X (6.5-9.0) A bright LPV, this is near a large bright parallelogram. Note a little triangle just E, of 7.3,7.2 and 7.1. Maxima of this star are given in the appendices.

V533 (7.0-8.0) A red star near ζ Serpentis. A large triangle of 6.5, 7.3 and 8.2 to the North can be used to estimate it.

V1010 (6.1-6.8) A chart is provided for this bright eclipsing binary.

Clusters and Nebulae

IC 4665. A really magnificent sight in binoculars, which will show over a dozen stars. An easy naked-eye object.

NGC 6633. Another marvellous cluster. A rather V-shaped group of many stars with an irresolvable glow behind them.

M.12 (NGC 6218). A globular cluster, of which class there are many in Ophiuchus. It lies between two 6m stars.

M.10 (NGC 6254). Another globular. A nebulous point in 6x30's.

M.19 (NGC 6273). Yet another globular, as are M.9, 14 and 62. Try them all.

ORION

What can one say about Orion? The swordsman of the sky overwhelms us with his brilliant stars whether we use either naked eye or optical aid. The whole constellation teems with interesting objects - and is one of those rare instances in which the stars in a constellation really are, at least for the most part, connected with each other. Many wisps of nebulosity, and not just in the region of the famous nebula, either. Search out the whole constellation for these delicate objects when you have a suitably transparent night.

Groups of stars

1. The "shield" repays sweeping, even with the smallest glasses.

2. λ (3.7) forms a wide triple with ϕ^1 and ϕ^2 , with a pretty little straight line of three close by. Some nebulous gleams here, which is in what is called a "T-association" - areas of active star formation with many young variable stars in evidence.

3. Small arc below ξ (4.4) terminating in the cluster NGC 2169.

4. 71. One of a bright semi-circle.

5. Brilliant trapezium of 21, 23, 25 and ψ . Many bright stars here.
6. Sweep along Orion's belt, noting another fainter and less regular line to the South.
7. The equally famous sword is one of the most beautiful parts of Orion; the powerhouse of the whole group, if you like. Note two small sprays near the Nebula, M.42.
8. 56, 59 and 60. Members of a bright but isolated group. 59 is an easy double star, and 56 is orange.

Wide doubles

π^1 . A wide triple lies closely SW, of 6.1, 6.8 and 7.6. See if you can see any colour in the brighter stars.

72. A 5th-mag star with two yellow associates. Small group to the N.

π^3 The *lucida* of the shield has two attendants which form a right-angle.

π^5 . Forms a contrasted wide pair with the red star 5 Orionis.

ζ. Between this star and gamma Monocerotis is a wide triple of 6.0, 6.2 and 7.1.

Close doubles

OΣΣ58. Magnitudes 6 and 9, separation 89". One of a quadrangle.

Σ627. Easy to find, but hard to split at only 21".

23. One of group (5), this pair appears to me both white, though it is supposed to be yellow and blue. What do you think? Distance is 32".

δ. An easy pair, more definitely coloured, with a reddish secondary.

Σ747. A beautiful object near the Orion Nebula. 36" apart and both white.

Σ855. A more testing object of mags 6 and 7, 29" distant.

Variable Stars

U (6-12) A red Mira star which can become quite prominent, and which is very easy to find and estimate near maximum (see appendix). It is close to a bright trapezium, led by the 4.5m star χ^1;
the other stars are of 5.8, 6.0 and 6.6.

W (5.9-7.7) This deep red star is well-served with comparisons, including 21 (5.5) and a 6.3m just S. of rho. Directly between π^5 and π^6 is a fainter star of 7.9m. Use small glasses on this variable, otherwise it appears too red.

RT (8.1-8.9) A chart is supplied for this red variable.

Fig. 29: Irregular variable stars near the Orion Nebula

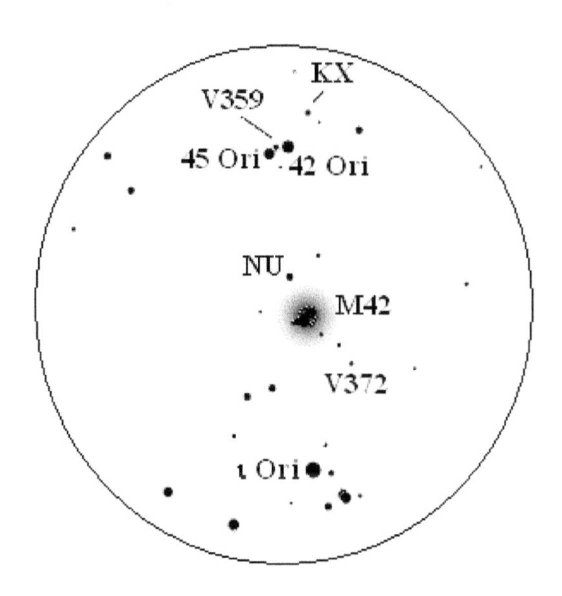

BL (6.3-6.9) This red star on the border with Gemini is one of a quadrilateral. Use the Northern members of 6.5 and 6.8 for comparisons.

BQ (6.9-8.9) There is a flattened trapezium just to the NE of 7.6, 8.6, 9.4 and 8.7. A good star to follow, with a decent range in light.

KX (6.9-8.1) This, along with NU, V359 and V372 below, appears on the chart. All these stars are eruptive variables, associated in some way with the Nebula; young, and recently-formed out of the vast cloud of gas they inhabit.

NU (6.5-7.3) See above.

V359 (6.9-8.1) See above.

V352 (7.4-8.9) As a relief from nebular variables, try this red star, which is in a line of four. West to East, they are: 7.1, 8.4, 8.6 then our star, which has a 9.1m just to the North.

V372 (7.8-8.5) See above.

Clusters and Nebulae

M.42(NGC 1976). The great Orion Nebula is easy with the naked eye, and binoculars of whatever size, but the larger the better, give an impressive view. Move the binoculars slowly around, as motion may well enable you to detect more detail than simply gazing. The diagram under KX above shows the area in close-up and includes several bright representatives of the class of Nebular

variables. One has to be careful in this area, as there are so many variable stars here that you never know if your comparison stars are variable themselves!

M.43 (NGC 1982). Visible as a condensation around NU Orionis.

NGC 2169. A cluster visible as a misty patch.

PEGASUS

Another superb group for the binocular owner, and far more interesting than could be guessed at by the eye alone, especially in its Northwestern border with Cygnus. A common test of naked-eye acuity, or the clearness and darkness of the sky (or both) is to see how many stars you can make out within the square.

Groups of stars

1. Wide group of the red stars 63, 72 and 73, plus 64 and 67.
2. 78 (5.0) has a fine region to the S.
3. 32 and η . Between these there lies an amorphous group of 12 stars.
4. Arc of eight stars between ϕ and γ.
5. The region of υ (4.6) is worth perusal with small glasses.
6. An interesting area around 2 (4.8) and 12 (5.5).
7. The telescopic pair Σ2841 is one of a large, diverse group.
8. ε . Note the tiny inverted V to the NW that points to a 6m star.
9. Fine, singular group of coloured stars - 55, 57, 58 and 59. The second of these is slightly variable, and is also known as GZ Peg.
10. Smaller group of 34, 35 and 37. Incidentally, all of these are telescopic double stars.

Wide doubles

π. Forms a pretty pair with 27.

61. A 6.3m star with a 7m companion. Making an isosceles with 60 and 61 is another double, but closer and fainter.

40 and 41. A wide, sixth-magnitude pair.

Close doubles

3. An easy coloured pair, white and blue, 39" apart.

P.271. An eighth-mag equal pair, separated by 141 seconds.

P.272. Two 7m stars 131" apart.

P.274. Another seventh-magnitude pair, slightly wider at 185".

P.275. Inside the square of Pegasus, this is a harder object of 7.9 and 8.4, and separated by 63".

Variable stars

TW (7.0-9.2) A chart is supplied for this red variable; a star with a persistent secondary period (i.e., the mean magnitude itself varies over a longer period of time than the main variation).

TX (7.7-9.0) A star of similar type, with an 8.7m companion. It makes a long triangle with a 7.0 and 8.7, the latter to the N.

AG (6.0-9.0) This star erupted in the late 19th century, and is nowadays content to fluctuate between 8th - 9th mag, but needs watching just in case. It is one of a small cross whose other stars are

7.6, 8.0 and 8.5.

GO (7.1-7.8) An easy star to find and estimate, since there is a small vertical line of three (7.0, 8.5 and 7.8) closely E. A bit of a shame that the range is so small, though.

Clusters and Nebulae

M.15 (NGC 7078). A brilliant globular cluster. In 20x70s, this is my favourite among these objects. Note a strange little Y to the South.

PERSEUS

After Cygnus, this is probably the finest group in the Northern sky for owners of binoculars, of whatever size. Its clusters are especially notable, several being in fact visible with the eye alone. Brilliant sweeping along the borders with Camelopardus and Cassiopeia.

Groups of stars

1. α (1.8). Lies in a glorious low-power field (see below).

2. Fine large group of λ, μ, 48 and b. Many fainter stars and little gleams in this area.

3. 43 (5.5). Beautiful sweeping around this star.

4. Sweep the triangle bordered by θ , ι and κ.

5. 20 (5.3). Fine sweeping for small bins around this area.

6. Small, rather square group closely E. of ε (3.0)

Wide doubles

29 and 31. A superb wide pair near Mirfak.

σ . A lovely wide triple, the other stars being 5.5 and 6.0m.

o. A star with a faint companion SE.

49. Forms a wide threesome with 50 and a third star which is actually a telescopic pair.

ε and ν . Making an equilateral with these is a 6m star with a wide, equal pair lying to the SE. A nice object for small glasses.

55 and 56. Another wide pair in a fine field.

Close doubles

η . This is said to be red and blue. The companion was equally said to be 8m, but it always looks at the very least, 9th-magnitude to me. A very hard one this, even for small telescopes, never mind binoculars!

OΣΣ44. A fine, both-white pair 58" apart.

OΣΣ47. These are 75" apart, with a third faint star forming a triple.

57. A lovely yellow and purple pair. Mags 5 and 6, distance 114".

Variable stars

T (8.3-9.3) A red star in a crowded field. Note a beautiful clustering nearby.

X (6.0-6.6) A hot white variable, made up of the main star plus an exotic neutron star. X Per is a strong emitter of x-rays because of this, and is readily seen in the smallest glasses. It makes an equilateral with ζ and a 6.1m star, with another of 6.6 closely NW of zeta.

TT (7.6-9.0) This red star has a wide pair of 7.4 and 8.4 NW, which are useful as comparisons.

XX (7.5-8.8) Another red variable, for which you can use two stars of 7.8 and 8.2, Northeast of the nearby 4 Persei. This and the preceding star lie close to a well-known red Long-Period Variable, U Persei, which reaches the 8th magnitude at maximum.

AD (7.7-8.4) Worth finding for its colour and also the fact that it is a member of the double cluster, as are several other red variables in the area, such as SU, PR and KK below. These have not been individually shown, as the area is too crowded. Just look for red stars near the clusters!

IZ (7.7-8.9) An eclipser between two phi's; phi Per and phi Cas. Difficult because of a close, faint companion.

KK (6.6-7.6) I watched this star for about two years in the mid-seventies without seeing it move very much from 7.9m. Similar behaviour is typical of many small-amplitude red variables. In the double cluster, so even if you have scant luck with its variations, you will at least get a good view of the cluster!

Clusters and Nebulae

NGC 869 / 864. The magnificent double cluster, one of the most beautiful sights in the whole Northern skies. Use as large a pair of bins as you can get hold of and you will not be disappointed. Note the presence of several red stars. Very easy with the naked eye as a brightening in the Milky Way.

M34 (NGC 1039). Another splendid cluster. 15 stars are seen in 10x80's, and I note that it lies in a long, irregular pentagon. Easy in 6x30 as well.

NGC 1528. A fine sight in average binoculars, and partially resolved in large ones.

Mel 20. A brilliant group of bright stars around Mirfak. (*Mel* is short for Melotte - Messier and the Herschels didn't bag all the clusters in the sky!)

Cr 29. Named from the catalogue of another cluster specialist, Collinder, this is a fine group of faint stars near the double cluster.

NGC 1245. With 10x or 16x50, you will see this as a misty spot.

NGC 1342. A fine cluster, which binoculars will partially resolve.

PISCES

A faint, dull group to the eye, but with some fine fields and pairs, especially in the North of the constellation near Andromeda.

Groups of stars

1. 82 (5.0) lies in a rich region.
2. Small isolated diamond of 35, 36, 38 and 41.
3. Beautiful bright Y composed of $\psi^{1,2,3}$ and χ .
4. The circlet. An asterism formed by γ, θ, ι, λ, κ, 7 and TX, the last of these a deep red star.
5. The group formed by 27, 29, 30 and 33. The latter two are red stars.
6. Small triangle of 73, 77 and 80. Many fainter stars around here.
7. ξ (4.8). Note a bright group W of this.
8. Small, bright assemblage including 1, 2 and 3. A close pair lies NE of the latter.
9. 23h 15m, -02°. Small faint Y, with a tiny line of three SW.

Wide doubles

τ . Two distant companions lie to the SW.

54 and 55. A conspicuous wide pair in a fine field.

χ Note a wide double of 7.0 and 7.2 to the SW.

ρ and 94. A notable low-power pair.

103 and 105. A similar but slightly fainter pair near the dim galaxy M.74.

κ . In the circlet, this has an orange attendant of 6.4m.

Close doubles

ψ^1 . A fine, roughly equal pair of the 5th magnitude. Dist. 30".

77. This star has a 7m comes 33" away.

ζ . A close (24") double of yellow and purple.

ΟΣΣ19. The distance here is 68" and the mags. are both 7.

Variable stars

Z (7.0-7.9) The little Y of 8.7, 7.8, 7.6 and 8.4 closely NW of this red variable make good comparisons.

RT (7.6-9.0). A fainter star in the same field, and for which the same comparison stars can be used.

TX (5.0-6.0) The stars in the circlet can be used on this deep red star. 22 Psc (5.8) is useful when TX is faint.

PISCIS AUSTRINUS

A compact group whose leader, Fomalhaut, can become quite bright from Europe and the southern parts of Britain. The rest of the constellation, however, needs altitude to reveal the numerous fine binocular double stars here.

Groups of stars

1. Bright, singular group that includes the 5m stars μ and τ.

2. λ (5.4). Fine sweeping north, and towards eta (5.4).

3. Large bright arc extending from ε to β .

4. δ . Lies in a fine field, with stars of assorted brightnesses.

Wide doubles

μ This has a companion of 6.4, also tau nearby.

δ . A degree north is an orange pair of 6.1 and 6.5.

Close doubles

h5356. A fine, easy pair of mags. 6 and 7 that are 85" apart.

P.211. Tucked away in the South of the group is this rather faint pair of magnitudes 7 and 8. Distance is 78".

P.212. A bright (6.3m) star with a 9m neighbour 235" away. On a dark night, owners of powerful bins may also be able to divide this star into 9.1 and 9.3m stars 78" apart.

P.214. A much easier object, both yellow and 7th-mag. Distance 142".

P.216. Easy to find near β , this is a good star for average glasses. Mags are 5.7 and 7.3, 91" apart. Both stars are orange.

P.217. An unequal pair of 7.4 and 8.8, distant by 108". It is the southern member of a line of three bright stars. Note the red tint of the Northern one.

Variable stars

V (7.7-9.1) This lies agonisingly close to Fomalhaut, between that star and 21 PsA (6.0). SW of V, and in line with it, are two stars of 9.2 and 9.0, with another of 8.5 South of the brighter of these.

Another star NW of the variable is of magnitude 7.6.

SAGITTA

A small but distinctive group containing some beautiful fields. It actually extends some distance east and west of the four little stars which form the arrow itself.

Groups of stars

1. Fine bright group of α, β, δ, and ζ . Note the curving line stretching from delta and running through...

2. ζ . Beautiful sprinkles around this star.

Wide doubles

ζ Aql. 3° North of this star is a notable wide pair of 5.7 and 6.4.

2 and 3. These make a similar fine wide double.

9. Note a tiny 9m. equilateral triangle closely NE. Near the cluster M.71.

15. A 5.9m star with a 6.8 companion.

Close doubles

ε . A beautiful coloured double 90" apart, and of mags 6 and 7.

Hh630. A harder pair of magnitudes 7 and 9. Distance 29".

Variable stars

U (6.4-9.0) An eclipsing binary; one of a small rectangle whose other stars are of 8.2, 8.3 and 8.5m.

BF (8.0-9.0) A red star in a very dense field. A chart is supplied, which also shows two other red variables; HU Sge, a small-amplitude red irregular, and a more interesting but fainter star, X Sagittae, which varies between magnitudes 7 and 9.7, so that you will need larger-than-average glasses to catch it at minimum.

SAGITTARIUS

The richest region of the entire sky, the centre of our Galaxy, lies beyond the glowing star-clouds of the Archer. Best seen from the USA or Southern Europe, the most notable part of the group to the eye is the large, symmetrical collection of second- and third-magnitude stars affectionately called the "teapot" because of its shape.

Groups of stars

1. 18h 00m, -22°50'. Small oval of brightish stars.
2. Small bright group, including 24 (5.7).
3. μ (4.0). Beautiful fields between this star and gamma Scuti.
4. o and π. Two members of a beautiful large and bright arc.
5. ω, 59, 60 and 62. Beautiful bright cross.

6. Large, regular arc, directly east of η .

Wide doubles

7 and 9. Both of these are closer pairs. 7 is the best - 6.9 and 8.5, distance 35". In the Lagoon Nebula, M8.

15 and 16. A third star is visible, thus making a small equilateral triangle.

o. A star with a 6m associate NW.

52. Forms a wide pair with 51, one magnitude fainter at 5.7.

53. In the same field, this is a closer pair of 6.1 and 6.2.

$\beta^{1,2}$. A fine, wide pair in a rich field.

Close doubles

GC25327. A close pair of magnitudes 5 and 7.

P.278. A lovely white pair of stars separated by 121". Mags are 6.5 and 7.0.

P.281. Though distant at 196" this object is quite faint - both 8th mag.

Variable stars

RY (6.5-14) A star of the unpredictable R Coronae type. It can be estimated at maximum with three stars of 5.6, 6.5 and 7.1 to the SE, and is closely North of a 7.4m star; when you feel the need

to use this star as a comparison, you will have to make this observation known, as RY Sgr will probably be entering one of its unpredictable fades. It also exhibits rather more regular, Cepheid-like variations with an amplitude of about 0.5m.

UX (7.6-8.4) A wide double of 5.8 and 7.2 points to UX, which lies a degree South of a 5th-magnitude star, closely SE of which is a fainter one of 8.0.

AQ (6.6-7.6) A chart is supplied for this deep red star.

V356 (6.9-8.0) An eclipsing binary near 29 (5.4), with a star of 8.2 between the two serving to note when V356 is at or near minimum.

V505 (6.4-7.6) A star of the same type, with a period of only 1.25 days. It is exactly halfway between 61 Sgr and a 6.2m star, and there is a fainter one of 6.9m closely SW of the variable.

Note : there are several Mira stars in Sagittarius that are easily visible with binoculars at maximum. I have provided predictions for the following examples:-R, RR, RT, RU and RV.

Clusters and Nebulae

M.8 (NGC 6523). The beautiful Lagoon Nebula, visible with the naked eye. It has an attendant cluster, but needs good altitude for an impressive view.

M.21 (NGC 6531). With 8x30's, you will see several bright, and many faint, stars before an irregular nebulosity.

M.24 (NGC 6603). Not so much a cluster as a rich star-cloud. Twenty stars are visible with 10x80. A bright elliptical glow in small binoculars.

M.22 (NGC 6656). A very bright globular cluster, looking rather like a comet, with a noticeably brighter centre in 8x30s.

M.25 (IC 4725). A dense cluster of several stars beyond which can be seen a nebulous glow.

M.17 (NGC 6618). The Omega or Horseshoe nebula. Imre Toth, using 10x80's, sees this as "Triangular, with apex to the South and the brightest part to the North... bright grey colour".

M.18 (NGC 6613). An open cluster appearing to the same observer as a bright grey nebulosity, rather whiter in parts.

M.20 (NGC 6514). The famous and photogenic Trifid Nebula has a faint greenish tinge, with several stars involved in it.

M.23 (NGC 6494). A nebula visible as a bright diffuse streak.

M.54 (NGC 6715). A moderately bright globular in a fine field near zeta Sagittarii.

M.55 (NGC 6809). Another globular, much brighter, but more isolated.

NGC 6530 Many stars are visible in this cluster, which lies in a really beautiful region.

NGC 6716 Again, some stars can be resolved, before a white glow.

SCORPIUS

Like the previous constellation, this is a magnificent group in every possible way. Also alike insofar as you need altitude to appreciate its best objects!

Groups of stars

1. π (3.0), is one corner of a large, bright lozenge.

2. Bright quadrilateral formed by ξ , 11, ψ and χ .

3. The "sting" is extremely beautiful with any optical aid. Its brightest star is λ (1.7).

Wide doubles

2 and 3. A wide but rather unequal pair in group (1).

ω . A bright pair in a brilliant field.

$\mu^{1,2}$. A similar beautiful pair.

ζ^2. This star has two distant companions. Amazing field.

Close doubles

ν . Two stars separated by 41".

22. A closer, fainter pair near Antares. Mags 5 and 7.

P.289. This is a good test object for average-to-large glasses, of magnitudes 5.7 and 7.7 but only 24" distance.

P.291. A beautiful equal pair, 96" apart.

P.292. Again equal, but 2 magnitudes fainter, the 8th-mag stars being 88" apart.

P.293. Rather closer at 65 seconds, but slightly brighter.

P.294. This seventh-magnitude pair is wider (162") and easy.

P.295. A fainter object, 217 seconds apart.

Variable stars

SS (7.5-9.5) An irregular variable of the orange type K. A chart is supplied.

SU (8.0-9.4) This rather faint red star makes a right-angle with two stars of 5.9 and 6.5. Between the latter and SU lie two useful stars of 8.5 and 9.0.

AK (7.8-9.3) A nebular variable near mu. A chart is supplied.

BM (6.0-7.9) Though brighter, this is difficult to estimate as it is in the brilliant cluster M.6.

FV (8.0-8.7) An Algol-type star, this lies between two bright stars (5.5 and 6.0). Closely SE of the variable is a small, neat triangle of 8.0, 8.3 and 8.5.

V393 (7.4-8.3) Near the wonderful cluster M.7, this eclipser is one of a small right-angle whose other stars are 7.4 and 7.6m.

V453 (6.8-7.3) Another eclipsing star, this time of the less-common β Lyrae type. A fifth-magnitude star nearby forms a triangle with two to the E., of 6.8 and 7.0.

V856 (6.8-8.0) A nebular variable, and a wide double. Its companion is of 6.7m and is found directly to the N. A useful little right-angle lies just SW of the variable. Its members are of 7.0, 7.9

and 8.0m.

Clusters and Nebulae

M.4 (NGC 6121). A large globular cluster, quite bright and conveniently close to Antares.

M.6 (NGC 6405). A beautiful coarse cluster that binoculars will partially resolve into about a dozen stars.

M.7 (NGC6475). This is one of the most superb of all binocular objects. Easily visible with the eye, your bins should reveal around 20 stars here.

NGC 6231 A cluster which streams out towards ζ , this is yet another showpiece of the Scorpion.

NGC 6322 This appears as five stars in a Y-shape before a misty patch.

SCUTUM

Although a small group, this contains some interesting objects in addition to the brilliant star-cloud easily visible to the naked eye. Many of the more recent constellation figures had double-barrelled names, such as Columba (formerly Columba Noachii, or Noah's Dove) or Sculptor (ex Apparatus Sculptoris, the Sculptor's Tools). Scutum was another of these, its former name being Scutum Sobieskii, named after a Polish nobleman. (That's a nobleman from Poland, not somebody who got rich quick from selling furniture-buffing products). It is shown on the map with the nearby Aquila.

Groups of stars

1. Long trail extending from the variable R, to the 4th-mag. α .
2. 18h 51m, -04°45'. Group of 8-9m stars in the form of a 7. A small line stretches from here to 7 and 8 Aquilae.
3. γ (4.7). A fine region E., including a red star.

Wide doubles

ε . A star with a fainter companion to the SE.

γ . This star has a bright associate closely E of it.

Close doubles

Σ2391. A faint, rather hard pair. Magnitudes 6 and 8, 38" apart.

Variable stars

R (4.5-8.4) A well-observed RV Tauri star which has been studied by amateurs for many years. It has recently emerged that this star displays chaotic behaviour, and those amateur observations, made by members of bodies such as the AAVSO, BAA and so on, are proving very important - a good example of how one branch of study feeds into another, this time amateur astronomy providing food for the mathematicians. Why not have a look yourself - you'll get a bonus view of the "Wild Duck" as well. R Scuti is one of a little quadrilateral (other stars are 6.1, 6.7 and 7.1m).

S (7.0-8.0) A deep red star; one of a flattened pentagon whose other stars are of (E to W) 7.0, 7.0, 7.3 and 7.9.

RZ (7.9-9.0) I have provided a chart for this eclipsing binary.

Clusters and Nebulae

M.11 (NGC 6705). A bright cluster visible as a hazy triangular patch. In small instruments, it well-deserves its nickname of the Wild Duck - it really does look like a flock of ducks flying in a V formation, with a brighter orange star near the centre.

SERPENS

The only double constellation in the sky, made up of Serpens Caput (the head) and Cauda (the tail) though objects are named as though the group were one. The idea is that Ophiuchus, who is struggling with the snake, has pulled the poor thing in half! That aside, Serpens contains many fine fields, notably in the E. In terms of star naming, both halves are treated as being one constellation, but each half is given its own chart in the maps at the end for clarity's sake.

Groups of stars

1. Sweep the "head" of ι, β, γ and κ. The last of these is reddish. There is another fine region just W, where there are no less than 8 stars bearing the Greek letter τ !
2. The area of 4, 5 and 6 is worth scrutiny with larger glasses.
3. ζ . This makes a neat triangle with two fainter stars.

4. η (3.4) is in a rich field of bright stars.

Wide doubles

ψ . This 5.8m star has two fainter associates.

47. A red star with two fainter companions, one of which is FQ Ser, slightly variable.

ξ . A bright pair, though separated by three mags.

64. An attractive, nearly equal double, both 6m.

Close doubles

Σ1919. Rather hard for binoculars, with a distance of only 25". The mags are 6 and 7, and this is said to be a coloured pair. How do they appear to you?

θ . Not normally considered to be a binocular double because of its rather small separation of 22", this is a superb equal pair which I have to say I find reasonably easy with 10x50s.

Variable stars

τ^4 . (5.9-7.0) A red star in the "tau's" and well-served with comparisons; τ^1 is 5.5, τ^2 is 6.1 and has two neighbours of 6.7 and 7.6.

R (6.7-13.4) A Mira star easily found between β and γ . Its period is very nearly a year to the day, and I have supplied predictions for it. R Ser is one of a regularly-spaced line of three (other stars are 7.3 and 7.4) and its redness at maximum will distinguish it.

Clusters and Nebulae

M.5 (NGC 5904). A large, blazing globular near the 5m star 5 Serpentis. A real bunch of fives! With small glasses, a bright nebulous star.

IC 4756 This is a really splendid cluster, appearing as a double row of stars of assorted brightnesses. The southern branch is more noticeable; a good object to try and draw.

M.16 (NGC 6611). Long-exposure photographs of the Eagle Nebula have become practically *de rigueur* in space documentaries, science-fiction movies and the like. The Hubble Space Telescope recently provided breathtaking pictures of this object as a giant stellar nursery, inside which one could lose the entire solar system with no difficulty at all. With giant telescopes it really is impressive of course, though the binocular observer will have to be content with rather less. Using 10x80 glasses Imre Toth says of it "the North edge is the brightest part, silver-grey in colour". There is also an attendant cluster, some stars of which you may see.

SEXTANS

A small, faint group below Leo, with few objects of interest for the observer using binoculars. Included in the chart for Hydra.

Groups of stars

1. ε (5.4) is in an attractive region of fairly bright stars.
2. 25 (6.1) lies in an area scattered with wide pairs.
3. 36 (6.3) Note a tiny tick-shaped group just East of this orange star.

Wide doubles

17 and 18. An attractive red and blue object, good for small bins.

25. About 1° south is an equal (7.0m) wide double.

41. Note an 8th-mag companion SW. A fine field with large glasses.

Close doubles

9. A faint but wide pair. Mags 6 and 9, distance 52". Primary is red.

Variable Stars

RT 7.9 - 9.0. A red semi-regular with a 96 day period, that can be found 2 degrees North of the 3.6m star λ Hydrae. A degree South of RT are two useful stars of 8.4 and 9.0 mags, while pointing to these from the East is a wide pair of 6.0 and 8.0.

TAURUS

A splendid group, containing two spectacular clusters in addition to a host of interesting objects of all types. Its leader, Aldebaran, the eye of the bull, is a suitably angry red, with the Hyades cluster as a sparkling background. Taurus has everything for the binocular observer; beautiful wide double stars, interesting variables, and several notable clusters.

Groups of stars

1. Fine, brilliant group that includes μ (4.3) and 47 (5.0).

2. Small oblique cross of ω , 51, 53 and 56. Note a small circlet to the East.

3. Not far away is another singular bright group of κ, 67, υ and 72.

4. The Hyades. Note especially the numerous wide pairs and triples around this area. A good first object to draw; begin with the two faint lines marking the arms of the V, then add the stars. Note that another arm extends southwards from Aldebaran, making the famous V into a not-so-famous italic N!

5. Brilliant, long, meandering line beginning at 120 and ending at 134.

6. Slightly SE of 12 (5.8) is a tiny "cluster" of four stars.

Wide doubles

21 and 22. A white, unequal pair in the Pleiades.

27 and BU. BU is Pleione, in the Pleiades, and is said to be one of the few single stars that are noticeably purple in colour, though it always looks white, or at best blue, to me.

37. This forms a wide pair with 39.

$\theta^{1,2}$. A beautiful naked-eye pair in the Hyades.

$\sigma^{1,2}$. Another beautiful wide pair. A third star is visible.

10. This has a fainter companion to the N, which is in fact a small-range variable, V711 Tau.

Close doubles

Σ I 7. A 7th-magnitude pair 44" apart.

φ . Difficult, because of the magnitudes of 5 and 9. Distance is 53".

62. Another unequal pair, this time of magnitudes 6 and 8 with a distance of 29".

88. Easier though still unequal; mags 4 and 7, 69" apart.

τ . Distance in this case is 63" and the mags are 5 and 7.

Σ I 12. An easy pair of 6m and 7m, 78 seconds apart.

OΣΣ64 This 8th-magnitude pair are 76" apart. A fine field.

OΣ118 The distance is the same, but the stars rather brighter (6m and 8m).

Variable stars

Y (6.8-9.2) This variable forms a triangle with two stars of 5.9 and 7.2, the latter with a 7.9m neighbour. Three other stars of 8.1, 8.6 and 9.4 lie just to the West. A satisfying star to observe, its deep red colour aiding identification.

TT (8.1-8.8) South of this deep red star is an elegant little horizontal Y of 7.5, 8.9, 9.0 and 7.2.

TU (5.9-9.2) Another deep red star. I have provided a chart.

BU (5.0-5.5) Pleione is a hot irregular variable known as a shell star, though its range of only half a magnitude and its slow changes do not make it the most suitable type of variable for amateur observers. Use 16 (5.5) as a comparison star.

HU (6.0-6.8) An eclipsing binary for which you can use a wide pair to the southeast (6.6 and 7.2.)

Clusters and Nebulae

M.45. The Pleiades are visible in their true glory only with binoculars, which are also useful for revealing the faint nebulae that reflect the light of the hot stars nearby. The area around Merope (23 Tau) is especially bright, though as with all nebulae, any haze or mist will ruin the view. I once did an observational project on the appearance of the Pleiades nebulosity and came up with the conclusion that it was by no means standard from one night to the next. This does not of course mean that the nebulosity actually changes, however - just the observing conditions!

The Pleiades, by their beautiful appearance, have awed all those who watch the sky since time out of mind: and are known by names as diverse as the cultures which bestowed them, witness the following selection of epithets - a bunch of grapes, speckles of dust, the little turkeys, the little

Fig. 30: The Crab Nebula

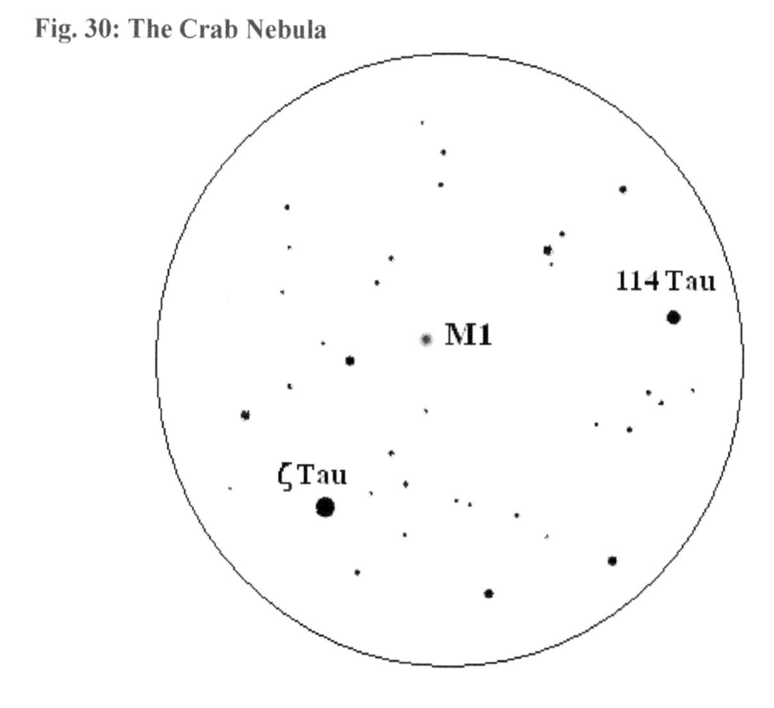

nanny-goats, or the hen and chickens; but we probably know them best as the *seven sisters*, in Greek myth the daughters of Atlas. By name they are Merope, Alcyone (the brightest Pleiad) Celaeno, Electra, Taygete, Asterope and Maia. Beautiful names for beautiful stars.

M.1 (NGC 1952). This is the Crab Nebula, remnant of a star that went Supernova several thousands of years ago and seen in 1054 by Chinese astronomers, among others. It is very interesting from the astrophysical point of view and indeed it has been said that there are two sorts of astronomy - the ordinary sort and Crab Nebula astronomy! It is a testing object for binoculars so I have provided a chart to help you find it.

NGC 1758. A large cluster of faint stars, quite impressive in large glasses, where its brighter members are seen to form a kind of S-shape.

NGC 1647. This is a large, fairly bright cluster that reveals some of its 30 stars to the binocular.

TRIANGULUM

A small but readily-spotted constellation containing some good fields, notably near the borders with Perseus. It is included in the chart for Aries.

Groups of stars

1. 01h 45m, +32°. Small isosceles triangle with a group of seventh- and eighth-mag stars to be seen just to the North of its southern member.

2. Sweep within the trapezium formed by 6, 10, 11 and 12.

3. 15 (5.6). Good sweeping SW of this red star, including the LPV R Trianguli (see below).

Wide doubles

γ. Together with 7 and δ this makes a fine, wide triple for small glasses.

12. A star with three fainter acolytes, one of which is 13 (5.9).

15. Lies in an area of many wide pairs.

Variable stars

R (5.8-12) 15 Trianguli nearby makes a good comparison for this popular variable when at maximum, which occurs on average every 267 days. Predictions for R Tri are included in the appendices.

W (7.5-8.8) A red star in the same field. It lies between two comparisons of 7.0 and 9.0. If R is visible, why not observe them both together. Do you see any difference in their colours?

Clusters and Nebulae

M.33 (NGC 598). A large, isolated galaxy, visible to binoculars as a round spot, brighter in the centre than at the edges. On one occasion I have glimpsed it with the naked eye. Its light is spread out over a large area, so it looks very pale.

URSA MAJOR

The most famous constellation in the sky is much larger than is popularly thought, extending far beyond the "seven stars in the sky", the Plough, Big Dipper or Charles' Wain as it is variously known. The last of these names is interesting, referring not to King Charles as many think, but quite the opposite - the Churl's Wain or peasant's cart! Be that as it may, the Plough has been used since time immemorial to help sea-voyagers and lost way-farers navigate. An old mariner's rhyme runs:-

Where yonder radiant hosts adorn the Northern evening sky
Seven stars, a splendid glorious train, first fix the wandering eye
To deck great Ursa's shaggy form, those brilliant orbs combine
And where the first and second point, there see Polaris shine.

Groups of stars

1. Small, bright triangle of 2, π^1 and π^2. Fine sweeping between this group and the regular triangle of ρ , and $\sigma^{1,2}$. Note a tiny tick-shaped collection of stars a degree SE of π^2.

2. Large, bright fan of stars including 38 UMa.

3. Sweep the bowl of the Plough, noting that the nearby 43 (5.8) is just North of a small faint line, rather angular in shape.

4. Beautiful bright trail stretching from Mizar to κ Bootis.

5. 51 (6.1). Note that a star 2° to the Southeast is the southern member of a regular figure-seven shape, and also has a close pair to its own South-East in turn.

Wide doubles

16. This has a companion NE which is in turn a close pair - Struve 1315, a 7th-magnitude double some 25 seconds apart.

67. Binoculars show one, possibly two, companions.

ζ and 80. This is of course Mizar and Alcor, the well-known naked-eye pair. You can even see Alcor when it is low in the sky, and Mizar itself is also double (2.1, 4.2; 14") and has been split with a good pair of 12x40 binoculars. You may in addition see another star between Mizar and Alcor called at one time *Sidus Ludovicianum* or "Ludwig's Star", named regrettably not after Beethoven, but by a loyal subject of King Ludwig, who thought he had discovered a new star! (The subject that is, not mad King L).

55. A 4.9m star with a 7m companion.

Close doubles

Σ1495. This lies between the pointers, Dubhe and Merak, and is 36" apart with a bright star to either side.

OΣΣ99. A wide pair in a barren area. Mags 6 and 8, distance 80".

OΣ199. A rather hard object; mags 6.6 and 8.3, distance 230".

P.283. Though the companion is quite faint at 8.3, this is a wide pair (187")

P.284. A beautiful object; both stars mag 6 and orange. Easy at 221 seconds.

65. Another fine sight, similar to the previous star, though closer (63").

P.285. A difficult double of 6.7 and 8.2, 98" apart.

P.290. Difficult again; mags 5.4 and 8.2, though twice as wide as P.285.

P.287. Rather easier at 146" and magnitudes 6.3 and 7.6.

Variable stars

Y (7.6-9.5) A red star not far from Alioth, the *lucida* of the Plough. It lies in a flattened triangle of 8.0 and 8.9 and has an easterly companion which is slightly variable.

Z (6.6-9.1) This well-observed and interesting star is in a faint little X whose mags are (from N to S) 8.8, 8.7, 8.6, 8.8. A brighter star of 7.2 lies between Z and Megrez, the faintest star of the dipper, though some believe this star may have faded somewhat over the course of centuries.

RY (7.0-8.0) An easy star to find, one of a small Y whose other stars are of 6.9, 7.4 and 7.8m.

ST (6.4-7.5) This bright red star is ideal for the beginner, though rather hard to find initially. It is the most Northerly of a distinctive, small vertical line of three whose southern member is 6.9m.

TX (6.9-8.5) This eclipser lies near a star of mag. 5, with a companion of 8.1m. A degree North-East of the bright star is another comparison of 7.3m.

VW (7.2-7.8) A red star, for which a little triangle some distance to the Northeast can be used (7.1, 7.7, 8.3). VW further lies between two 6m stars and has a southerly companion of 8.1m.

VY (6.0-6.6) A deep red star, good for the beginner with small glasses, though rather poor in light-range. Use the long triangle to the East of 6.1, 6.5 and 6.7. As with most stars of this class (the
red irregulars) it needs observing only once, or, in case of interesting behaviour, twice a month.

Clusters and Nebulae

M.81 (NGC 3031). A galaxy readily seen in average bins. You may also see its fainter neighbour, M.82. In 1993 a Spanish amateur discovered a Supernova in M.81 that has proved to be one of the most interesting explosions yet observed. Although 1993J as it was called never reached binocular visibility, this is still a good example of what an ordinary observer can achieve.

URSA MINOR

Some notable red stars, but really not much besides, at least for the observer using binoculars.

Groups of stars

1. 13h 36m, +77°. Small arc of five, with some fainter associates.
2. Sweep along the line of 4, 5 and β. They are all reddish stars.
3. η. Lies in a beautiful field, with some coarse groups of stars.
4. 15h 00m, +67°. Bright quadrilateral, of which the senior member is the variable RR UMi, a small-amplitude red star of magnitude 5.

Wide doubles

γ. This has a red companion, 11 (5.1).

β. A degree to the North is a bright, wide double.

Close doubles

π^1. Magnitudes 6 and 7, distance 31". Quite easy.

P.288. A faint, obscure pair of 7.5 and 8.4. Separation 210".

Variable stars

V (7.4-8.8) Found near group (1), this is the southern member of a small isosceles whose other stars are 7.8 and 8.1m.

VIRGO

A large group with some good fields and several nice variables. Its most notable claim to fame is the vast collections of galaxies which inhabit it, though only a very few of these are binocular objects.

Groups of stars

1. Sweep from σ towards ζ .
2. Long arc, extending from 95 (5.5) to 72 (6.1).
3. ι . This lies inside a pretty triangle, with a fine cross to the W.
4. 61 (4.8) is the centre of a large, bright collection best seen with small bins.

Wide doubles

27 and ρ . An unequal, wide pair.

64 and σ . South of these stars is another wide 6m double.

Close doubles

Σ1740. A close, equal double of the 7th magnitude, and 28" separation.

Variable stars

R (6.0-11.5) A mira star with the short period of 150 days. I have followed this star to mag. 10 with a 7x50 finder - You will never know quite what your optics can do unless you push them from time to time! Predictions are supplied.

RT (8.0-9.0) A rather isolated and dim star.

RW (7.0-8.2) There is a small triangle closely E of 6.6, 7.4 and 7.5, with an 8.4 inside it. RX, one degree away, is a similar type of star, but with a small range of only half a magnitude.

SS (6.8-8.9) An interesting star to follow, this is one of a vertical arc whose members are, from N to S, 7.3, 7.6, SS and 8.3. Observe once every 3 weeks.

SW (6.5-7.7) Again there is a vertical arc, this time North of the variable. Its stars are of (again N to S) 7.2, 7.6, 7.7, SW.

BG (8.1-9.1) This, together with its westerly neighbour of 8.4, is the most northerly of a triangle of wide, faint pairs. The E. pair is of 8.4 and 8.9, and between BG and 109 (3.8) are two other stars of 8.3 and 8.7.

FH (6.9-7.5) Though of small range, two stars near 59 Vir, of 6.9 and 7.2, are useful as comparison stars for this variable.

Clusters and Nebulae

M.49 (NGC 4472). A galaxy which appears as a faint gleam of light.

M.104 (NGC 4594). The "Sombrero Hat" galaxy is visible as an elliptical blur. The many other incredibly distant galaxies in this constellation are mostly very faint, but you could try sweeping slowly and carefully around the bowl of Virgo (that is, the area roughly bounded by δ and ε virginis and beta Leonis) for a glimpse of some of them.

VULPECULA

A really fine little group in the Milky Way, unfortunately rather shapeless. In fact there used to be two constellations here; the other was predictably Anser, the Goose - it is no longer officially in the sky, but presumably in the Fox's tum instead! Charted with Sagitta.

Groups of stars

1. 4 (5.3) is the leader of a brilliant little group known as the "Coathanger" from its shape - which may look more realistic in an inverting optical system such as many finderscopes are. In ordinary binoculars, you will have to conceive of an upside-down coathanger!
2. 18,19 and 20. 19 is surrounded by three faint stars, while some members of the open cluster NGC 6885 may be seen around 20. Very fine field.
3. 29. This is surrounded by beautiful faint groupings.

Wide doubles

α . This has a companion of 6.0m. Additionally, the interesting open cluster NGC 6800 lies nearby.

16. Another star with a 6th-mag. attendant.

32. Two, maybe three companions may be seen here.

Close doubles

OΣΣ181. Right on the border with Lyra, these stars are both of 6.5m. A fine object.

Σ I 48. Another equal 6th-magnitude pair, but thirteen seconds closer at 42".

Variable stars

V (8.1-9.4) Most good binoculars should be able to follow this interesting member of the RV Tauri class, easily located close to the 5.5m star 27 Vulpeculae. The variable lies exactly midway between this star and a faint comparison of 9.3 magnitude situated half a degree Northwest of 27 Vul. A brighter star of 8.6 magnitude forms an isosceles triangle with V Vul and 27. Observe this variable once a week, since its period is only about 75 days, and it often shows unpredictable behaviour.

SV (7.5-9.4) A golden-yellow Cepheid with the long period for these stars of 45 days, which by the period-luminosity law (i.e., the longer the period of a Cepheid the more luminous it is), means that here we have a very powerful star indeed. Not far away is a similarly-coloured star, S Vul, which is a small-amplitude semiregular variable, not so very dissimilar in fact from the Cepheid variable close by, other than the fact that its variations are less, and not quite so predictable.

DY (7.0-7.7) Just to the North is a star of 7.5, while directly North of 33 Vul is a brighter comparison of 6.7. A red variable.

FI (7.6-9.1) This is one of a small triangle, whose other members are 7.5 and 8.7m.

Clusters and Nebulae

Cr.399. This is the Coathanger, mentioned above. It was near here in 1976 that George Alcock, the famous amateur, discovered another of his many novae. This was no flash in the pan, however; for many years he had been watching the skies, learning the patterns of hundreds of stars along the Milky Way, an approach now copied by the rest of the world's amateurs involved in the discov-

ery of Novae.

He used for these purposes not a powerful telescope, but binoculars - proof again that you do not need to spend large amounts of money on flashy-looking pieces of high-tech equipment to make a mark in the world of Astronomy - though you do need dedication, devotion and sometimes a bit of patience!

M.27 (NGC 6853). The dumb-bell nebula, this is readily visible as a large grey spot near the 5m star 14 Vulpeculae. You might like to search out this constellation for other open clusters not discussed above, chief of which is probably NGC 6885.

The Southern Constellations:

ANTLIA

A rather dull, uninteresting group to the eye, though with some fine sweeping in its southern reaches. On the same chart as Pyx-is.

Groups of stars

1. ε (4.6). Some fair areas for sweeping around here and towards η
2. α (4.4). Another star in a fine area.

Wide doubles

$\zeta^{1,2}$. Another wide pair (both 7.4m) lie to the East.

Close doubles

P.12. A wide, easy 6th-magnitude pair. Distance 172".

Variable stars

U (5.7-6.8) A chart is supplied for this easy red variable.

RR (8.0-9.2) Another red star, found from the wide double ψ Velorum. 2° North is a little foursome of 6.4, 8.3, 8.4 and 8.5, though little else.

APUS

A small but easily-noticed constellation near the South Pole of the sky. Unfortunately, few items of binocular interest. Included on the chart of South-polar constellations (Apus, Chamaeleon, Mensa and Octans).

Groups of stars

1. Small triangle of β, γ and δ .

Variable stars

θ (5.2-7.0) This red star lies not far from α , and between the two are three stars of 6.2, 6.2 and 6.0. Epsilon Apodis (5.2) may also be used.

ARA

A small, though brilliant group in the milky way, containing several fine clusters.

Groups of Stars

1. α (3.0). Beautiful sweeping from here to the bright pair beta and gamma.
2. 17h 20m, -63°. Small group of assorted magnitudes.
3. σ (4.6). Numerous scattered collections of stars around here. Brilliant region.

Wide doubles

η . A fine, wide pair lies closely North.

μ . Between this star and λ is P.29 (6.2, 8.1, 186"). The primary is a small-amplitude variable, V626 Arae.

Close doubles

P.69. A faint pair of mags. 7.3 and 8.0, 165 seconds distant.

P.70. Both these stars are orange, and are 283" apart. A third star is visible.

P.92. A rather unequal pair of 5.9 and 8.2, distance 153". Fine field.

Variable stars

R (6.0-6.9) An eclipsing binary, well-suited to small glasses.

V713 (8.2-9.0) A small-amplitude red star. It is one of a long trapezium of magnitude 7, whose two northernmost members possess a couple of useful stars of 8.4 and 8.9 closely South, but on the whole the range in magnitude is rather small for effective observation.

Clusters and Nebulae

NGC 6193. Binoculars will resolve some of this cluster's stars. It is also immersed in a gaseous nebula. Brilliant sweeping.

**Fig. 31: NGC 6397, a large Globular Cluster.
Field size is 3°**

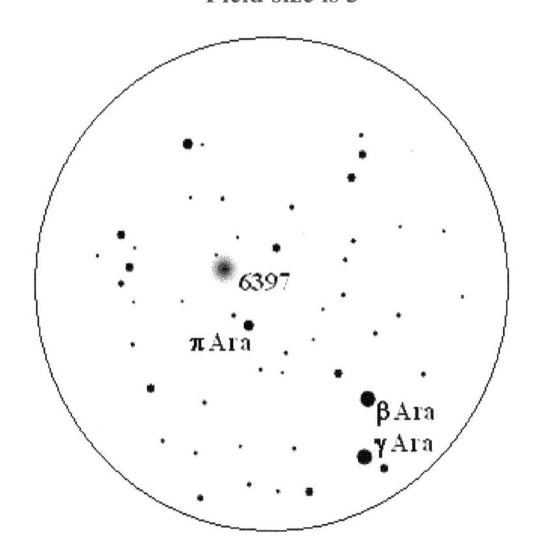

NGC 6397. A large and bright globular cluster, visible as a silvery round glow. It is one of the closer objects of this type, "only" 2300 light-years away. Use the chart here to find it.

CAELUM

A small group which, though visible from most of the USA. needs high altitude to show its few interesting objects to any advantage. It shares a map with its more interesting neighbour Columba.

Groups of stars

1. Sweep the area bounded by α and δ Caeli and α Horologii (3.8).

2. A fine large group North of α, containing some small doubles. The most northern star of this group is a wide orange pair, both magnitude 7. Two degrees South of this is another pair of 6.0 and 7.5, again both orange.

Wide doubles

γ. A 4.6m star with a 6m companion that is a small-range variable, X Caeli.

CARINA

This magnificent group, together with its "shipmates" is laden with some of the sky's most resplendent and interesting objects. The constellations of Vela, Puppis and Carina form together Argo, the ship of Jason in Greek mythology; and the three groups are treated together as one enormous constellation insofar as they share the Greek alphabet between them, a relic of the time not so long ago when the ship Argo had not been subdivided.

Groups of stars

1. Canopus. The sky's second brightest star by apparent magnitude (and one of the most powerful stars known, to boot) lies in a beautiful bright field.

2. 08h 25m, -55°. Magnificent group of about a dozen stars.

3. Large triangular group, including ε (1.7) c and f, this last also known as V344 Carinae, a small-amplitude variable.

4. Prominent quadrilateral of ι, g, h and the variable N Velorum. A superb part of the sky.

5. 10h 00m, -60°. Brilliant field of stars that includes the marvellous open cluster NGC3114.

6. s (4.1). Around this star we have one of the most brilliant areas of the entire sky. This region is so crowded with stars that it is hard to tell the difference between a true cluster and simply a rich star-field. Many of the stars here are in some way connected with the amazing "Keyhole" Nebula which dominates the area.

7. θ (3.0). Lies in a magnificent clustering of many brilliant stars.

Wide doubles

e^1. This forms with e^2 a bright wide double in a strangely sparse region. Not far away, b^1 and b^2 are a similar, but wider, pair.

g. A 4.2m star with a 6.4m neighbour.

x. Closely North is a fine wide pair of magnitude 6. Near the open cluster NGC 3532.

y. This star has two seventh-magnitude companions.

s. Note a bright, wide pair to the North.

Close doubles

u. Another brilliant object; magnitudes 3.9 and 6.6, distance 155". Have a look at the colours of these stars.

A. One degree South is a small right-angle which includes the beautiful pair I157 (6.6 and 6.8, 128 seconds apart).

I. A bright, wide double of mags. 4 and 6 separated by 250" of arc.

Δ60. Rather harder at 6.0 and 8.2, and 41" distance.

h3984. A faint pair, both white and 107" apart.

h4000. Wider as well as brighter; magnitudes 6 and 7.

P.26. An unequal object of 6.1 and 8.3 and 229" distance. The bright star is also slightly variable (QY Car).

P.27. Again unequal and wide; 4.9 and 7.3, 253 seconds apart.

P.35. A faint, isolated pair of 7th and 8th mags. Distance is 170".

P.37. This beautiful pair of 6.3 and 6.5 is separated by 61".

P.38. Another good target, less equal but wider at 172".

P.116. These stars are of magnitude eight, and are separated by 165 seconds.

P.119. More equal this time; mags 7.4 and 7.7. Distance 146". A fine high-power field.

P.122. Another 7th-mag double, with orange and yellow components.

Variable stars

R (4.6-9.6) One of the few Mira stars always visible in binoculars, and also circumpolar from most Southern latitudes. I have provided a chart for this rewarding star.

S (5.7-8.5). Another Mira variable, this time of spectral type K, rare among these stars. Its short period of 150 days means you need to estimate it twice a

month rather than the usual once. It is easy to find near the 3rd-mag star q Carinae. The best comparisons are a neat little triangle of 6.5, 7.0 and 7.6 between S and q above. A fainter star of 8.0 lies just south of S.

RR (7.3-8.5) A red semi-regular for which a chart is supplied.

VY (6.9-8.0) This Cepheid star has a period of 19 days and lies in a brilliant field between the bright stars x Velorum and u Carinae. The best comparisons all lie in a little line to the E, and they are of (S to N) 8.1, 7.0, 6.3 and 7.2. A good star for average glasses, plus the added attraction of one of the most awe-inspiring regions of the sky.

AC (7.8-9.0) A red variable; again a chart is provided.

AG (7.1-9.0) A highly massive, powerful star of which there are many in this area. It is an eruptive variable which you should look at on every clear night, or at the very least once a week. I have again supplied a chart.

BO (7.2-8.5) This red star lies not far from eta (see below) and is included on its chart.

BZ (7.5-9.2) Another red variable near an eruptive object, this appears on the chart for AG above.

CK (8.0-9.2) A fainter star, though readily found close to the bright star s Carinae. Two other stars lie in a regular line to the W, of 9.0 and 8.9m, and another of 7.9m lies to the North.

IW (7.3-9.0) A fascinating RV Tauri star with a period of only 68 days, although the mean magnitude of this period itself varies over a 4-year cycle, and large glasses may be needed to catch it around minimum light. This interesting star is included on the chart with R above.

IX (7.4-8.3) A red variable in a nebulous area. The best guide is a wide, bright pair of 6.1 and 6.7 on the edge of the eta Carinae nebula. IX has a companion of 7.0m, and a little line runs to the South of these wide pairs; its stars are of 8.3, 7.5 and 8.2m.

QX (6.6-7.2) This is an easy Algol-type object, though of rather small amplitude. It appears on the chart for RR Carinae.

V341 (6.2-7.1) This red star lies in the variable nebula IC 2220, and makes an equilateral triangle with a 5.8m star to the South and the lovely open cluster NGC 2516, which is bordered by a bright line of three (5.6, 5.7 and 6.4). Another line of three terminates in the 5.8m star above and their mags. are 7.3 and 6.7. Worth looking at for the nebula and cluster alone!

η (-0.8 - 7) This is one of the most amazing objects in the sky, and its minimum magnitude of seven rather belies its explosive past. Originally catalogued by Halley of comet fame as a fourth-magnitude star, eta Carinae underwent a unique series of fluctuations until in 1838 it reached magnitude -1 and thus became the brightest star in the entire sky with the exception of Sirius! A supermassive, superluminous star at a great distance, that now (2008) wobbles around the fifth magnitude, the improved Hubble Space Telescope revealed symmetrical tangerine-coloured bursts emanating from this star! Of course, you won't see this with bins, but you will see the many nebulous flecks which pervade the general area. One gets the impression that this whole area, with its titanic stars and exotic nebulae, is a part of the sky completely unlike any other; it should always be watched closely, for these highly-massive stars will not last long, and we could well see a supernova here any day. Naturally, I have had to supply a chart, though note that, because of the star density, this is a close-up view.

Clusters and Nebulae

NGC 2516. A marvellous sight in binoculars, resolving into a beautiful spray of stars before a white gleam.

NGC 3114. Another brilliant cluster, visible with the naked eye in fact.

NGC 3372. The Keyhole Nebula, around eta Carinae. Easily visible with binoculars and probably the most photogenic of the whole class of gaseous nebulae. This whole area is dominated by the nebula and the many hot and powerful stars connected with it.

CENTAURUS

Another vast and interesting constellation which is partly visible from the US of A, though its richer southern reaches are best appreciated from South of the line. A really outstanding constellation for the binocular owner.

Groups of stars

1. 1(4.8) is one of a fine collection which also includes u (5.6).
2. 13h 10m, -42°. A long string of variously-coloured stars; from N. to S., they are 6.2 red, 5.7 orange, 5.8 yellow, 6.1 orange, another the same, and 5.3 yellow. The smallest glasses actually give the best view of this fine group, which is a good example to practice your magnitude estimation skills on.
3. Small but attractive group of i, g, k and h, also called 1, 2, 3 and 4. The second of these is slightly variable. In the far North of Centaurus, this little group has been seen from Southern England.
4. Sweep the triangle bounded by the bright stars η, ψ and b. Many beautiful coarse fields.

5. λ (3.3). Around this star lies the richest region in the whole constellation. Note especially a brilliant curving line, and a small, rather faint group just South of λ.

6. ζ (3.1). This forms a neat triangle with two 6m stars.

7. Omega Centauri. The area of the famous globular cluster is worth sweeping.

8. K (5.3) is in an area of many bright stars.

Wide doubles

I78. This telescopic pair forms a wide double with a red star of 5.7m.

h4563. Another telescopic double star, but one of a triple in binoculars.

C. A brilliant wide triple of magnitudes 5.4, 5.5 and 5.6. C^1, the faintest of these, is a small-amplitude variable, V763 Cen.

γ . This bright star forms a nice, wide pair with w (4.6, orange).

M. 1° N is a wide pair of 5.5 and 6.1, and M itself is very near the globular cluster NGC 5286.

Close doubles

J. A fine pair of magnitudes 5 and 6, 60" apart. The companion is slightly variable, V790 Centauri.

u. See if you can spot the double in the little triangle closely West of this star. Its magnitudes are 6.7 and 8.4.

P.127. An equal but rather faint pair, both yellow and 197" apart.

P.128. These stars are also both yellow and magnitude 8, though closer at only 71".

P.131. This is an eighth-magnitude triple of which the two closer stars are 133 seconds apart.

P.132. 192 seconds divide these stars of 7.1 and 8.1. A fine field.

P.134. The mags here are similar, but the stars wider at 253".

Variable stars

R (5.8-11). An interesting Mira star with variations characterised by alternate bright and faint minima. R Normae and BH Crucis are stars of the same type, and neither of them are very far away from this object. You can find this star easily, between the brilliant α and β in a triangle formed by beta and two bright comparisons of 5.0 and 5.3. A wide pair NE of beta is of 6.5 and 7.7, while R itself makes a little equilateral with a 6.7m star to the North and a small line of 8.4, 8.9 and 8.2 to the East.

S (6.0-7.0)A good star for small glasses, this red variable forms a triangle with σ and τ , and makes a similar pattern with two stars of 6.2 and 7.0.

T (5.5-9.0) This star has one of the shortest periods of any Mira-type variable - 90 days, although strictly it is a semi-regular variable.

Y (7.3-8.4) Another red star; a chart is again provided.

RR (7.3-7.9) An eclipsing binary with rather a small range, but easy to find near α and β and between two 5m stars. Closely E of the Northern one of these is a small cross of (W to E) 7.2, 8.5, 8.2, 7.8 and 8.2.

SZ (8.3-8.9) Another eclipsing binary nearby and with the same range.

V412 (6.5-8.5) An irregular variable between β and ε. A chart is provided for it, which also includes SZ above.

V744 (5.1-6.6) This easy object lies near the 4.7m star M Cen. Two stars of 6.3 and 6.5 lie between this and V744, which has a faint comparison of 7.2m just to the South.

V766 (6.2-7.5) Slightly fainter, though still a good star for those with minimal optical power, this is an eruptive variable that lies on the edge of the open cluster NGC5281 precisely between two good comparisons of 6.7 and 7.3.

V854 (7.0-14) This fairly recently-discovered variable is of the peculiar R Coronae Borealis type, and is one of the more active of its class (some of these stars can go for decades without doing anything unexpected!) This object was used in a recent study by amateur and professional astronomers which revealed what I have to say I had always suspected - i.e., that visual observers could be highly accurate in their results, provided that good conditions prevailed (i.e., suitable comparison stars with accurate values, a good spread of observers, etc). This is a good point at which to reiterate that the charts in this book are "for amusement only". If you decide to take this branch of astronomy seriously, you should join an accepted society such as the AAVSO or BAA and obtain official charts from them.

Clusters and Nebulae

NGC 3766. This fine cluster betrays several stars to binoculars.

IC 2948. A much larger cluster even nearer to λ. It is a wonderful sight in binoculars, with several doubles and triples being visible.

Fig.32: Open Cluster NGC5460. The field diameter is 6°.

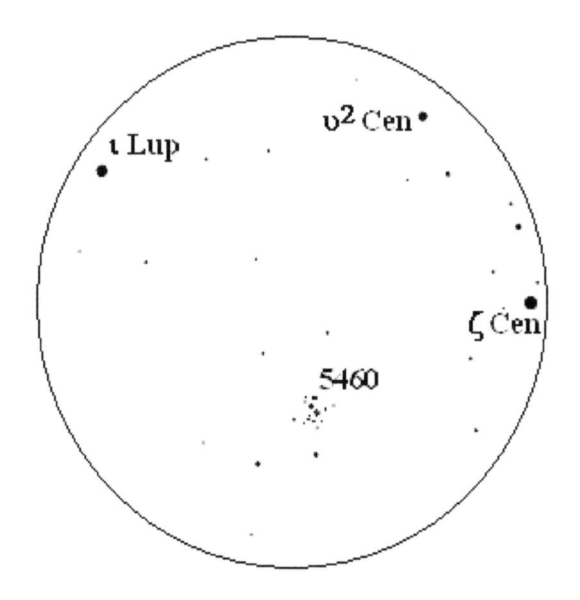

NGC 5460. A rather less "dominant" object, but still a beautiful sight, this is surrounded by delicate little groupings of stars. You can use the chart here to find it.

NGC 5662. Large glasses give a good view of this cluster, showing it to be beautifully symmetrical. Two tiny triples sandwich a wide pair, with a closer double between them.

NGC 5139. This is the brilliant, blazing globular called ω Centauri. It is one of the few globulars that are not spherical. What does its shape look like to you? Many other gleams of light, betraying fainter globulars, may be seen in the areas around alpha and beta.

CHAMAELEON

An indistinct little group not far from the pole, which contains several fascinating and newly-discovered nebular variables - all unfortunately far beyond the grasp of even large binoculars!

Groups of stars

1. α (4.1). A fine double-curved line of 7th- and 8th-mag stars lies closely East.

Wide doubles

η . This star has a companion which is variable, RS Cha (below)

δ . Another wide pair, separated by 5 arc-minutes.

ε . This has a closer but fainter neighbour to the NE.

Close doubles

P.156. A difficult pair, due to the close distance of 27" and the magnitudes of the stars (7 and 8.9). Definitely one for the large glasses.

P.157. Wider but fainter, this is an 8th-magnitude pair separated by 69 seconds.

Variable stars

RS (6.0-6.7) An eclipsing binary, whose dips in light occur every 1.7 days. It forms a wide double with η (5.6). Between the two is a fainter star of 7.7, while 2° West are two other stars of 6.8 and 7.3.

CIRCINUS

A small but rich group in the Southern Milky Way, with many beautiful high-power fields.

Groups of stars

1. δ (5.2) lies in a most beautiful field of bright stars.
2. 14h 46m, -66°. Delicate little group that includes five stars in a straight line.

Wide doubles

β . A star with a 6m companion. Fine field, on the edge of a large dark cloud in the Milky Way.

β and γ . Forming an isosceles with these stars is a fine 7th-mag pair.

Close doubles

Δ169. In a beautiful region, this pair is of mags 6.2 and 7.7, and 69" distance.

P.87. Fainter and wider, at 113" and magnitudes 7.7 and 8.1.

COLUMBA

A neat group below Lepus, with some fine fields and interesting double stars. Though visible from the USA, it feels more at home in the far southern groups.

Groups of stars

1. β (3.2) Sweep around here, and towards epsilon. Beta Columbae is sometimes called by its proper name of *Wezn*, which is the same as the star Delta Canis Majoris, and means "weight", the idea being that the stars in

question never climb very high above the horizon, and so look weighed down!

2. η (4.0). Lies in an interesting area.

3. 06h 35m, -37°. Beautiful bright group, of which the northern member is a wide double.

Wide doubles

$\pi^{1,2}$. A fine, wide pair in an attractive region.

Close doubles

θ . This has a 7m neighbour 279" away. Note a neat little triple to the NE, and a wide, rather faint pair NE of this.

GC 7735. Mags here are 5.9 and 8.3. Distance 74".

h3740. The separation of this pair is at most binoculars' limit of 25".

h3857. A beautiful object, both stars yellow and in a fine region. Magnitudes are 5.7 and 6.7, distance 65".

P.158. This fine yellow pair of magnitude 7 lie in a neat little group on the border with Lepus. Distance is 86".

P.167. Two 8m stars separated by 153". Lies in a straight line with the next two objects.

P.170. The mags are similar to P.167 but the stars are closer (66").

P.171. Mags are 6.2 and 8.7, so rather difficult for most bins. However the separation is wide at 175".

P.172. Difficult for the same reason (7.4 and 9.0) and even closer at 59".

P.175. An equal, wide double. 194" separates the orange stars, though their colour may not be evident in small binoculars.

P.177. A neat 7th-magnitude pair 66 seconds apart.

Variable stars

RV (8.5-9.5) A faint red star only suitable for large glasses. Note a wide unequal pair SE, and a tiny faint line NW, of 9.4, 9.6 and 9.7.

CORONA AUSTRALIS

Also called *Corona Austrina*, this is a well-named little group which, like its boreal counterpart, extends beyond the crown itself. Richer though fainter than Borealis.

Groups of stars

1. The arc running from δ to γ is a fine sight, notably in the region of α . In this area are several diffuse nebulae, both bright and dark, and a great many faint nebular variables.

Wide doubles

$\eta^{1,2}$. Note a third (7th-mag.) star forming a wide triple.

ζ This has a 6m companion to the NW.

Close doubles

κ . A difficult pair of 6.0 and 6.6, only 23" apart.

P.15. Easier at 222". Mags here are 5.5 and 6.7.

P.152. A wide but unequal pair; 6.6 and 8.5, distance 164".

P.153. Much easier this time, this is a fine yellow double of 7.1 and 7.5 separated by 138".

Clusters and Nebulae

NGC 6541. Binoculars show this globular cluster as a fuzzy grey spot very close to the 5th-magnitude telescopic double star h5014.

CRUX

Although the smallest constellation in the sky, Crux crams into its 100 square degrees a noble cluster, a large dark nebula, several doubles, an interesting variable and two first-magnitude stars!

Groups of stars

1. Fine sweeping from λ Cen to ε (3.4). This is the star that one of the old astronomers wanted moved, as it spoiled the symmetry of the cross!

Wide doubles

θ^1. This makes a beautiful wide pair with θ^2, slightly fainter. A brilliant field.

η . Again in a splendid part of the sky, this has a 7m companion to the South.

ι . Between this star and the "Jewel Box" is a fine pair, both white.

Close doubles

β. A brilliant pair for large glasses (the primary rather overpowers its companion). Mags 1.6 and 5.1, both blue. Distance 85".

γ . A wider but more unequal coupling, of 1.6 and 6.7, separated by 111".

h4548. Difficult because of the faint (9m) companion 53" distant.

P.85. This is an unequal pair of 6.6 and 7.8, 160" apart.

Variable stars

AO (7.3-8.8) A chart is supplied for this red variable.

BH (7.2-10) A peculiar Mira star like R Centauri, a sort of cross between the Mira stars and the RV Tauri objects; it shows a double period, with alternating bright and faint maxima, and is followable for nearly all its range by average binoculars. I have supplied a chart.

Clusters and Nebulae

NGC 4755. A cluster of many stars, and beautiful in large glasses. Called by Sir John Herschel the "Jewel Box" because its variously-coloured stars suggested to him "a superb piece of fancy jewellery".

β . Just South of this star, and extending over a sizeable portion of the whole group, is the famous "Coal Sack", a giant dark nebula blotting out the light of goodness knows how many stars behind it. It occupies an important place in Aboriginal and Bushman folklore.

DORADO

A group which would be unremarkable were it not to contain most of the large Magellanic Cloud. Though the Latin name of the constellation would seem to suggest a goldfish, Dorado actually represents a swordfish. Needless to say, it does not resemble one. The European observers who gave the Southern Constellations their modern names were obviously possessed of wild imaginations. Must be those sweltering southern climes!

Groups of stars

1. Sweep from ζ to κ , noting the fainter star between them, with a tiny triple nearby.

2. δ (4.5) lies in a fine field of bright stars.

Wide doubles

θ . Guide star to two wide doubles.

Close doubles

η^2 . Difficult due to the faint *comes*. Mags 4.9 and 8.1, 272" apart.

P.148. 50 seconds separate these stars of 7.1 and 8.5m, which are so close to the previous object that they effectively form a double-double.

Variable stars

R (4.8-6.6) A readily-observable star even with the smallest glasses, this is a semi-regular variable.

S (8.3-11) I have included this star for its interest value; it is an amazing eclipsing system with a period of 40 years, whose combined light-output is something like one million times that of the Sun! It is in the Magellanic cloud, in the cluster NGC 1910. Don't forget when you look at this star that it is in another Galaxy!

SN 1987. In February 1987, not far from the Tarantula Nebula, there appeared, in the Large Magellanic Cloud, the brightest Supernova seen since Kepler's Star in 1604. SN 1987, as it was called, turned out to be a most peculiar object. The obscure progenitor star which gloried in the designation Sk-69°202 was not an old red star (the usual precursor of a Supernova) but instead a blue object.

Some idea of the power of a Supernova may be gained by considering that an explosion 15,000 light-years away (that is, far more distant than any star we can see with the naked eye) would produce a star in our skies with a visual magnitude ten times brighter than Sirius! Since its explosion, the Supernova has thrown off a ring of gas (caused by its rapid rotation) and has faded back into visual - though not astrophysical - obscurity.

Clusters and Nebulae

NGC 2070. This is the great "Tarantula Nebula" close to the Supernova site. Seeing this object in binoculars, it is incredible to think that it lies in another galaxy, 200,000 light-years away. What it must look like at close quarters is a true inspiration for the astronomical artist!

ERIDANUS

This continues the section described under the Northern groups; the far South of this constellation is best seen from the Antipodes.

Groups of stars

1. e (4.3). Fine sweeping to the North-East. Note a large bright triangle with a much smaller group close by.

2 .The area of the fourth-magnitude stars g, f and h is worth scrutiny.

3. 02h 02m, -55°. Pretty angular line of 5 seventh-mag stars.

Wide doubles

θ and e. Forming a large right-angle with these, and to the Northeast of theta, there is a small triangle of 6th- and 7th-magnitude stars, closely followed by a wide triple.

χ . A star with a sixth-mag distant companion.

Close doubles

P.5. A pleasant pair of 7.2 and 7.6, 179" apart.

P.149. Another attractive little double of 7.0 and 8.1, close to P.5. Separation 52".

P.150. Wider but more difficult because of the 9m companion. 102" apart.

P.151. About a degree and a half S. of the 4th-magnitude star s Eri, this is a good object of 7.4 and 8.0, with a separation of 92".

FORNAX

A fine group for sweeping, visible in its entirety from the US of A, though needing a high altitude to show its rather faint stars. Several good doubles and many faint galaxies, though these latter are generally beyond the range of binoculars.

Groups of Stars

1. 01h 50m, -39°. Pretty group of assorted brightnesses.

2. η (4.7). Guide to rich area, notably to the N.

3. ι^1 (5.8) is another star in an interesting region.

4. α (3.9). Note the fine group of brightish stars which stretches from here through ε to β (4.5).

5. χ^1 (6.3) lies in a pretty X-shaped group.

6. σ . A fine area of many bright stars.

Wide doubles

$\eta^{2,3}$. Rather too wide for most binoculars.

Close doubles

χ^1. An easy double of 6.2 and 7.2, 140" apart.

P.179. An equal 8m pair separated by 185".

P.180. Very similar to the preceding object, this makes a neat little triangle with nu and pi.

P.181. Quite near to P.180, this is a yellow pair of 7.3 and 7.7, 289" apart.

P.182. Two 8m stars separated by 130 seconds of arc.

P.184. Another 8th-magnitude couple, but this time only 64" apart. Lies 1°S. of β.

Variable stars

X (8.0-9.2) Lying halfway between ω and γ^1, this can be identified from the rather singular line running from W. to E. Closely North of X is a little triangle of 8.6, 8.9 and 9.1. A red variable.

GRUS

A fine, conspicuous group with some good sweeping, and prominent wide double stars. Incidentally, the crane it represents is the avian, not mechanical, variety!

Groups of stars

1. γ (3.2) is one of a bright Y, with a wide pair at its branch.
2. φ (5.5) also lies in a bright, attractive field.

Wide doubles

$\mu^{1,2}$. A 5th-magnitude pair with a further star making a wide triple.

$\delta^{1,2}$. A brilliant wide double of the fourth magnitude.

$\sigma^{1,2}$. A similar, but rather closer object with both components of magnitude 6.

ν . Not far away, this has a 7m attendant to the North.

Close doubles

Δ249. A close, difficult object of 6.5 and 7.4. Distance just 27".

P.139. A fine seventh magnitude pair 167" apart. Both orange.

Variable stars

π^1 (5.4-6.7) The slightest optical aid will show this star of the rare spectral type S, which indicates a high proportion of the element Zirconium in the star's atmosphere. The variable makes a large triangle with the bright star Alnair (alpha Gruis) and a 6.2m star, and it has a companion in π^2 (5.8) just to the East.

HOROLOGIUM

Often thought of as a poor group, this actually contains several interesting objects; unfortunately, the lack of bright stars here makes it a bit difficult to find some of them. Included on the Eridanus chart.

Groups of stars

1. η, ζ and ι are members of a large diamond-shaped group. The region around the first of these stars is especially fine.
2. 04h 03m, - 45°. Large collection of faint stars, best seen with powerful binoculars.
3. ν (5.4). Sweep from here Southwards to the neighbouring group, Hydrus.

Close doubles

P.6. Difficult because of the faintish companion; mags 5 and 8, 71" apart.

P.17. A fine triple star of mags. 6.1, 7.5 and 8.0. The separations are 83 and 93 arc-seconds.

P.18. Wider but more difficult. The faint comes is 143" away, and of 8.7m.

Variable stars

R (4-14) This star has one of the largest ranges of any variable, though a normal maximum is magnitude 6. It forms a long triangle with a 6.1m star 1° to the South, and a 7.1m object to the W., much closer. Additionally, the stars of group (1), all of 5.3m, can be used. Approximate dates of maxima are given in the appendices.

V (7.3-8.4) A deep red variable for which a chart is supplied.

HYDRUS

A polar group with some fine fields and wide pairs, but not much else.

Groups of stars

1. Long, and appropriately sinuous line extending from the wide double β in a southerly direction.
2. 01h 50m, -77°. Large group of about 10 seventh-magnitude stars.
3. 03h 23m, -77°. A smaller but similar concentration.
4. 03h 20m, -70°. Large, regular hexagon.

Wide doubles

$\pi^{1,2}$. Equal wide double. There are several other fainter wide pairs in Hydrus, notably in the areas around epsilon and zeta.

INDUS

A dull group to the eye, but revealing many interesting objects when examined with binoculars, fitting for its purpose of honouring the original inhabitants of the Southern hemisphere.

Groups of stars

1. 21h 05m, -46°. Splendid large group of twelve stars of magnitude 8 and brighter, plus many fainter ones.
2. 21h 17m, -50°. Several groups arranged in lines. Note a beautiful little group of six bright stars to the South.
3. ε (4.7) is situated in a beautiful coarse field. One of the closest stars to the Sun incidentally, only 11 light-years away.
4. ι (5.2). This is the chief star of a large Y, from which a long, faint trail extends to just North of μ (5.2).

5. o (5.5). One of a neat little triangle.

Close doubles

P.32. An easily-found attractively equal pair 138" apart.

P.33. A fine triple star with the two fainter members of 7.0 and 7.9 being 153 and 191 seconds from the 6.8m primary.

P.39. This is one of the members of group (5), and is a good object for medium-sized bins. Magnitudes 7.3 and 7.5 and distance 181".

P.137. Though nearly equal in brightness, both these stars are quite faint at magnitudes 8.2 and 8.4. Distance here is an easy 238" of arc.

Variable stars

T (5.7-7.4) An excellent star for the beginner. I have provided a chart.

LUPUS

A fine milky way constellation which, though strictly wholly visible from parts of the USA, has been included among the Southern groups. Excellent low-power fields.

Groups of stars

1. ι (4.1) lies in a brilliant field of stars.

2. Large bright arc extending from δ (3.4) to ϕ^1(3.6).

3. Bright Y that includes θ , a fine wide triple star.

4. π, λ, ε, μ and κ . Together with e, these form a brilliant wide group.

Wide doubles

$\tau^{1,2}$. A beautiful 4th-magnitude double in a rich field for small binoculars. One degree South is a fainter and slightly fainter pair.

β . A bright star with a sixth-magnitude companion.

δ . This is another bright star with a 6m attendant.

ν^1. Similar in appearance to the two previous stars, but fainter. Beautiful field with small glasses.

γ. This has three eighth-mag companions. One of a bright triangle.

Close doubles

κ . Only suitable for large glasses, this is made up of 4.1 and 6.4m stars a mere 27" apart.

μ . Even more difficult at 24" and mags of 4.4 and 7.2, though a wider pair lies closely South.

ι . Much easier at 4.1 and 7.6. Separation 166 seconds of arc.

ζ . Another hard pair to split, because of the magnitude difference. The stars of 3rd and 7th mags. are 72" apart.

Δ192. A beautiful equal pair separated by 35".

Δ178. The distance here is the same, but the stars are not quite as similar in brightness (6.4 and 7.1m).

P.58. An attractive equal pair of magnitude 7 separated by 319".

P.62. Rather more testing. Magnitudes 7.2 and 8.4, 151" apart.

Variable stars

FQ (8.0-9.0) This red variable in conveniently found between two stars of 8.1 and 9.2 which all lie in a straight line just North of the 5th-mag star η Lupi.

Clusters and Nebulae

NGC 5822. A cluster visible in small glasses as a large gleam of light.

MENSA

A small, totally unremarkable constellation containing part of the greater Cloud of Magellan. No star brighter than magnitude 5 makes it hard to find anything!

Groups of stars

1. β (5.3). Fine sweeping around this star, which lies between us and the Magellanic Cloud - though of course immeasurably nearer.

2. ξ (5.8). Note the regular curve sweeping away from this star. A smaller line emanating from delta converges with it.

Close doubles

P.143. A faint but wide pair of 325" distance. You will need a good atlas and large glasses to find the stars of 8.5 and 8.9, and if you do find it, you may notice a third star making a tiny isosceles
triangle.

P.144. Far easier and more impressive, this is a pair of seventh-magnitude yellow stars separated by 247 seconds of arc.

Variable Stars

TZ (6.2-6.9) A bright eclipsing variable with a period of 8.5 days. Both xi Mensae (5.9) and a star of 6.7m, about 1° South can be used to estimate TZ.

MICROSCOPIUM

A fine group for the binocular owner which, though visible from the USA, needs altitude to enable its faint stars to be seen properly. Shown on the chart for Grus.

Groups of stars

θ . An area of good sweeping, especially to the North.

Wide doubles

ζ . This 5.4m star has a slightly fainter companion closely North.

Close doubles

ι . Quite an easy object of magnitudes 5.1 and 7.7, 170" apart.

Δ236. This is the central star in a line of three between η (5.6) and iota. A good object for binoculars, with a separation of 58". Mags are 6.5 and 6.9.

Variable stars

T (6.3-8.0) A chart is supplied for this semi-regular variable, which lies in the Northern reaches of this little constellation.

MUSCA

A small but very interesting group near a rich region of the Milky Way.

Groups of stars

1. ε. Beautiful areas around this red star, which is slightly variable. Note especially a little group about a degree to the South.
2. 12h 12m, -74°. A long line of 7m stars.
3. 13h 33m, -66°. Tiny Y-shaped collection of seventh-magnitude stars, closely North of which is a brighter star with a coarse trail emanating from it.

Wide doubles

γ. A degree S. is a fine wide triple of 6.0, 7.0 and 7.0. Very close by is the Globular Cluster NGC 4372, which you may spot as a rather faint smudge of light. An interesting field.

ζ^2. A star with two distant companions.

δ. This 3.6m star has a bright neighbour to the North.

Close doubles

η. Owners of average glasses might like to try this pair of 5th and 8th magnitude stars, separated by 60 arc-seconds.

P.145. A lovely seventh-magnitude pair, both stars blue and 164" apart.

P.146. Not far away, this is a slightly closer object of magnitudes 6.4 and 8.4. Large glasses may show a third star of 8.6m.

Variable stars

BO (6.0-6.7) A good star for small bins, which will show its red colour very well, though its amplitude is rather small. A chart is supplied in the appendices.

DZ (8.3-8.9) A small-amplitude Algol type eclipser, this is the Eastern-most member of a little triangle closely NE of a magnitude 5 star. The other members of the triangle are of 7.9 and 8.3 magnitude, with a small threesome of 8.1, 8.7 and 8.9 to the West.

Clusters and Nebulae

NGC 4372. A large, rather dim globular cluster near gamma.

NGC 4833. This is a smaller and brighter object of the same type. Use the chart here to find them.

Fig. 33: Two globulars in Musca. Field size is 5 degrees.

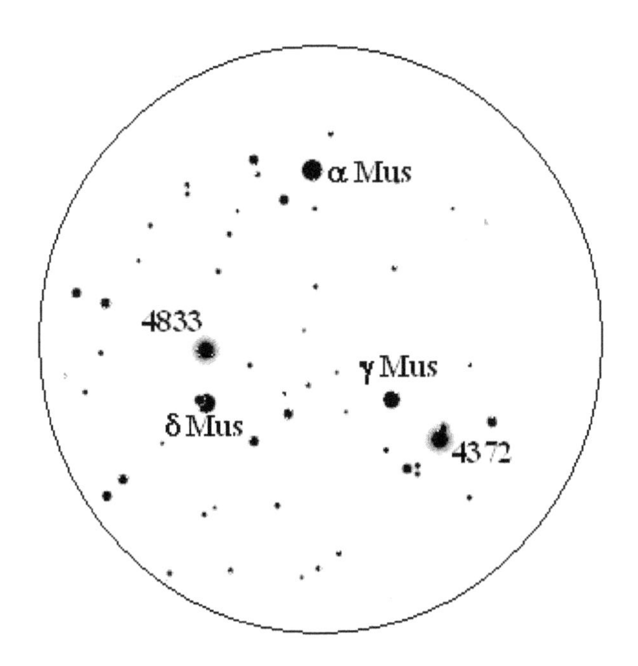

NORMA

A small, but extremely rich, constellation in the Milky Way; few really bright stars - but hordes of faint ones.

Groups of stars

1. κ (5.1). Brilliant region of many small groups and clusters. Sweep South to the slightly brighter i1

2. 15h 44m, -52°. A field of fine groupings and lines of stars.

Close doubles

ε . A difficult test for good, large binoculars, the mags are 4.8 and 7.5 and separation only 25".

P.28. Much easier at 114" and magnitudes of 6.7 and 7.9.

P.61. Though rather faint, these 8m stars are quite easy at 106".

P.64. A lovely equal pair of magnitude 7, 216" apart. There is another pair closely to the East of this, though slightly less equal in brightness.

Variable stars

R (5-12) Another star like R centauri, and though the minima of this star will be out of reach with normal bins, you can easily cover the rest of its changes in light. To find it, locate the 5m star η . Our star is just 5° away, and is the last in a regular bright line ending just N of eta. The other members are of 6.3, 5.8 and 6.0m. The second of these has a 6.8 South and a 7.5 North, while R itself makes a tiny triangle with two stars to the S, of 7.3 and 8.9. Predictions for this star's maxima are included.

S (7.0-8.0) A Cepheid variable with a 10-day period, this lies in the centre of the beautiful open cluster NGC 6087. Therefore, the area is actually too crowded to give a meaningful description!

Clusters and Nebulae

NGC 6067. A fine open cluster visible with small binoculars, in a little triangle of bright stars.

NGC 6087. Many members of this magnificent cluster can be seen with binoculars. Note the contrast in the richness of the field to the South as opposed to the N.

NGC 6167. A gleaming open cluster, close to an unequal double.

H.10. A large, brilliant group of stars not far from NGC 6067, with many doubles and triples seen in average glasses.

OCTANS

An unremarkable group whose sole claim to fame is that it contains the South Celestial Pole. The "official" S. Pole star, sigma Octantis, is too faint to be employed for any useful purpose.

Groups of stars

1. ω (5.9) and rho (5.7) are members of a pretty arc of stars.
2. 08h 00m, -84°. Fine V, extending over a sizeable area.
3. γ^1 (5.1) forms a singular group with the slightly fainter γ^2 and γ^3 .
4. 18h 20m, -80°. Long Y of brightish stars.
5. ν (3.7). A fine area between this, Octans' brightest star, and nu Indi.

Wide doubles

$\pi^{1,2}$. These stars, both of 5.6m, are rather too wide for most binoculars, but small field glasses should bring out their charm.

Close doubles

P.96. An obscure 8th-magnitude pair 202 seconds of arc apart.

PAVO

An impressive constellation, well-befitting its noble original. Packed with interesting objects of all descriptions, notably several interesting binocular variables. Good fields, especially in the Northern parts.

Groups of stars

1. π (4.4). Fine fields around this star.

2. θ (5.9). Another star in an attractive area.

3. 19h 20m, -63°. Beautiful field of tiny stars.

4. 19h 42m, -66°. Superb field, extending over several degrees to the North.

Wide doubles

μ^1. This forms a fine wide pair with μ^2. Delta Pavonis, in the same field, is also a wide double as well as being a naked-eye variable star.

Close doubles

ν . The primary is 4.8 and the secondary 3 magnitudes fainter, at a distance of 99" of arc.

σ . A good binocular double of 5.5 and 7.1. Distance is 135".

P.30. This is a fine pair, even for the smallest glasses, both 7m and yellow. Easy at 178".

P.94. Faint but equal, separated by 180". Near the following star.

P.95. Less alike in brightness (6m and 8m) and slightly closer at 163".

P.98. A lovely seventh-magnitude pair, white and orange in colour, and 142 seconds of arc apart.

Variable stars

S (6.7-9.3) A rewarding star to observe of the "SRa" class - that is, a semi-regular variable but with a persistent period. Many red variables spend long periods changing little, or not even varying
at all, but this isn't one of them. A chart is supplied for it.

X (7.9-9.3) Another red variable, shown on the same chart.

Y (5.7-8.5) This is also a good star for binoculars, with a large range of light variation, and again a chart is provided.

Z (7.7-9.1) This variable makes a triangle with two other stars of 6.7 and 7.6. Z itself lies at the centre of an equilateral triangle of 7.4, 7.9 and 8.4.

KZ (7.7-9.3) The best guide to this red star is the fifth-magnitude o. This makes the sharp point of an isosceles triangle with KZ and a 6.8m star with an 8.3m attendant. On the opposite side from

these is another star of 9.0m.

OW (7.8-8.9) Quite easy to find near the fourth-magnitude pi, the field of this red variable is unfortunately rather devoid of useful stars.

Clusters and Nebulae

NGC 6752. A globular cluster visible as a large misty gleam behind a star of magnitude seven.

PHOENIX

Another of the Southern Birds, and a fine group to both the eye and binoculars.

Groups of stars

1. υ (5.2) Fine sweeping around here, especially SE and NW.
2. ϕ (5.0) lies in a beautiful region of coarse groups of stars.
3. GC 705. An orange star of magnitude 5.9. Note a tiny triple close by, and several long lines of faint stars.

Wide doubles

ι. This has a 6m companion to the S., and a neat little triangle Northwards.

Close doubles

P.142. Forming a long isosceles triangle with two bright stars, this is an attractive orange pair separated by 36", though quite faint.

Variable stars

S (7.4-8.2) This red variable is well-served with useful stars; three, just to the South, of 7.6, 7.7 and 6.7, and just East of the variable is an 8.9m star.

SX (6.5-7.5) A remarkable star to observe, this is a (very) short-period Cepheid of the type known as the RR Lyrae stars. In fact, these stars are also further divided on the basis of period, and this
is the prototype of these ultra-short period stars. The period of SX Phe is only just over 1 hour - so you can actually watch it brighten over the course of a few minutes! Take a look at it, for instance, between estimates of other stars, taking note of the time as exactly as you can. I have supplied a chart for this easy variable - though you might not always catch it on the rise, of course.

PICTOR

An apparently faint group to the eye, though like Camelopardalis in the North, it has many stars just beyond naked-eye visibility. A fine binocular group, included on the map for Dorado.

Groups of stars

1. $\eta^{1,2}$. A very wide pair, close to a beautiful little bright Y.

2. 05h 25m, -52°. Large triangle, including θ. Note a fine V of stars near ζ (5.2) and numerous wide groups.

3. β (3.9). Note an attractive group to the South, of which the southern member is double.

Close doubles

θ . A beautiful object of 6.3 and 6.8m, separated by 38" of arc.

P.19. Another attractive pair, both of 7.4m. Distance is 121".

P.21. A rather harder one this time, on account of the dim (8.0) secondary star. Separation in this case is 64 seconds.

P.109. A lovely bright double of 5.5 and 6.5, 197" apart. Easy with any form of optical aid.

P.110. Mentioned under group (3), the mags. here are 6.4 and 7.6 and the distance is 84".

P.112. Magnitudes of 6.0 and 8.6 make this a difficult object to split, though it is quite wide at 189" of arc.

Variable stars

W (7.5-9.3) This is a deep red star of spectrum N, though if it is faint, the light level will be too low for the cones of the eye to register much in the way of colour. A chart is given.

PUPPIS

A really splendid constellation which, though accessible from many Northern countries, seems to belong with Vela and Carina in the good ship *Argo*.

Groups of stars

1. Large, brilliant distorted hexagon with a bright central star. Includes m, and adjoins another rich area around p, k, 3 and ξ (3.5).
2. Large, rather amorphous group containing 9 and 10.
3. c (3.7). This red star is the centre of the brilliant cluster NGC 2451. A bright wavy line trails from here to zeta, which is one of the hottest stars in the sky.
4. Small but bright oval group including q and r, with a splendid group below r, and a wide double in the centre. r is slightly variable and is also known as MX Puppis.
5. Small group attending π including the wide double v^2, also known by a variable star name, this time NV Pup.
6. E(5.4) lies in a fine area of pretty groups.
7. F(4.2) is another star in a brilliant region.
8. H(5.1). This is the southern member of a large bright V.

Wide doubles

n. Note a fine wide pair to the South.

h3834. A telescopic double, but with two binocular companions.

GC 8960. This 5.1m star has a delicate wide triple a degree to the S.

2 and 4. An unequal but bright pair in a beautiful field.

k. A bright star with two companions, of magnitudes 6 and 8.

ξ. This bright object has a yellow neighbour of 5.3m.

11. Good glasses show three faint acolytes attending this fourth-magnitude star.

v^2. See above. Magnitudes 4.7 and 5.1. Beautiful. They form a threesome with the nearby third-magnitude star π.

d^1. This has a faint neighbour to the North. d^3 nearby is a more equal pair. This whole area is laden with clusters and sprinkles of stars.

Close doubles

Δ38. A bright but close pair: mags 5.8 and 6.9, distance only 26".

h4038. Very difficult, of the same separation as the previous star, but more unequal magnitudes.

h5443. Rather easier, though the companion is still rather faint at 8.0m. Separation is 214".

P.7. A faint pair 134" apart immersed in the cluster NGC2220.

P.9. A fine double, though rather disparate in magnitudes of 5.4 and 7.0 and 61" distance. Beautiful field.

P.10. Another splendid double, near r. Magnitudes 5 and 6, distance 67".

P.22. A attractive red and blue pair 191" apart. Equal magnitudes.

Variable stars

L^2 (2.6-6.2) This brightest example of the semi-regular variables needs only the slightest optical aid to show it even at minimum. The following stars, all shown in Nortons 2000, can be used as comparisons: σ (3.3), I (4.5), L^1 (5.0), C (5.3) and M (6.0).

RU (7.6-9.3) A deep red variable best found from the bright star ρ Puppis. It is the northern member of a little equilateral of 8.7 and 9.0m stars, and a small triangle to the SE which points to RU is

of 65, 8.1 and 8.7.

RY (7.0-8.0) This is only a probable range, as there is some doubt as to whether this star is actually variable. So you can try and see for yourself, I have provided a chart.

OT (8.1-9.1) This red star is the easternmost of a neat right-angle (other stars 7.7 and 8.1, between which are two fainter comparisons of 8.7 and 9.0).

QY (6.2-6.7) A peculiar variable which, although of small amplitude, is easy to find between 2 and 6 Pup. It is the North member of an attractive double whose other member is of 6.8m. Observe

it once a week.

Clusters and Nebulae

NGC 2477. Visible as a large misty spot attending b (4.5), and in a brilliant region.

NGC 2546. Large glasses will reveal several stars in this cluster, which is in a beautiful field.

M.47 (NGC 2422). Most binoculars will show about a dozen stars in this bright cluster, which does not appear as nebulous as most.

M.46(NGC 2437). A rather faint cluster for most binoculars.

M.93(NGC 2447). In 7x50s this appears as about half a dozen stars before a nebulous glow.

PYXIS

A small group, representing the compass of the Argo (I know the Greeks are supposed to have had a word for most things, but did they have one for "compass-box" even before the compass had been invented? No - *Pyxis* actually just means 'box') Several good binocular double stars and a notable eruptive variable. Shown on the map for Antlia.

Groups of stars

1. 08h 30m, -32°. Small T-shaped asterism that includes a red star.

2. Bright parallelogram that includes α and β .

Close doubles

P.186. A challenging object for large binoculars because of the faintness of the stars. Mags are 8.3 and 9.0, separated by 100". You may see a third star of magnitude 9 here.

P.187. Slightly easier, though still faint; mags 7.5 and 9, distance 122".

P.191. Again, rather difficult because of the faint companion, though a bit wider at 185 seconds.

P.192. Another faint pair, though more equal at 8.2 and 8.7. Readily swept up on a good night closely South of the 4.9m star λ Pyxidis.

P.193. A critical test for good-quality large glasses, this pair is of magnitudes 6.7 and 9.2 separated by 68".

P.194. It is refreshing to turn at last from these testing, faint objects to one not quite so taxing! This is a beautiful equal pair of magnitude 7 whose stars are 162" apart. Can you see the colours of these stars at all? It is readily found between alpha and a 5th-mag. star which is a telescopic pair called h4115, lying about 2° North of beta.

P.196. These stars of 7.6 and 8.6m are 152" apart.

P.197. Hard for average-sized binoculars, this pair is of magnitudes 7 and 9, and separated by 129 seconds.

Variable stars

T (7-14) This star is actually a species of nova, but one not content with exploding just the once! Its first outburst took place in 1902, and since then it has repeated this errant behaviour on several occasions. It forms a long triangle with two stars of 6.9 and 7.8m which make good comparison stars. Watch the area on every clear night.

RETICULUM

A compact, pretty little group containing some fine objects. Included in the map for Dorado.

Groups of stars

1. 03h 30m, -58°. Large group of assorted brightnesses.

2. κ (4.8). Note a curious-looking, rather square collection of faint stars to the North.

3. Symmetrical semi-circle that takes in both η (5.2) and θ (6.2).

4. β (3.8). Guide star to a large triangle to the South. The northern member of this is one of a small Y, and is near two wide pairs.

Wide doubles

ζ^2. This forms a striking double with ζ^1, only slightly fainter at 5.5m.

Close doubles

h3670. A pretty, if rather faint pair of mags. 6 and 8, 32" apart. Closely S. is the Mira star R Reticuli, which can attain magnitude 7 at times, thus making a temporary double with this star.

P.154. Quite a difficult pair, due to the faint (9th-mag) companion. Distance is 58 seconds of arc.

P.155. Easier and more equal in brightness, these 8m stars are separated by 162 seconds.

SCULPTOR

A group usually maligned as containing little of interest, though binoculars disclose some notable objects. It contains the South Galactic Pole, though most of the many galaxies visible here are not within binocular range. This is one of those groups with a curtailed name, for it was originally *Apparatus Sculptoris*. The sculptor since appears to have lost his apparatus!

Groups of stars

1. 23h 18m, -29°. Small, rather square group, with a bright stream going off to the North.

2. ζ (5.0) lies in a beautiful field, including a nearby small triple.

3. Issuing from the 5.7m τ is a fine, almost straight, line that runs towards α (4.4). Tau itself is in a fine bright group of stars.

4. 01h 36m, -37°. Fine large trapezium of 5.5, 5.6, 5.9 and 6.1. This group actually spills over into neighbouring Fornax.

Wide doubles

μ . This 5m star has a 6th-mag companion.

GC 33175. A closer pair, with another fainter star to the North.

S. Just South of this variable is a fine wide pair of 6.7 and 7.4, both orange in colour.

Close doubles

P.198. A rather faint pair of 7.9 and 8.5, but wide at 141".

P.199. Mags of 6.5 and 8.7 make this a more difficult object. Distance 105".

P.200. This and the next star are both hard objects because of the faint secondaries. Distance here is 147".

P.201. Slightly easier, this forms an isosceles with P.200 and the 4th-magnitude δ .

P.203. A rather faint triple star. The 8.1m primary has two attendants, both of 8.4m, at 166 and 227 seconds of arc distance.

Variable Stars

R (5.8-7.7) A chart is supplied for this semi-regular variable.

Y (7.5-9.0) Another red star, easy to find from a bright right-angle East of Fomalhaut (5.6, 6.5 and 6.6) with the Southern member of this triangle being the westernmost of a regular line of four whose other members are the variable, then 8.4 and 7.7 stars.

SW (7.3-9.3) This is a red supergiant with a good magnitude range but rather lacking in good comparisons. It lies in line with two bright stars (5.7 and 5.2) and just N. of the variable is a little line of three, the middle one of which is a small-range red variable, XY Scl. On either side of it are 7.1 and 8.2m stars.

Clusters and Nebulae

NGC 253. A large galaxy, visible in good glasses as an elongated blur.

NGC 288. A bright globular cluster, not far from NGC 253, the previous object.

NGC 55. A slightly fainter galaxy than 253. It can be found by the regular line of three 6m stars nearby.

TELESCOPIUM

A rather appropriate name for a constellation, I have always thought! There used to be a Telescope in the Northern sky for a time, called Telescopium Herschelii which was near the traditional group of Gemini and was meant to commemorate the discovery of Uranus in that part of the heavens. The southern version likewise lies near the Milky Way and contains several interesting objects.

Groups of stars

1. α (3.8) makes a large, right-angled group with ε and ζ . Note a faint, wide triple star between these two.

2. Bright quadrilateral including κ, λ and ρ . The area around κ is particularly well-populated.

3. η (5.2) is another star in a superb field, and one of a large group rather reminiscent of the "steep-roofed house" shape of the constellation of Cepheus.

Wide doubles

$\delta^{1,2}$. A fine, fifth-magnitude wide pair.

GC 25861. A 5.5m red star with a 6.3m white companion.

Close doubles

Δ227. Bright but close, the magnitudes are 6.1 and 6.7, and the distance just 27".

h5114. Wider but less equal; 5.6 and 8.3, distance 71 seconds of arc.

Variable stars

RX (7.5-9.0) A red star, found to the South of a bright, wide double, itself not far from another similar pair (β Sagittarii). The variable has a northerly companion of 9.0, with another of 8.3 making a little triangle.

BL (7.2-9.3) A faint eclipsing star, but with a good range. A bright star lies to the N., North again of which are two others of 6.3 and 7.2m. A fainter comparison of 8.6 makes a right-angle with BL and the nearby rho Tel.

HO (7.9-8.6) A star of similar type, but much smaller amplitude. It is one of a little quadrilateral whose other stars are of 7.7, 8.2 and 7.5m.

TRIANGULUM AUSTRALE

A brighter, more symmetrical and far richer group than its Northern counterpart. On the same map as Circinus and Norma.

Groups of stars

1. β (3.0) lies in a magnificent area, especially to the N.

2. α (1.9). In a flash of imagination, aviators christened this star Atria (Alpha TRIanguli Australis - get it?). Note two beautiful sprays; the brighter extends to δ (4.0) and the smaller, though containing more stars, sweeps out in the same direction. A similar arc runs through ζ (4.9).

Close doubles

h4809. A difficult pair, due to the magnitude 9 companion, 48" of arc away.

Variable stars

S (6.5-7.7) A Cepheid variable with a period of 6.3 days. The stars described under the following variable can be used, though take care not to confuse S with a slightly fainter neighbour of 8.2m.

U (7.6-8.4) A star of the same type, that lies between δ and a sixth-magnitude star. It lies in a beautiful little wandering line of five stars which run, in order from delta, 7.4, 6.9, variable, 7.7 and 8.4.

X (5.6-6.6) The smallest glasses can observe this deep red variable, for which I have supplied a chart.

Clusters and Nebulae

NGC 6025. A beautiful binocular object in a beautiful field, with many faint stars and a misty glow being visible.

TUCANA

A good constellation for the owner of binoculars, containing some fine doubles as well as two famous globular clusters. To be found on the same chart as Indus.

Groups of stars

1. γ (4.1) is one of a small parallelogram, two of whose stars are unequal wide doubles.

Wide doubles

η . Note a fine orange pair of 5.7 and 6.9 about two degrees away.

$\beta^{1,2}$. (Both 4.5, double). These form an imposing group with β^3 (5.2).

κ . Itself a telescopic pair, this has a more distant neighbour of 7.6m.

λ^1. Again a double, this has a distant binocular associate of 6.3.

Close doubles

β . A close but beautiful equal pair (see above) 28" apart.

λ^1. Very difficult again, at only 22". Magnitudes are 5 and 7.

P.76. An excellent binocular double formed by two orange stars of magnitudes 6 and 7, and 250 seconds apart.

Clusters and Nebulae

NGC 104. This is generally held to be the brightest and finest globular in the sky, though there is a bit of friendly rivalry from Omega Centauri for the honour. Also called 47 Tucanae (why, I have never been able to figure out), this is easy with the naked eye, and a brilliant, blazing object in binoculars.

NGC 362. Another bright globular, though rather in awe of its big brother above! The binocular shows this as a bright fuzzy spot, with the Small Magellanic Cloud as a brilliant backdrop.

VELA

A really awe-inspiring constellation from the binocular observer's point of view, this is full of wonderful fields of bright stars, a host of doubles, variables and clusters, as well as the brightest nova of the nineties.

Groups of stars

1. 08h 38m, -40°. Brilliant group of sixth- and seventh-magnitude stars.

2. d (4.1). Near this wide double is a stupendous clustering of many stars from the 5th to 8th magnitudes.

3. λ (2.2). Marvellous field, especially to the South and East.

4. q (4.1) is one of a brilliant V. It is situated right in the centre of an equilateral triangle of 6m stars.

5. m (4.6). One of a small arc that also includes u (5.3). It is near this star that the milky way is visibly broken to the eye. Of course, this is not an actual break, but is caused by large clouds of obscuring dark nebulae.

6. γ (1.8). The brightest star in Vela is one of the hottest and most luminous of the naked-eye stars and is in an incredible region, with arcs and streams of bright stars shooting out from it. To cap all this, it is also a wonderful double star!

7. o (3.7) is in another region of overwhelming splendour.

8. a (4.1). One of a brilliant cross of hot-looking stars. Indeed, the Northern observer gets the impression that this whole region of the sky is lit by stars more powerful, more numerous, hotter, and generally more exotic than his own, and constantly looks forward to the day when he can see them for himself (preferably from some silvery-sanded and palm-fringed shore!) The star is also a wide triple.

9. J (5.2). Beautiful field to the North.

Wide doubles

o. A brilliant wide triple of 3.7, 5.4 and 5.7, and chief star of the open cluster IC 2391.

Close doubles

γ . Large, good-quality bins, if possible with a good anti-reflective coating, will show this to be a beautiful pair of magnitudes 2.2 and 4.6, 41 seconds of arc apart.

t. Rather difficult; magnitudes 5 and 9, also 41" distant.

Y. This time, the magnitudes are the same as the previous star, but the distance is slightly greater at 50". Due to the way the early astronomers named the stars, we actually have two stars called Y Velorum; one of course is a variable star - this is the other one.

m. An unequal pair of 4.6 and 7.3, 129" apart in a good field.

P.13. Another wide pair, this time of 6.2 and 8.0, separated by 138".

P.14. Not far away, this is an attractive double of 7.1 and 7.5 with a separation of 155".

P.25. 185 seconds divide these stars of 5.9 and 7.7m.

P.121. A fine, easy object of 7.1 and 7.5, both yellow and 137" of arc apart.

Variable stars

S (7.7-9.5) You will need large bins to see this Algol-type star fall by nearly 2 magnitudes every 5.9 days. Closely North lies an eighth-magnitude variable called U Velorum which does not concern us because its range is too small (only 0.3m). However, just West of these two variables is a vertical line of N to S) 8.9, 8.8, 8.1 and 9.0.

X (7.3-8.6) This is a deep red variable readily found near a 6.3m star, which makes an equilateral with X and a 7.4m star. Closely West of X is a fainter star of 9.3m.

SW (7.4-9.0) A Cepheid with the rather long period for these stars of 23 days, easily found in a beautiful region half a degree East of the 5th-magnitude n Velorum. A small curved line just S has stars of 6.7, 8.1 and 7.4m, and just S of the central one is a fainter star of 8.7.

SY (7.4-8.7) A chart is given for this star as well as the next one.

WY (7.4-8.8) Another red variable, here made more interesting by the fact that it is not an ordinary red star, but a "symbiotic" binary subject to nova-like outbursts from time to time.

AI (6.4-7.1) A variable of the RR Lyrae type, with a period of only 2 hours! It is best found from a bright vertical line of three which lie between γ Velorum and ζ Puppis. The northernmost of the three has a distant neighbour of 6.4m which points straight to AI. The variable itself has a close accomplice of 7.4m directly South of it.

CV (6.7-7.4) An eclipsing binary in a crowded field. A chart is supplied.

Clusters and Nebulae

NGC 2547. A beautiful cluster with well over a dozen stars resolved in most glasses.

IC 2391. A brilliant group centred on o.

IC 2395. Yet another fabulous collection of bright stars that abound in this constellation.

VOLANS

A neat constellation to the eye on a transparent night, but otherwise dull, and rather unfitting to close the chapter on after the previous group. Appears on the map for Carina.

Groups of stars

1. 06h 54m,-70°. Small scattered group of bright stars.
2. κ (4.4). A pretty line of seventh-magnitude stars lies to the North.
3. Regular, sinuous line stretching from the telescopic pair L3846 Carinae to η Volantis (5.4).

Close doubles

κ . This very attractive double of 5.4 and 5.7 is interesting, as the fainter member has an 8th-magnitude companion at 40" distance. The main stars themselves are 65" apart.

P.147. An unequal pair (6.8 and 8.6) separated by 102" of arc.

CONSTELLATION CHARTS

These run parallel to the text as a rule; save for where more than one constellation appears in a chart, first the charts for Northern Constellations are shown, and then those for the South. All objects

mentioned in the text are identified. Not every 'ordinary' star is labelled. I thought it best to adopt this approach to avoid unnecessary cluttering. In the charts, where a close double star does not have a catalogue name (which is sufficient to identify it) duplicity is shown by a small line above or below the star's name, while with variable stars, the designation style is sufficient to show variability.

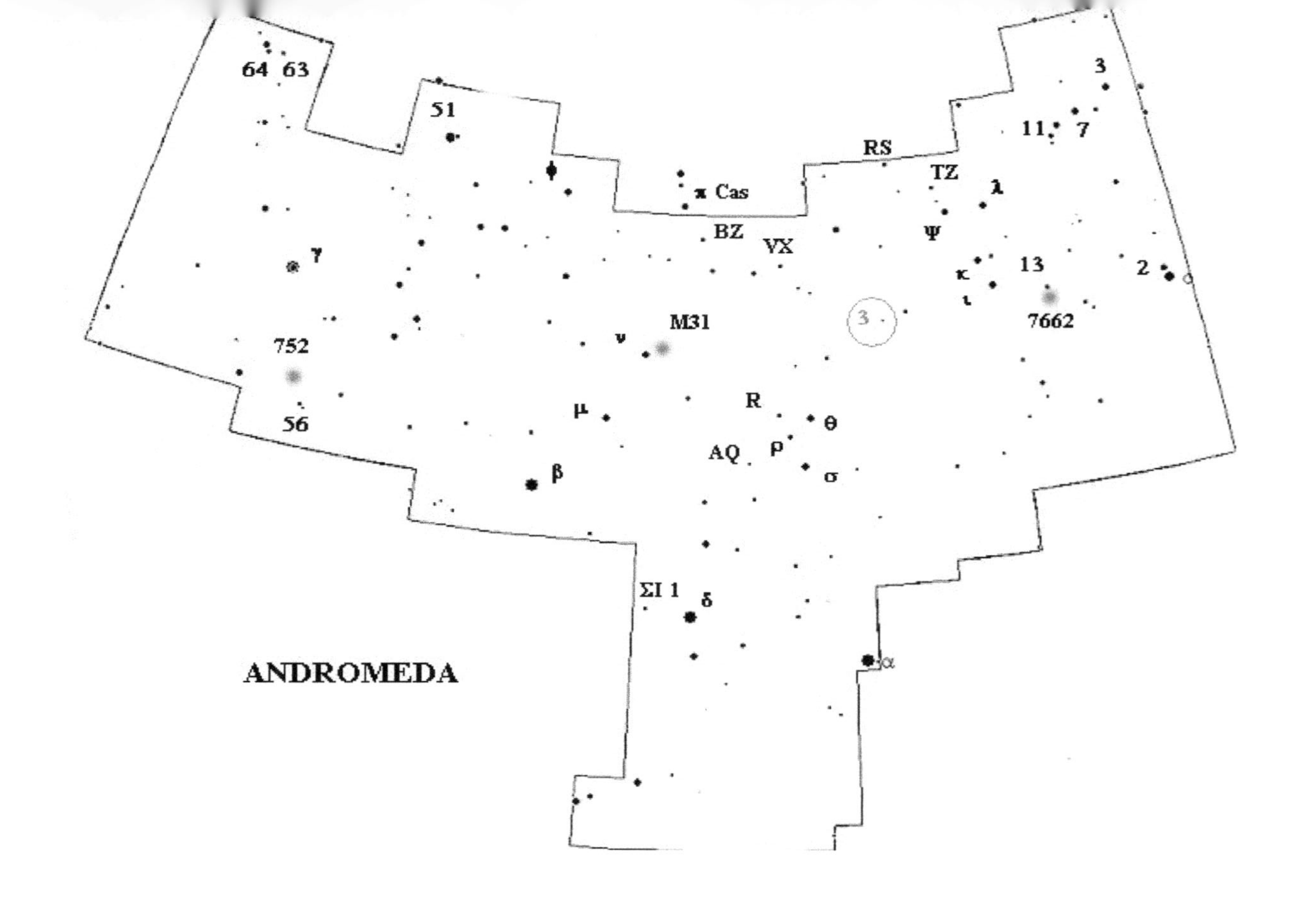
64 63
51
κ Cas
RS
TZ
λ
3
11
7
BZ
VX
ψ
γ
κ
13
2
ι
M31
ν
3
7662
752
56
μ
R
θ
AQ
ρ
σ
β
ΣI 1
δ
α
ANDROMEDA

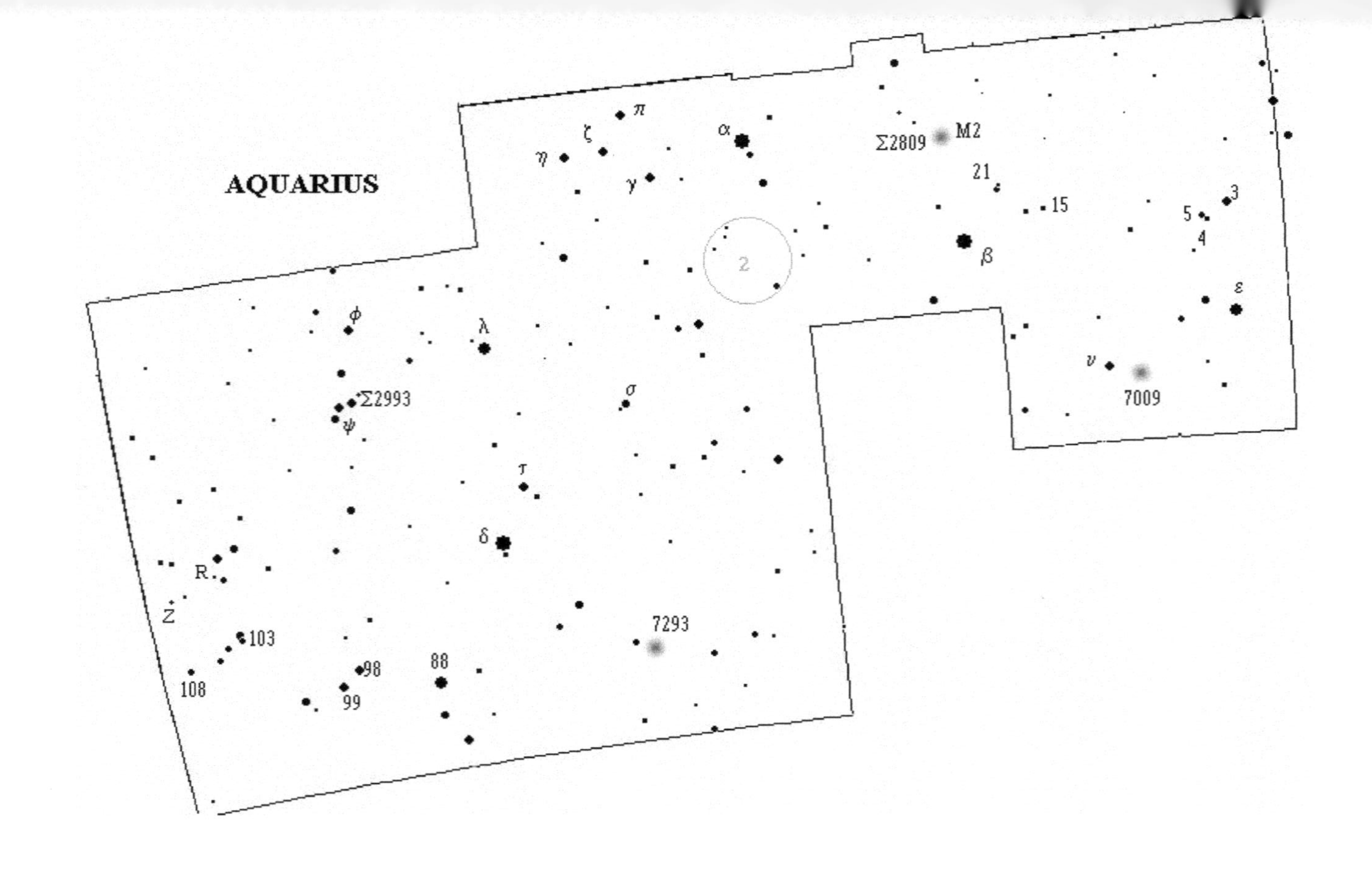
AQUARIUS
π
ζ
η
γ
α
Σ2809
M2
21
15
5
3
4
β
2
ε
φ
λ
Σ2993
ψ
σ
ν
7009
τ
δ
R
Z
103
108
98
99
88
7293

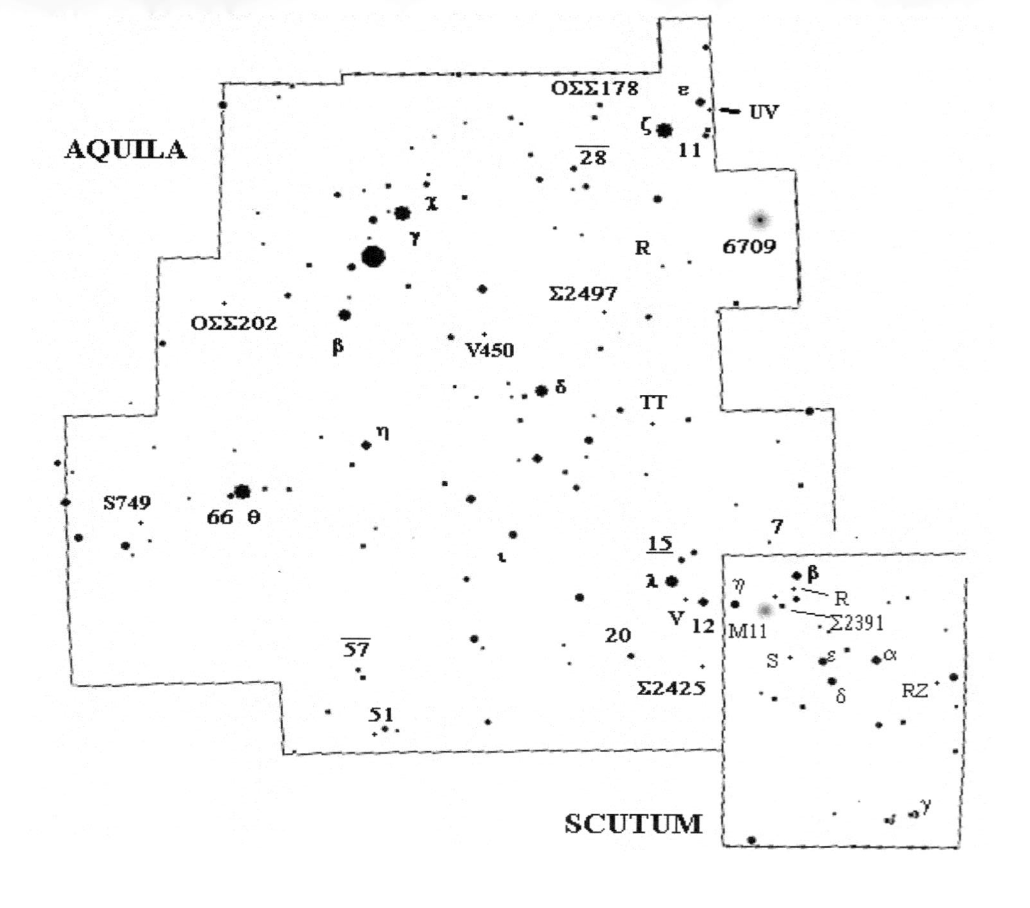
AQUILA
OΣΣ178
ε
UV
ζ
11
28
χ
γ
R
6709
Σ2497
OΣΣ202
β
V450
δ
TT
η
S749
66 θ
7
15
ι
λ
V
12
20
57
Σ2425
51
SCUTUM
β
η
R
M11
Σ2391
S
ε
α
δ
RZ
γ

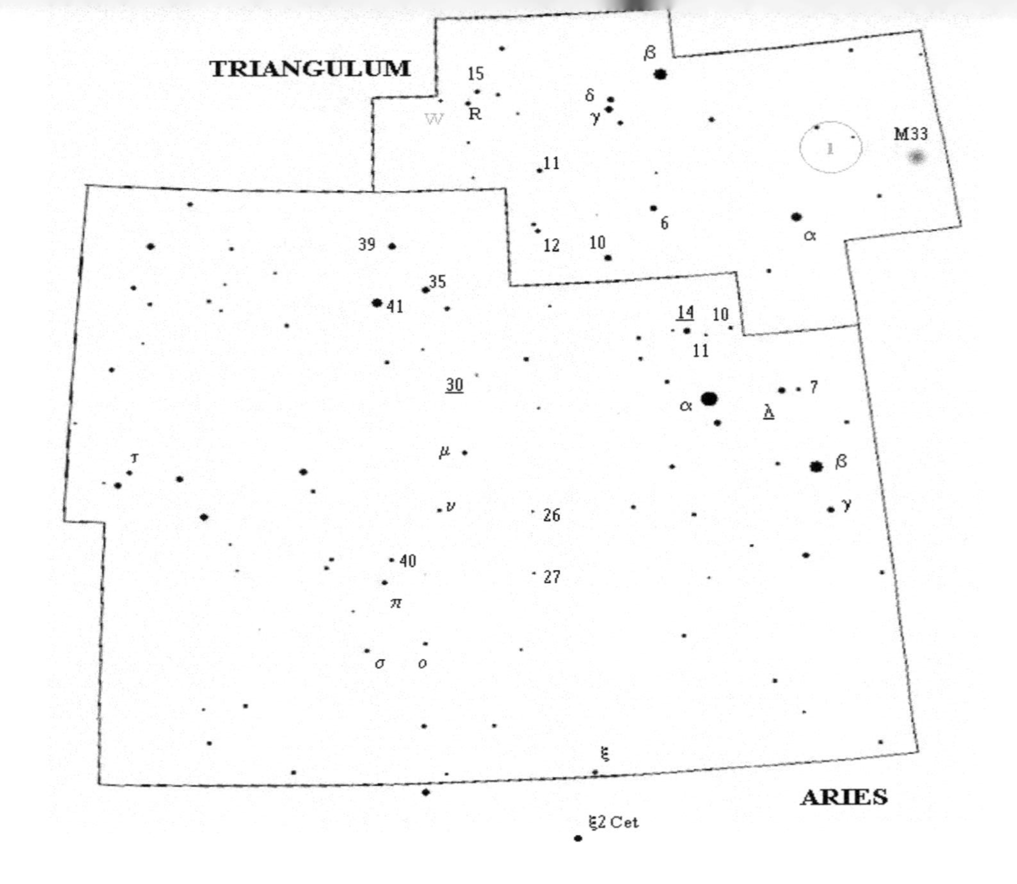
TRIANGULUM
15
R
W
δ
γ
β
11
M33
6
α
12
10
39
35
41
14
10
11
30
α
7
λ
μ
β
γ
τ
ν
26
40
27
π
σ
ο
ξ
ARIES
ξ2 Cet

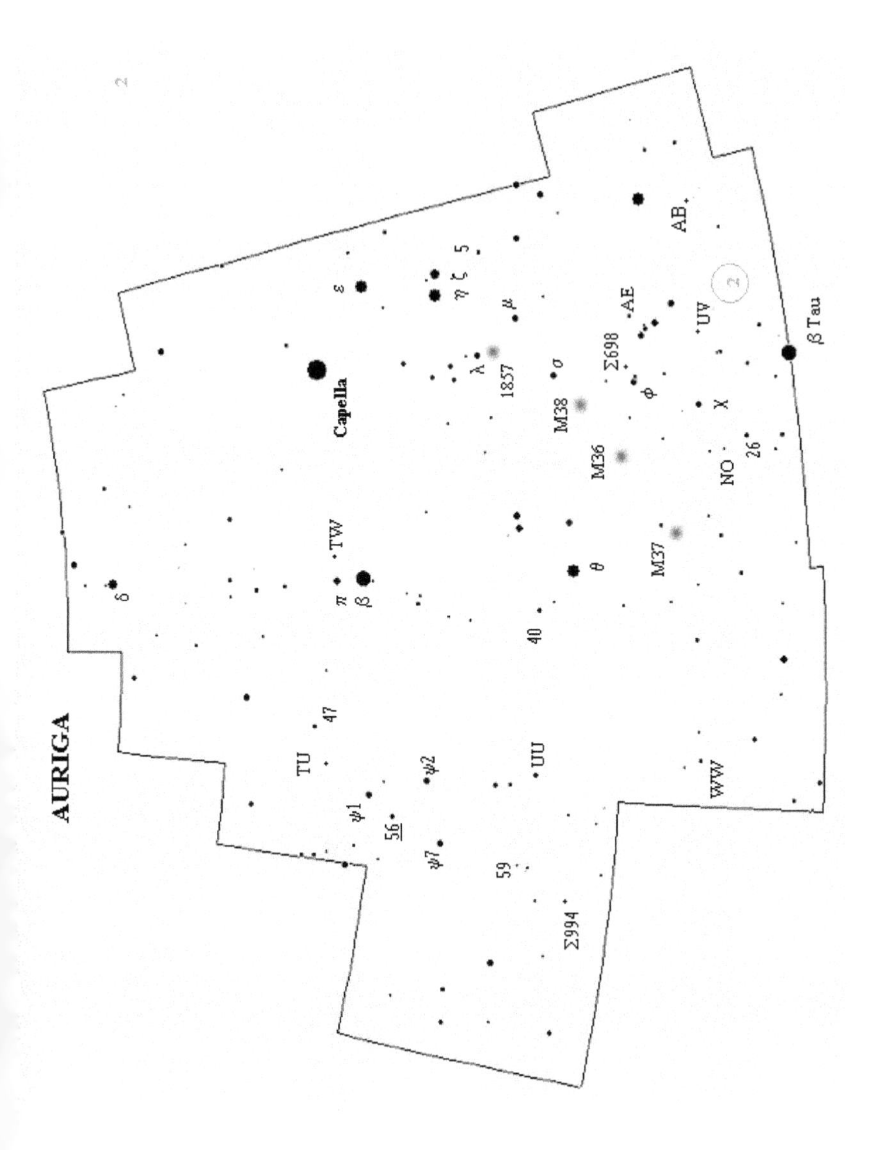
AURIGA
Capella
β Tau
ε
η
ζ
5
μ
λ
1857
σ
Σ698
AE
AB
UV
φ
χ
M38
M36
M37
NO
26
TW
π
β
δ
θ
40
47
TU
ψ1
ψ2
56
ψ7
UU
59
Σ994
WW

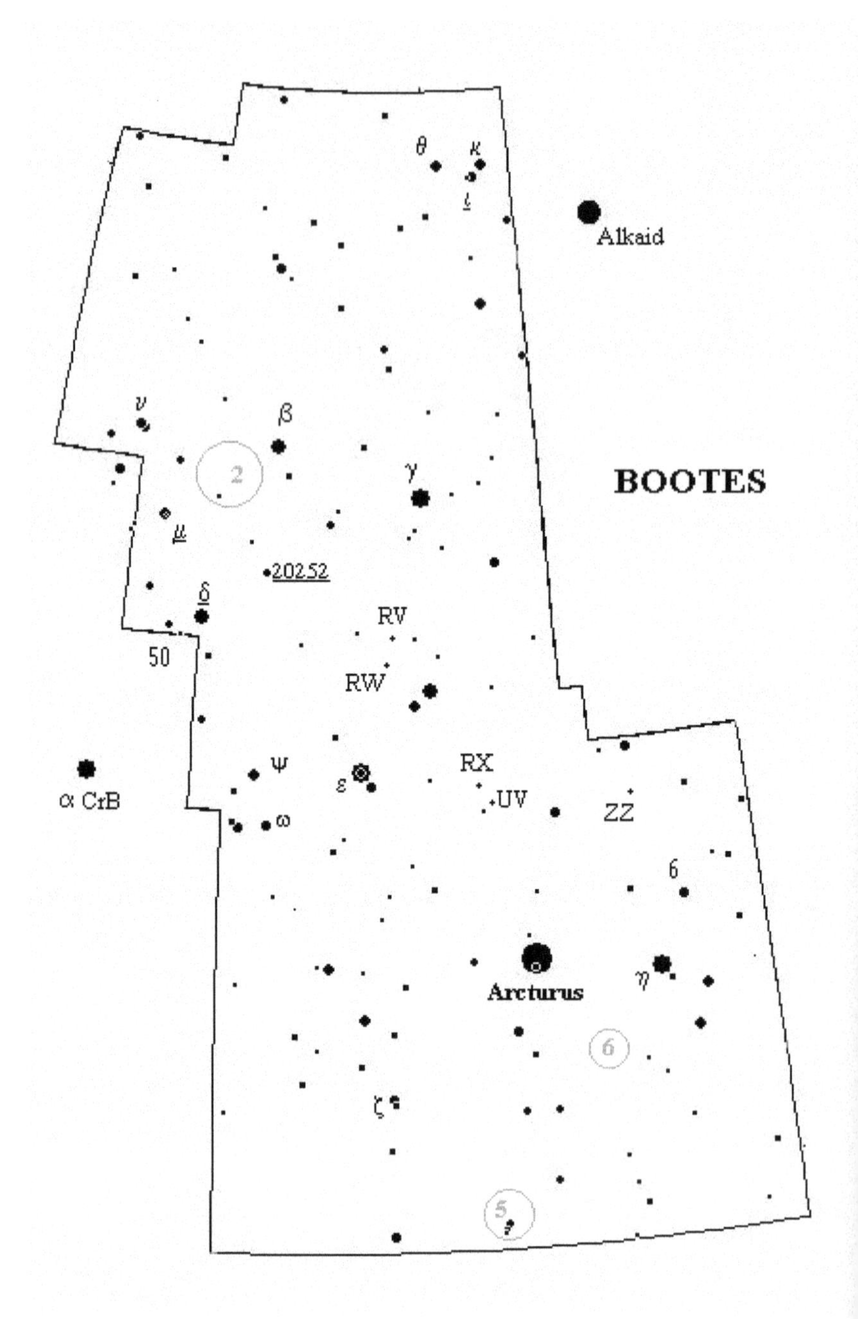
θ
κ
ι
Alkaid
ν
β
2
γ
BOOTES
μ
20252
δ
RV
50
RW
ψ
RX
α CrB
ε
UV
ZZ
ω
6
Arcturus
η
6
ζ
5

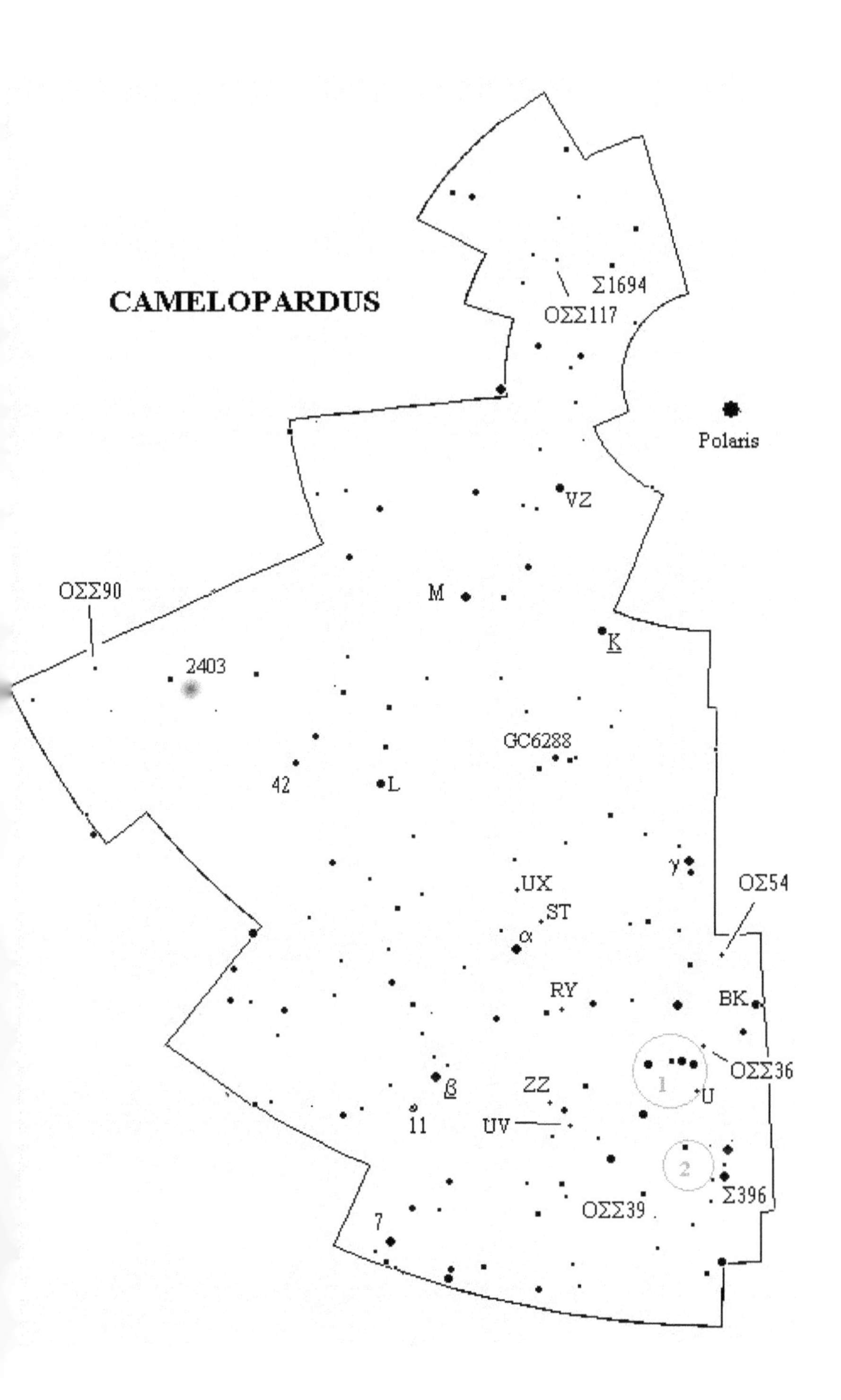
CAMELOPARDUS
Σ1694
OΣΣ117
Polaris
VZ
OΣΣ90
M
K
2403
GC6288
42
L
γ
UX
OΣ54
ST
α
RY
BK
OΣΣ36
ZZ
β
U
11
UV
Σ396
OΣΣ39
7

CANCER

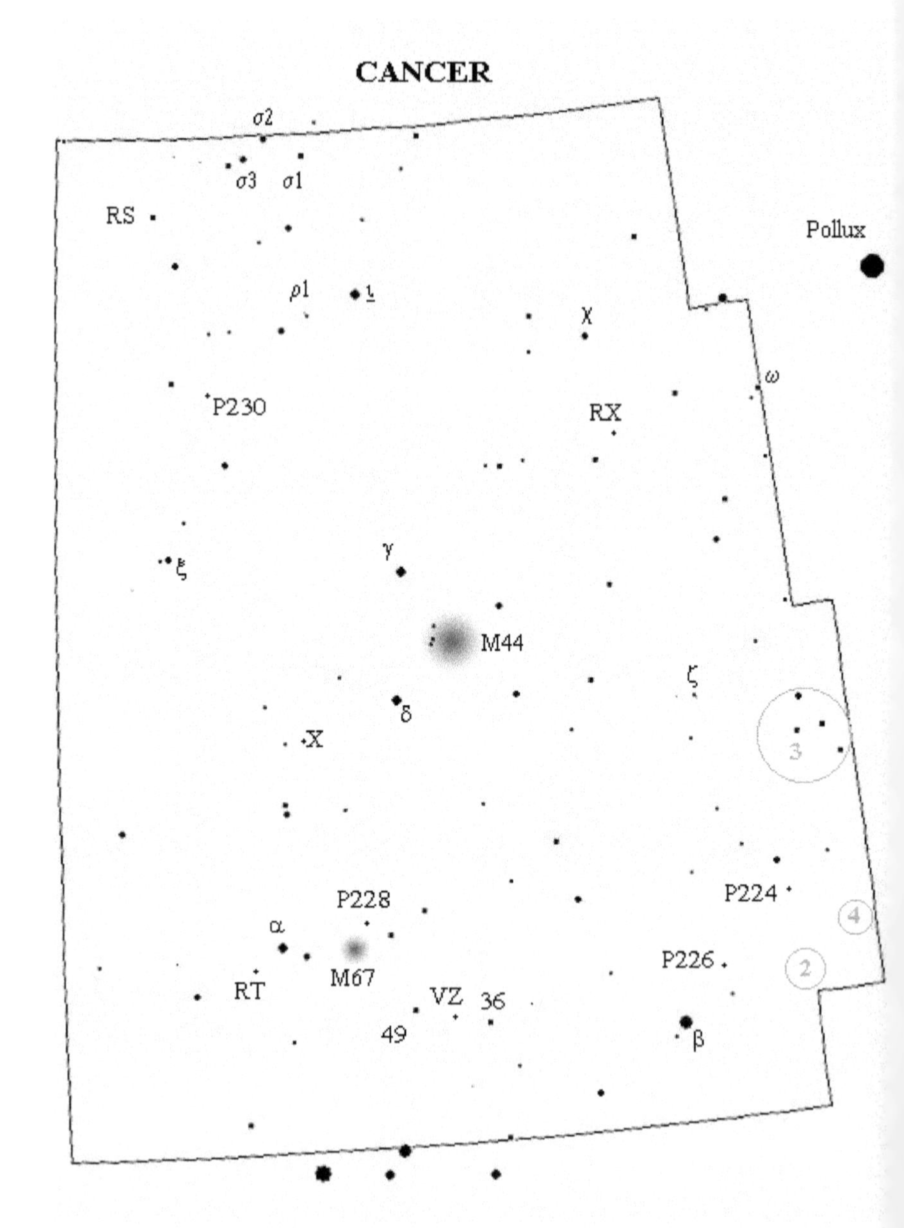

σ2
σ3
σ1
RS
Pollux
ρ1
ι
χ
ω
P230
RX
γ
ξ
M44
ζ
δ
X
3
P228
P224
α
4
P226
M67
2
RT
VZ
36
49
β

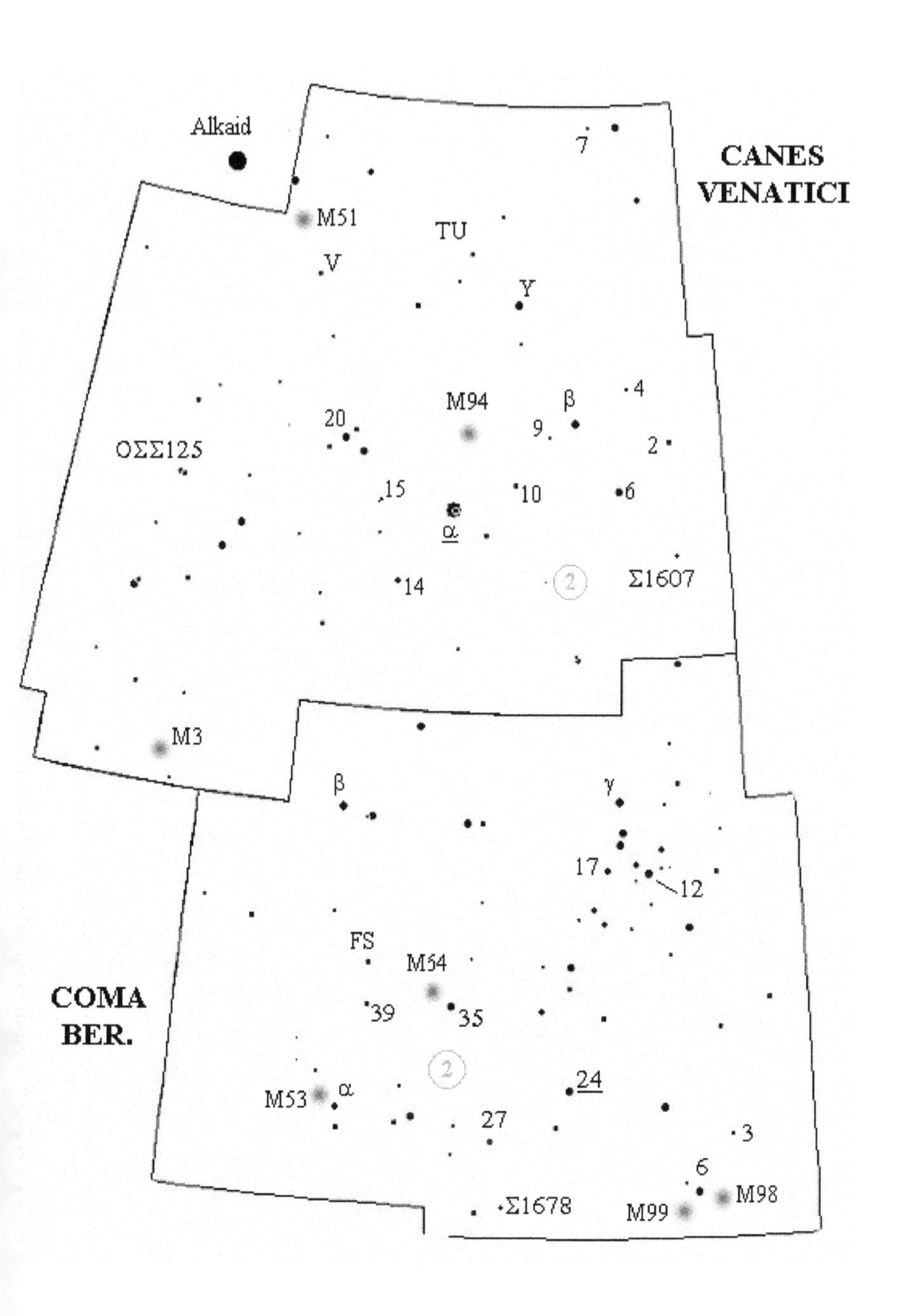

Alkaid
CANES
VENATICI
M51
TU
V
Y
7
M94
β
4
20
9
2
OΣΣ125
15
10
6
α
Σ1607
14
M3
β
γ
17
12
FS
M54
COMA
BER.
39
35
24
M53
α
27
3
6
Σ1678
M99
M98

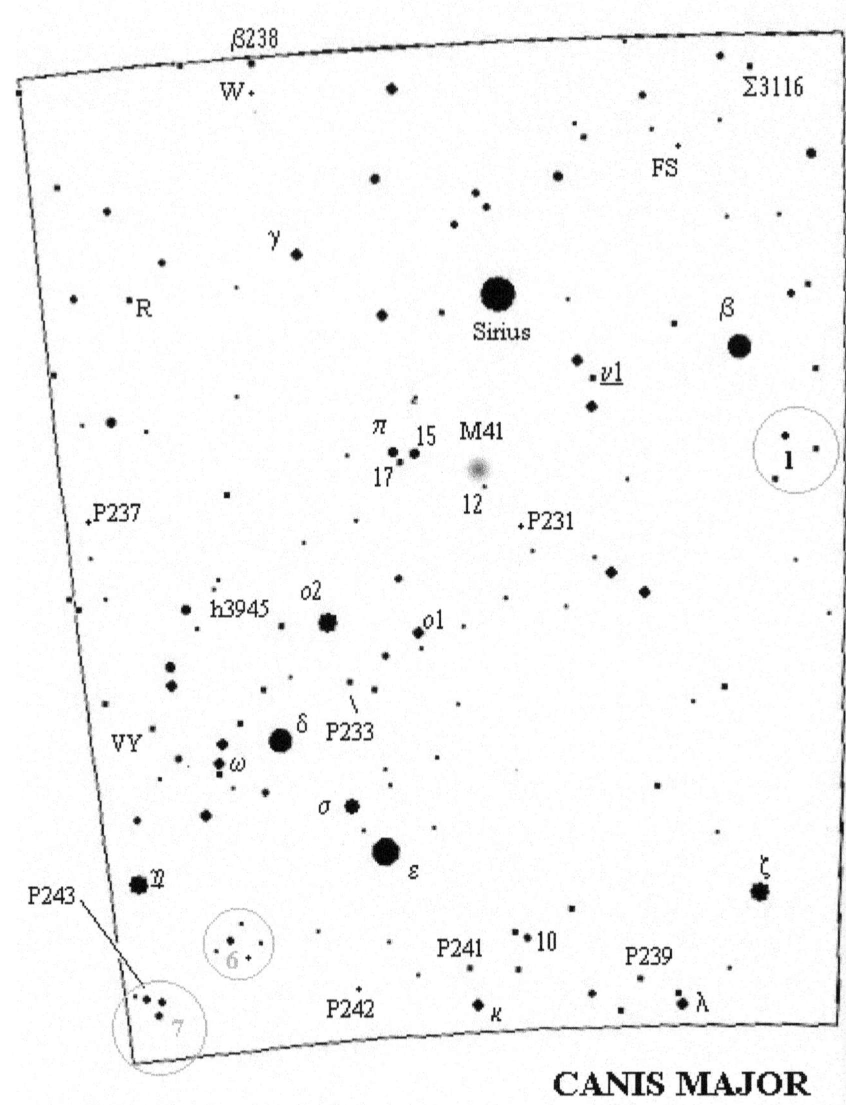
β238
W
Σ3116
FS
γ
R
Sirius
β
ν1
π
15
M41
17
12
1
P237
P231
o2
h3945
o1
P233
δ
VY
ω
σ
ε
ζ
P243
6
P241
10
P239
P242
κ
λ
7

CANIS MAJOR

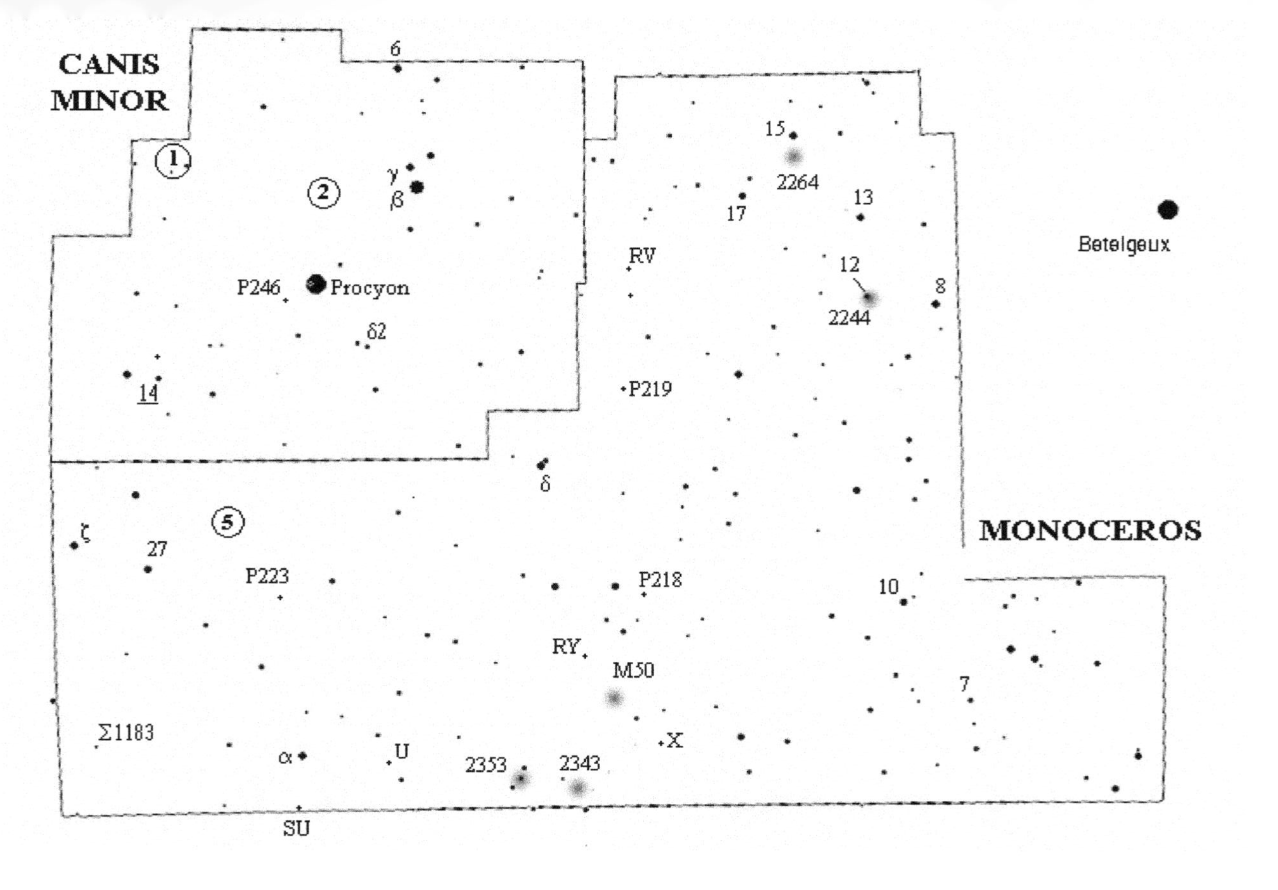
CANIS MINOR
6
1
2
γ
β
P246
Procyon
δ2
14
15
2264
17
13
RV
12
2244
8
Betelgeux
P219
δ
MONOCEROS
5
ζ
27
P223
P218
10
RY
M50
7
Σ1183
α
U
2353
2343
X
SU

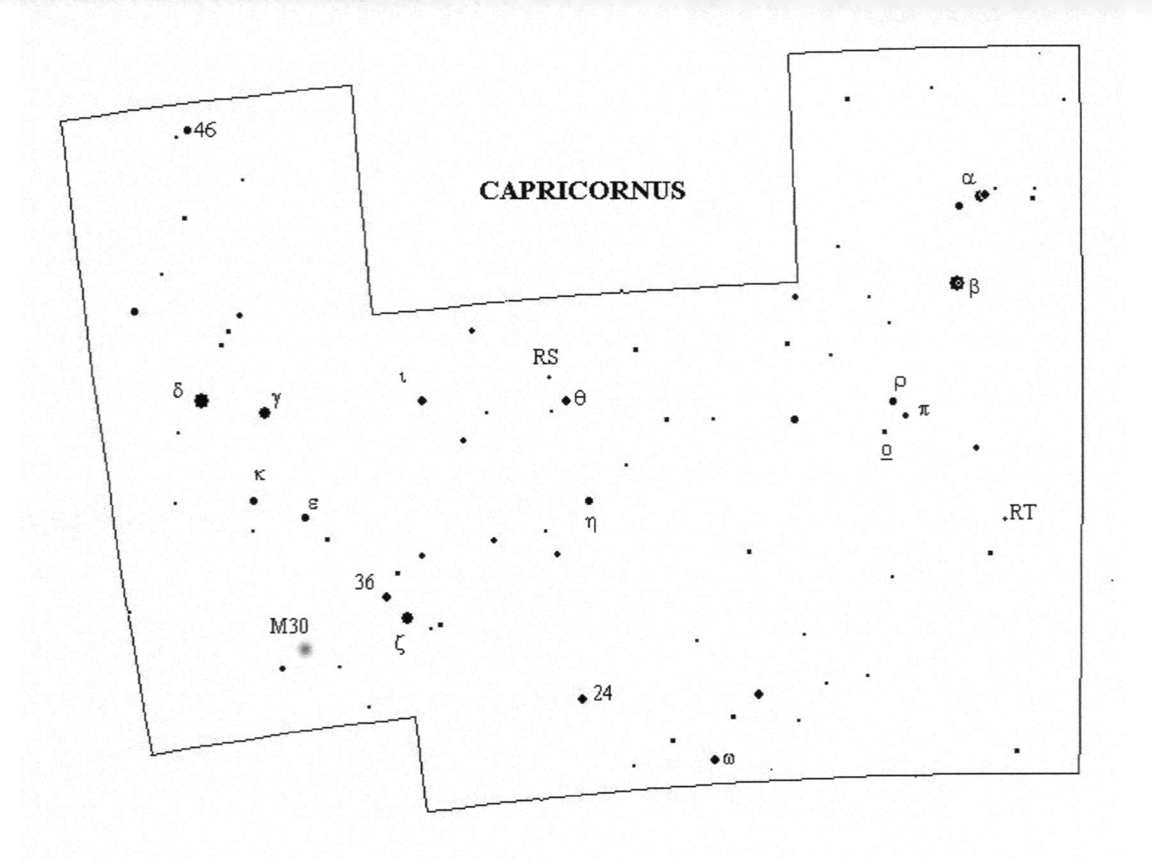
CAPRICORNUS
46
α
β
δ
γ
ι
RS
θ
ρ
π
o
κ
ε
η
RT
36
M30
ζ
24
ω

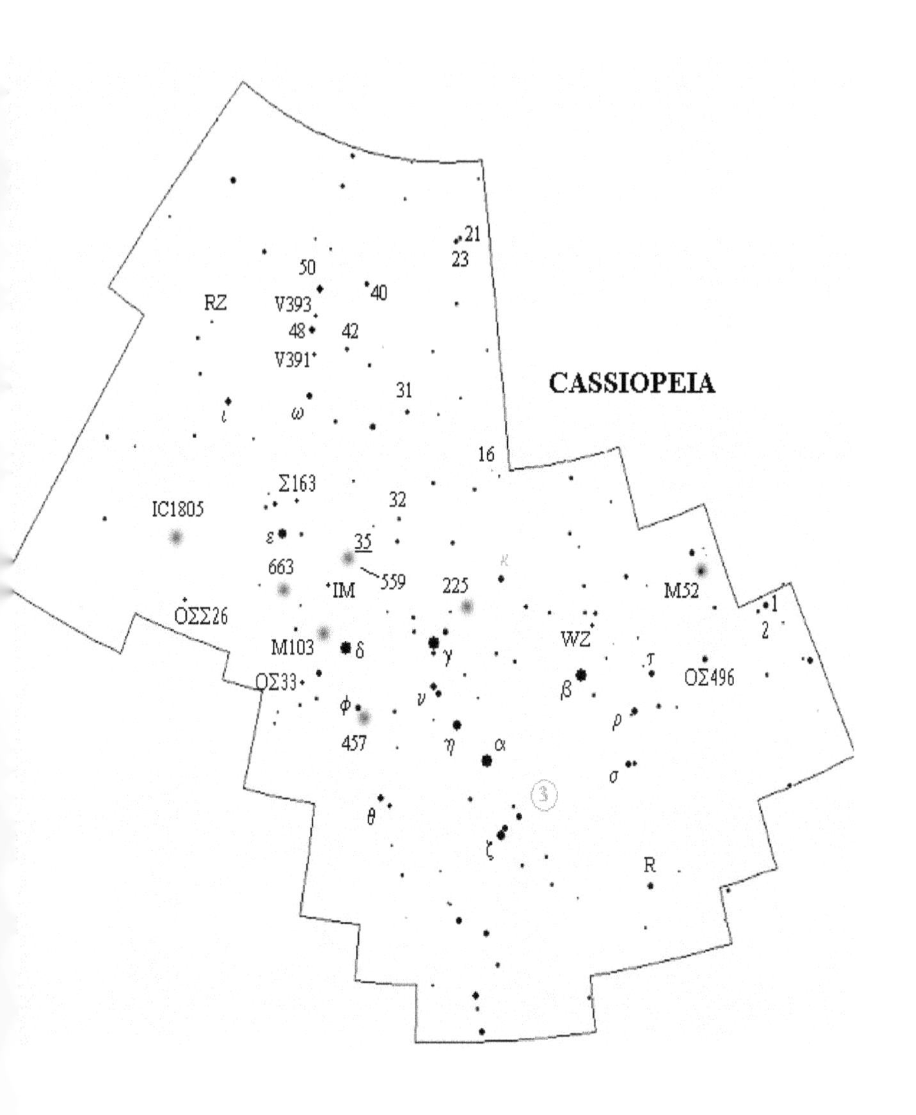
CASSIOPEIA
21
23
50
40
RZ
V393
48
42
V391
31
ι
ω
16
Σ163
32
IC1805
ε
35
κ
663
559
IM
225
M52
ΟΣΣ26
1
2
WZ
M103
δ
γ
ΟΣ33
τ
ΟΣ496
β
ν
φ
ρ
457
η
α
σ
θ
ζ
R

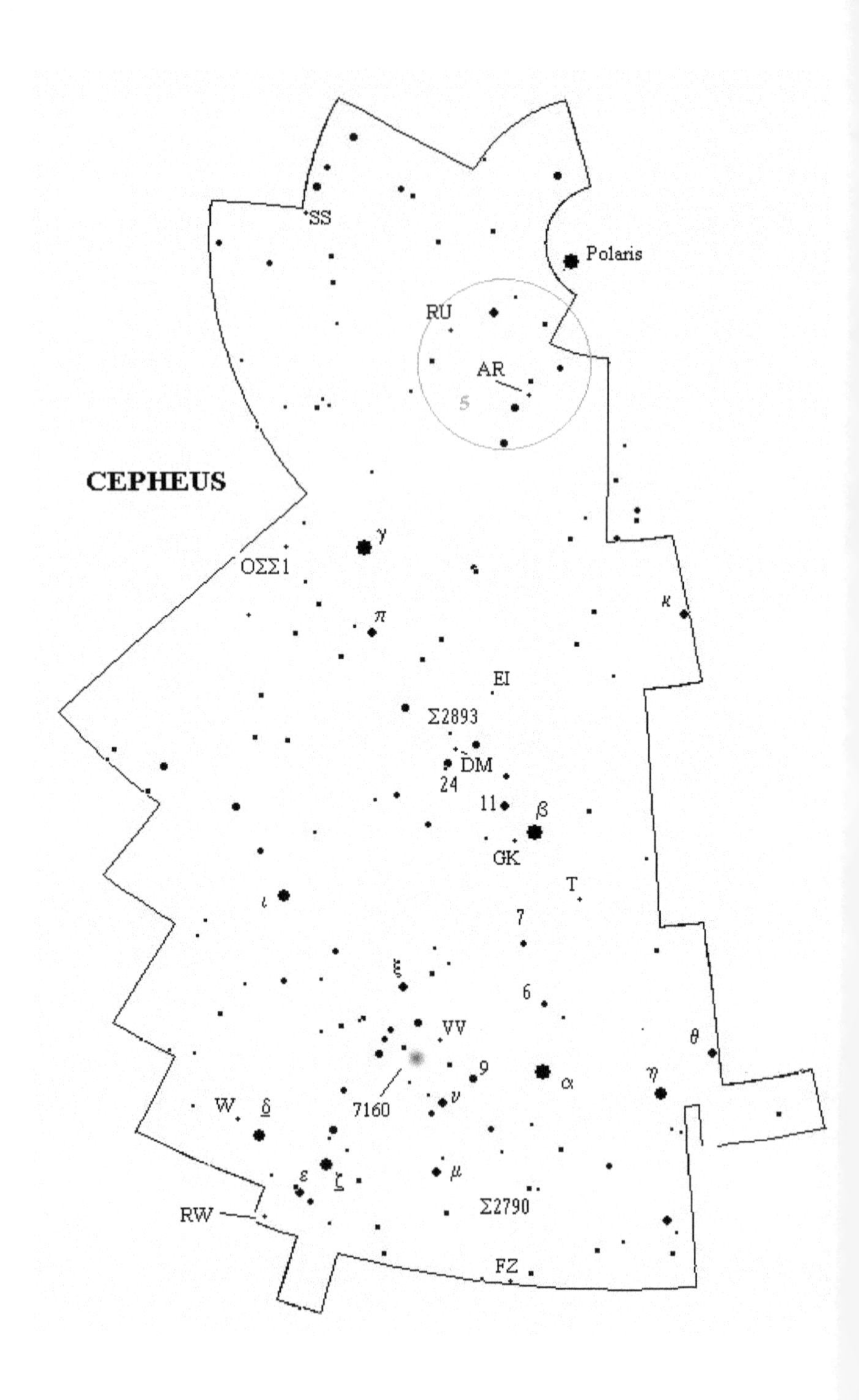
CEPHEUS
SS
Polaris
RU
AR
5
γ
ΟΣΣ1
κ
π
EI
Σ2893
DM
24
11
β
GK
T
ι
7
ξ
6
VV
θ
9
α
η
7160
ν
W
δ
ε
ζ
μ
Σ2790
RW
FZ

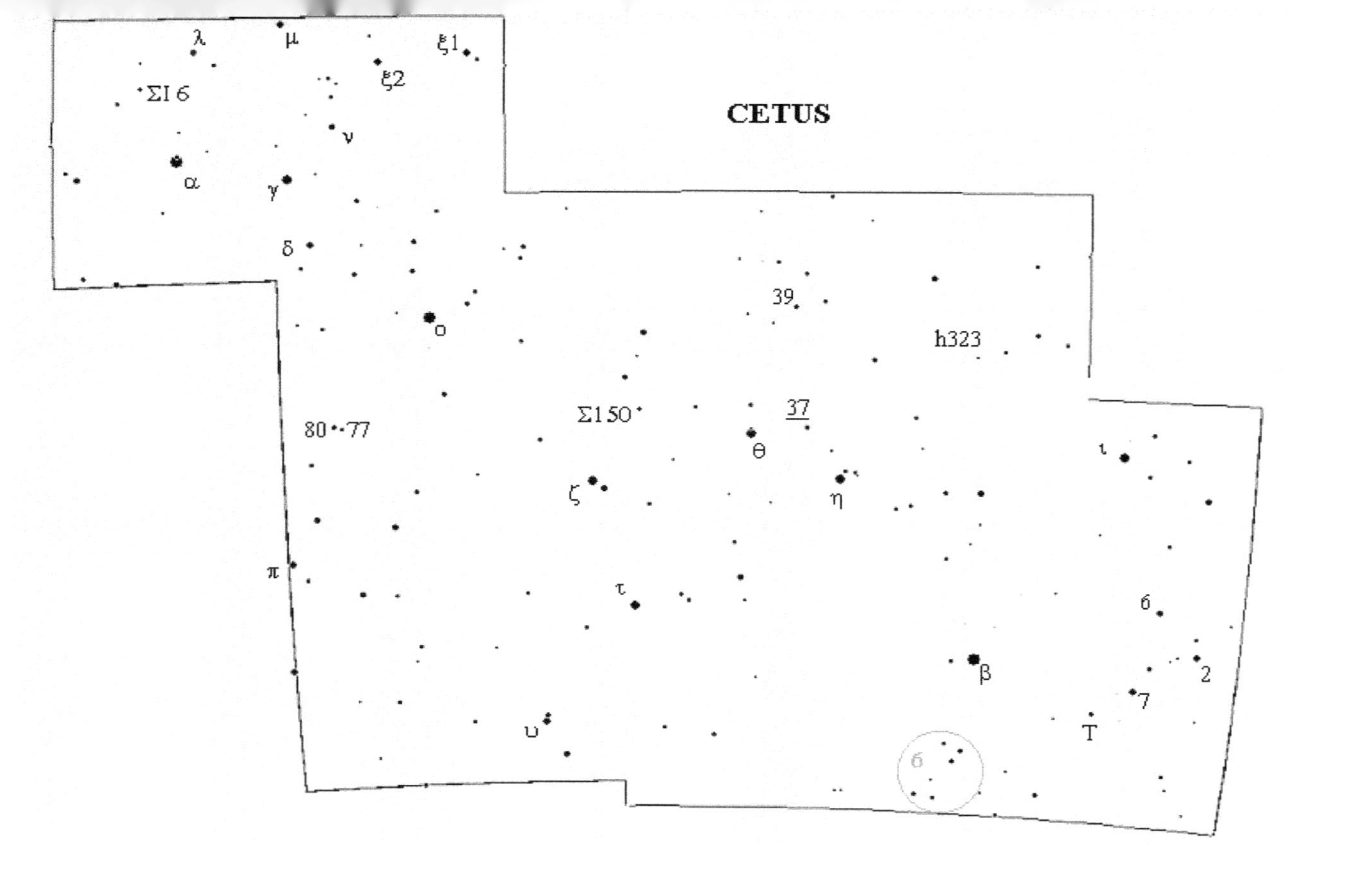

CETUS
λ
μ
ξ1
ξ2
Σ16
ν
α
γ
δ
39
h323
ο
Σ150
37
80
77
θ
ι
ζ
η
π
τ
6
β
2
7
T
υ

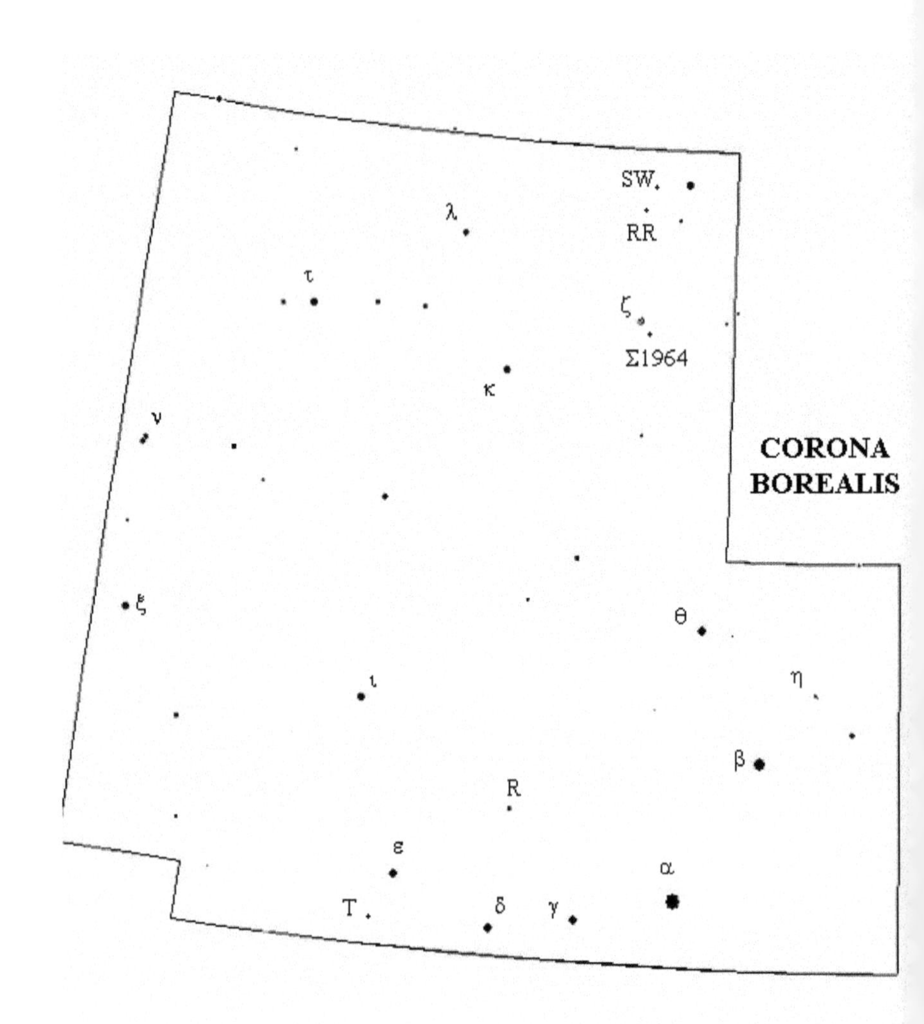
SW.
RR
λ
τ
ζ
Σ1964
κ
ν
CORONA
BOREALIS
ξ
θ
ι
η
β
R
ε
α
T
δ
γ

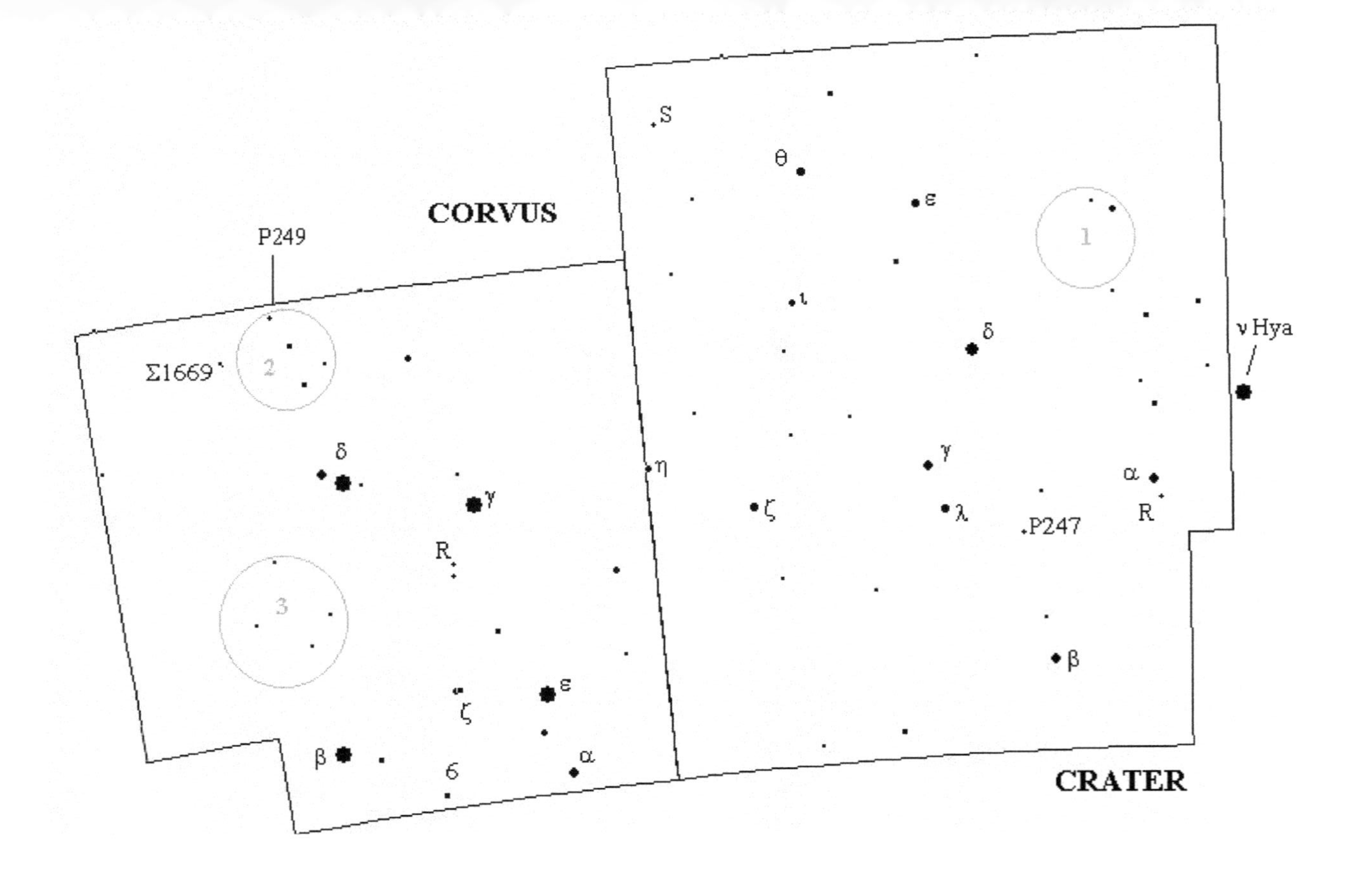
CORVUS
P249
Σ1669
2
3
δ
γ
R
ζ
ε
β
6
α
η
S
θ
ε
ι
δ
1
ν Hya
γ
α
ζ
λ
P247
R
β
CRATER

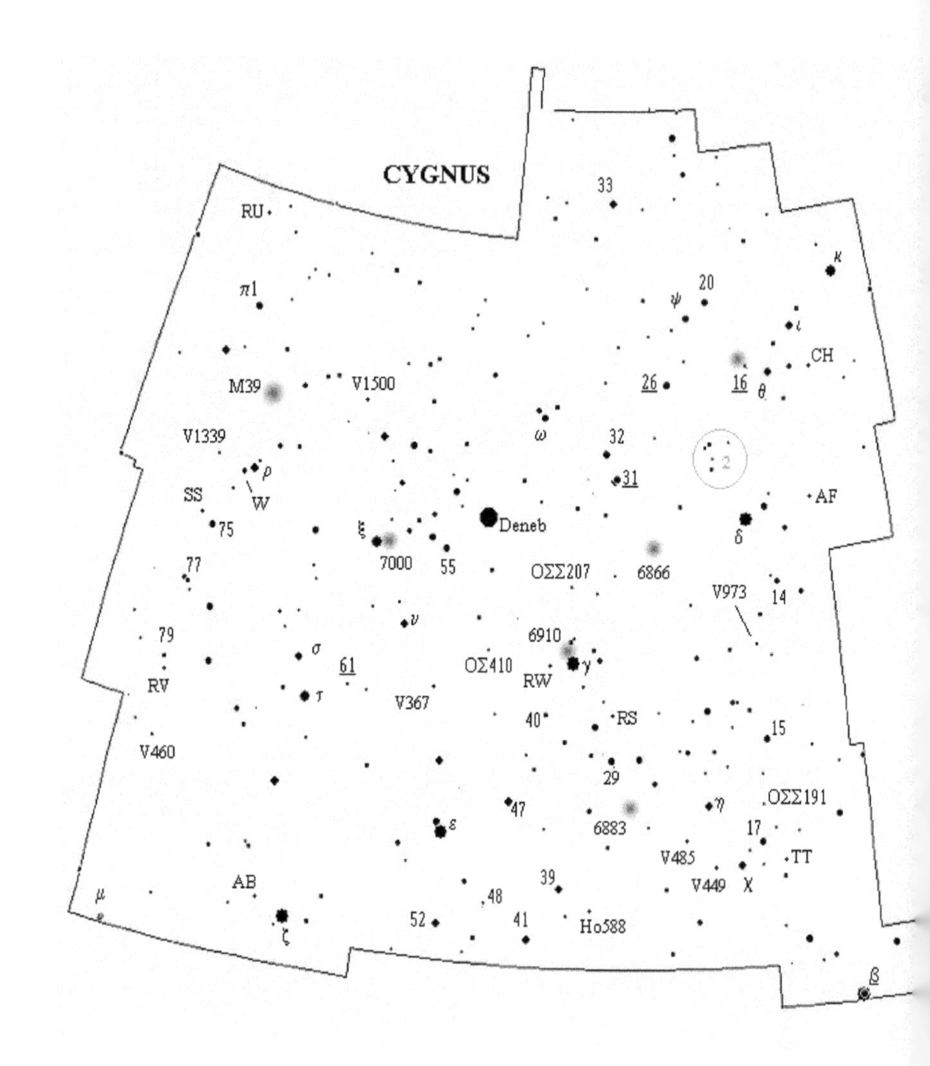
CYGNUS
RU
33
π1
20
ψ
κ
ι
CH
M39
V1500
26
16
θ
V1339
ω
32
ρ
31
SS
W
AF
75
ξ
Deneb
δ
7000
55
77
ΟΣΣ207
6866
V973
14
79
ν
6910
σ
61
ΟΣ410
γ
RV
τ
RW
V367
40
RS
15
V460
29
47
η
ΟΣΣ191
ε
6883
17
V485
TT
AB
39
V449
χ
μ
48
ζ
52
41
Ho588
β

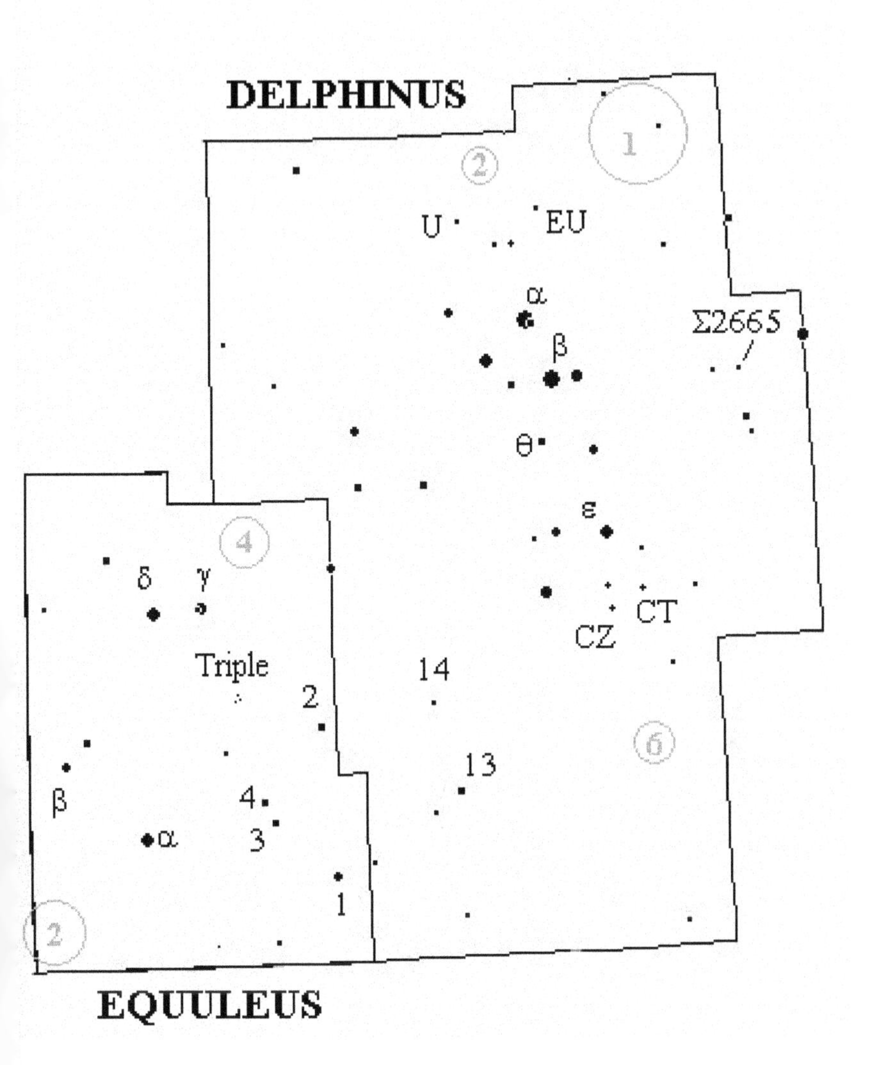
DELPHINUS
1
2
U
EU
α
Σ2665
β
θ
ε
CT
CZ
14
13
6
4
δ
γ
Triple
2
β
4
α
3
1
2
EQUULEUS

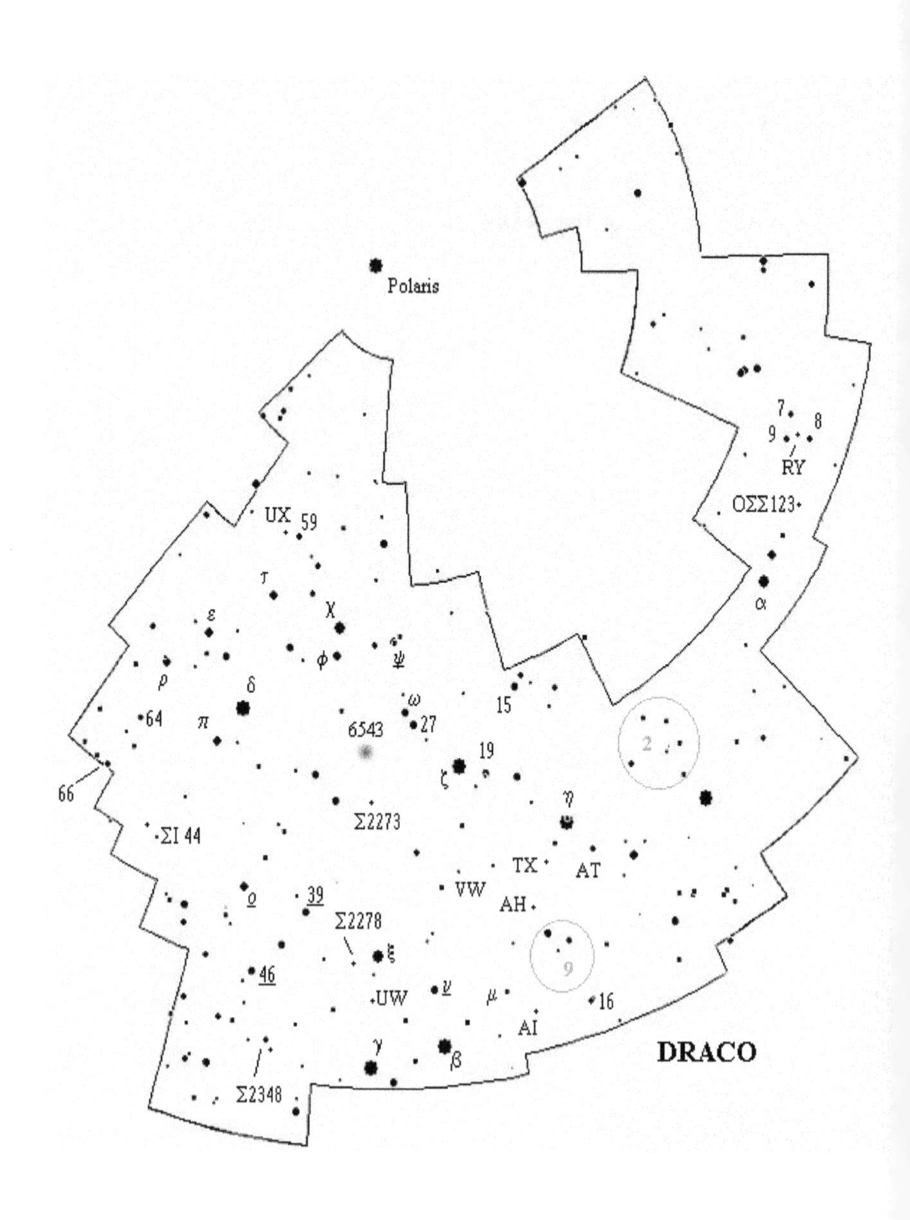
Polaris
7
8
9
RY
OΣΣ123
α
UX
59
τ
ε
χ
ϕ
ψ
ρ
δ
15
ω
64
π
6543
27
19
2
66
ζ
η
Σ2273
ΣI 44
TX
AT
VW
AH
o
39
Σ2278
ξ
9
46
UW
ν
μ
16
AI
γ
β
Σ2348
DRACO

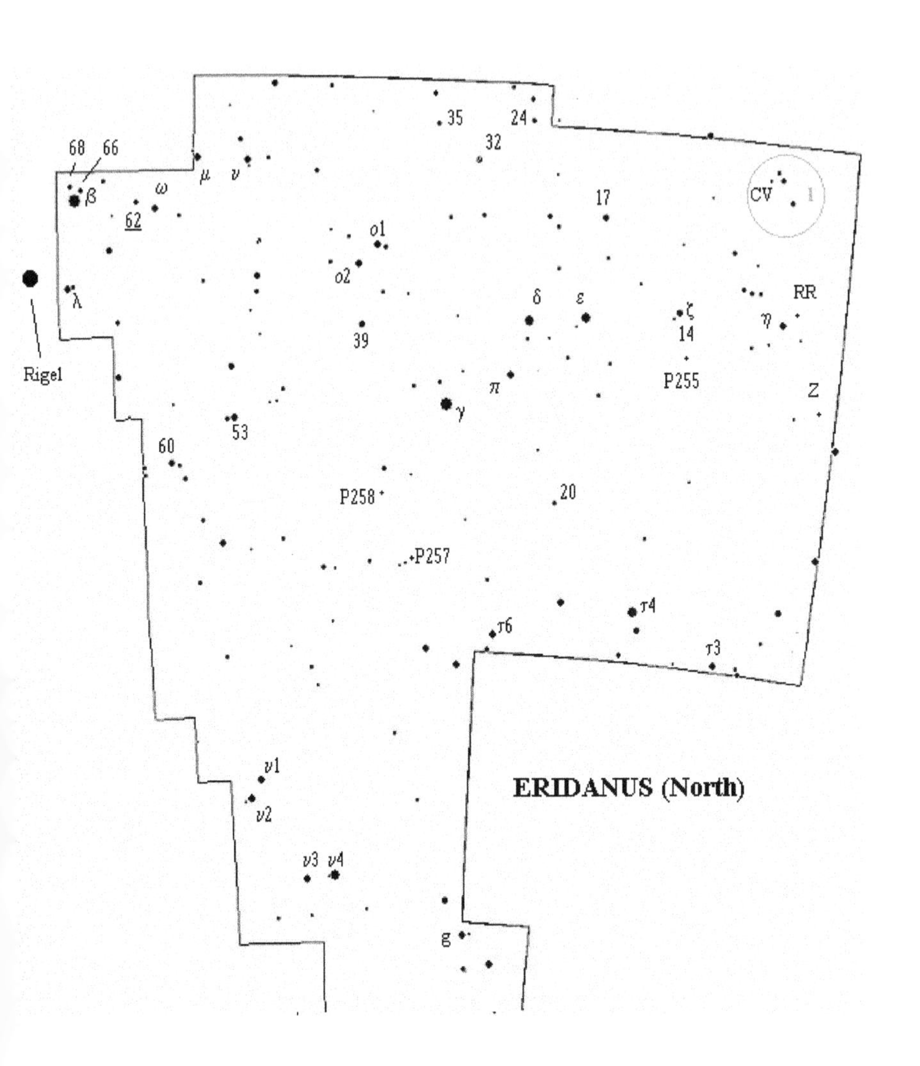
68
66
β
ω
62
μ
ν
35
24
32
17
CV
o1
o2
λ
Rigel
δ
ε
ζ
14
RR
η
39
P255
π
γ
Z
53
60
P258
20
P257
τ4
τ6
τ3
ν1
ν2
ν3
ν4
g
ERIDANUS (North)

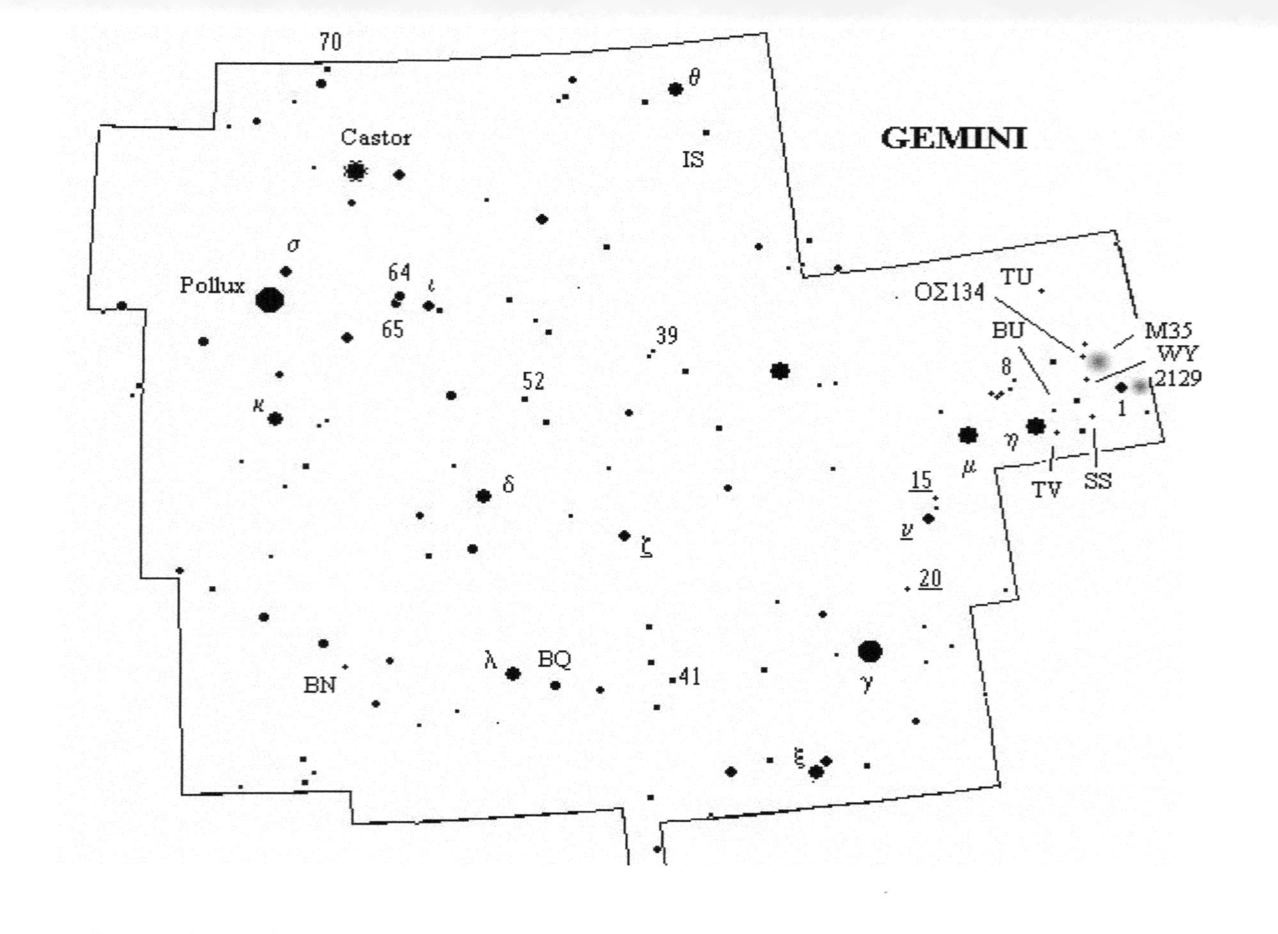
GEMINI
70
θ
Castor
IS
σ
Pollux
64
ι
65
39
52
κ
δ
ζ
OΣ134
TU
BU
M35
WY
2129
8
1
η
μ
TV
SS
15
ν
20
λ
BQ
41
BN
γ
ξ

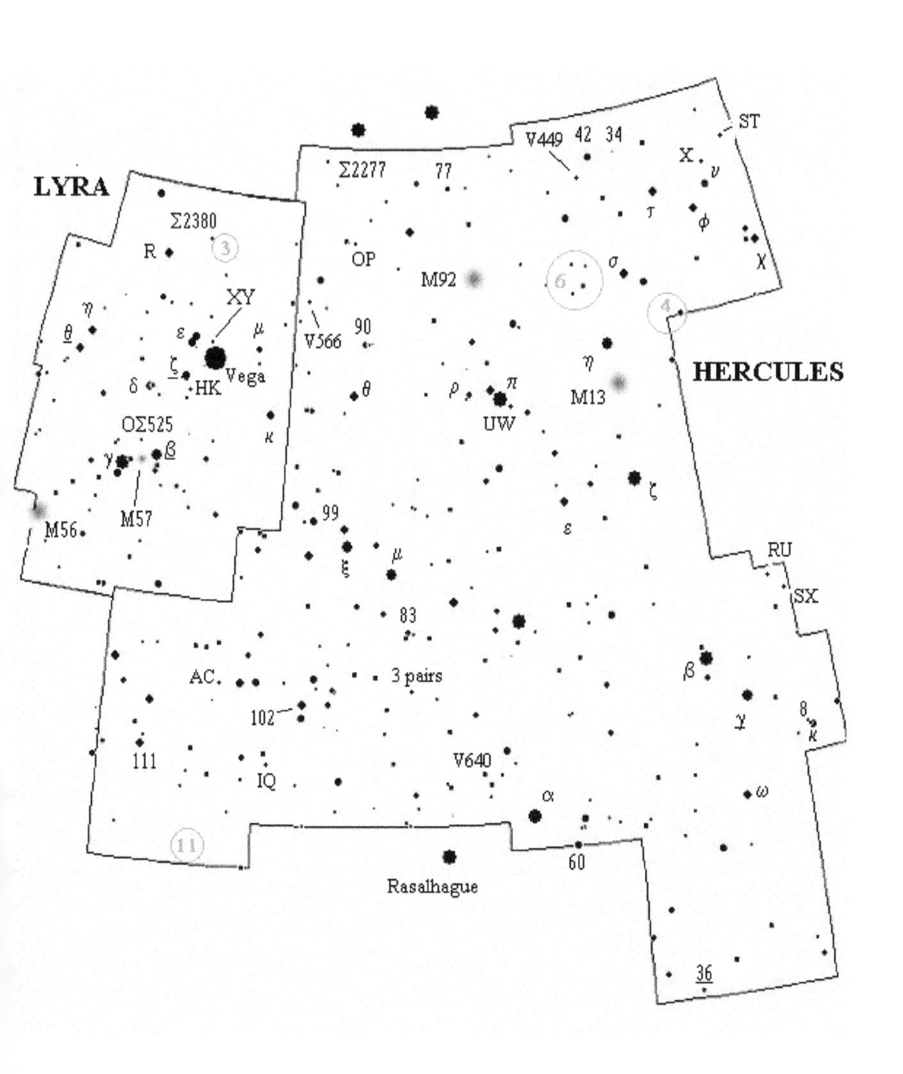
LYRA
HERCULES
Σ2277
77
V449
42
34
ST
X
Σ2380
R
OP
M92
XY
V566
90
Vega
HK
M13
UW
OΣ525
M57
M56
99
RU
SX
83
AC
3 pairs
102
111
IQ
V640
60
Rasalhague
36

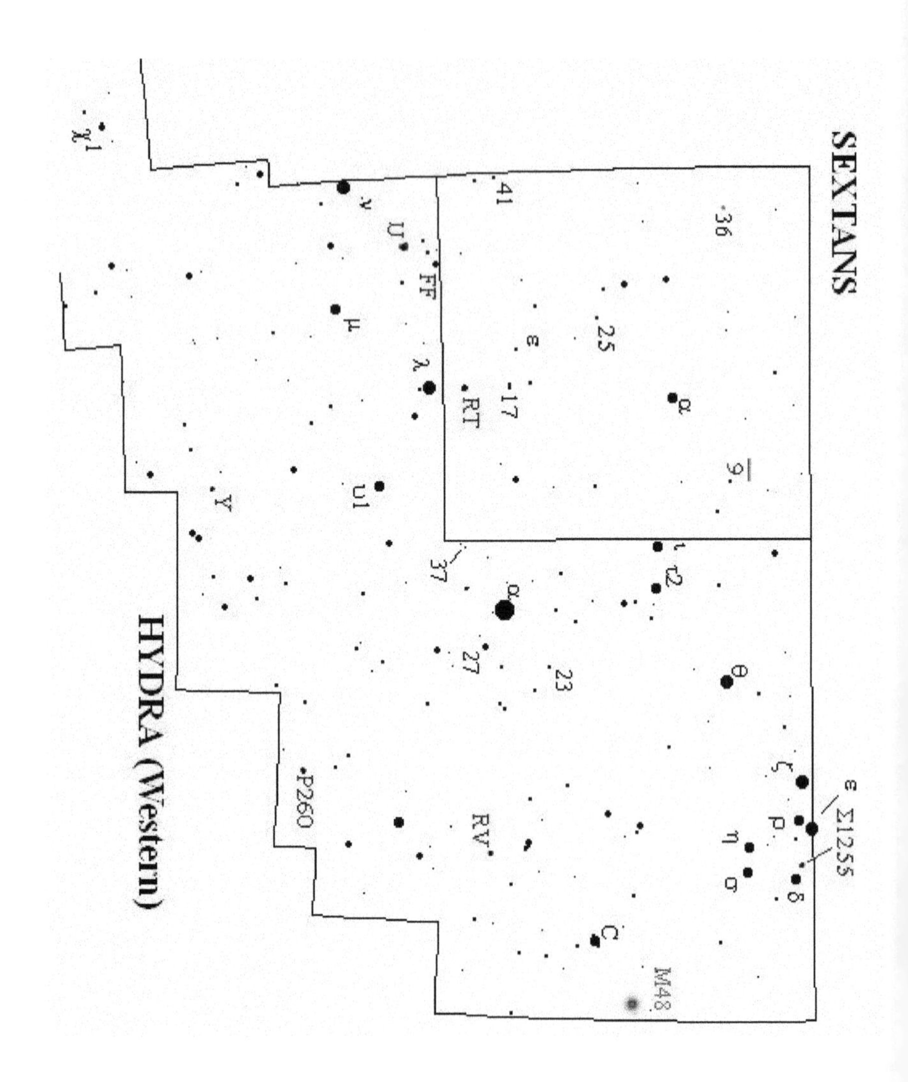

SEXTANS
HYDRA (Western)
Σ1255
M48
P260
RV
RT
FF

HYDRA (Eastern)

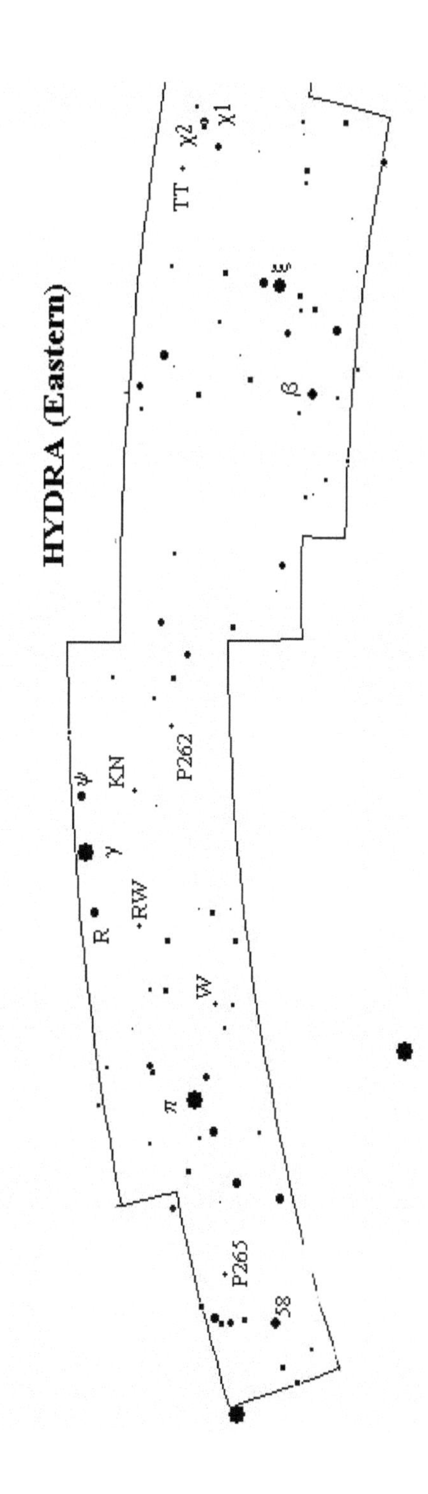

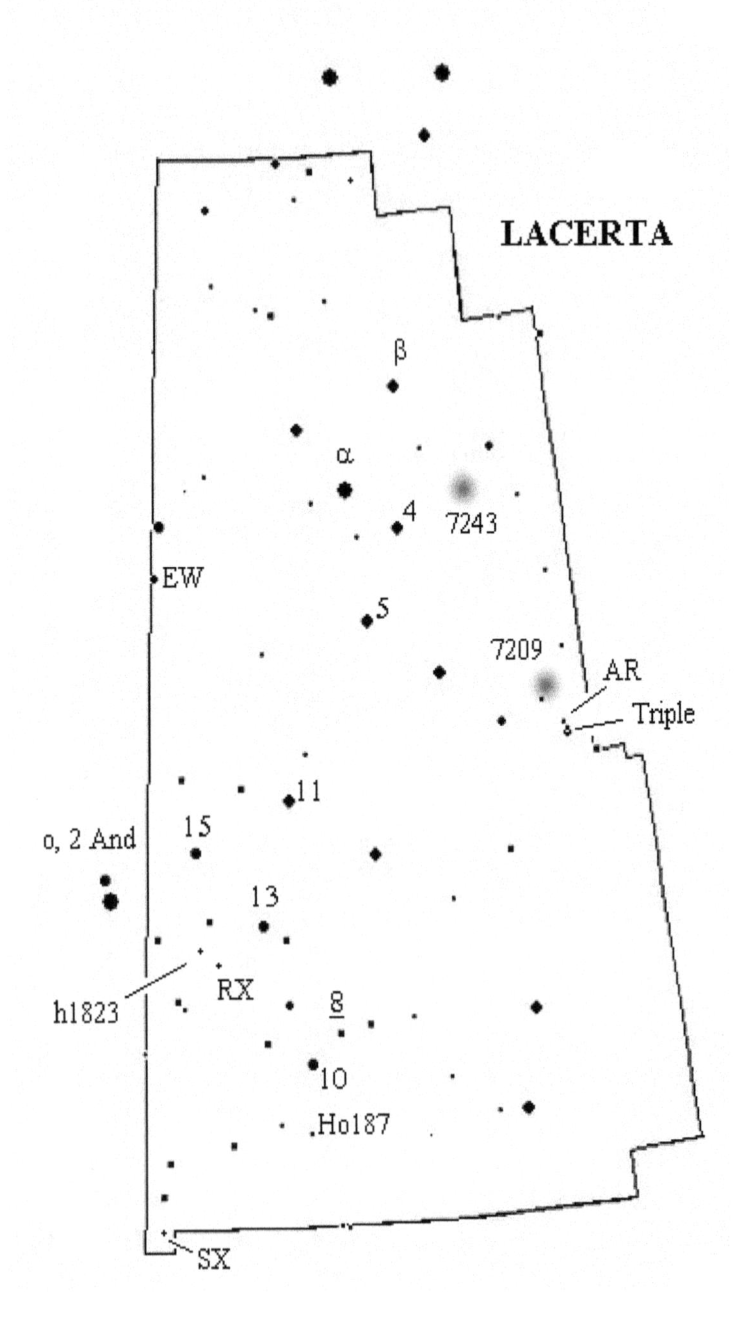
LACERTA
β
α
4
7243
EW
5
7209
AR
Triple
11
15
o, 2 And
13
h1823
RX
8
10
Ho187
SX

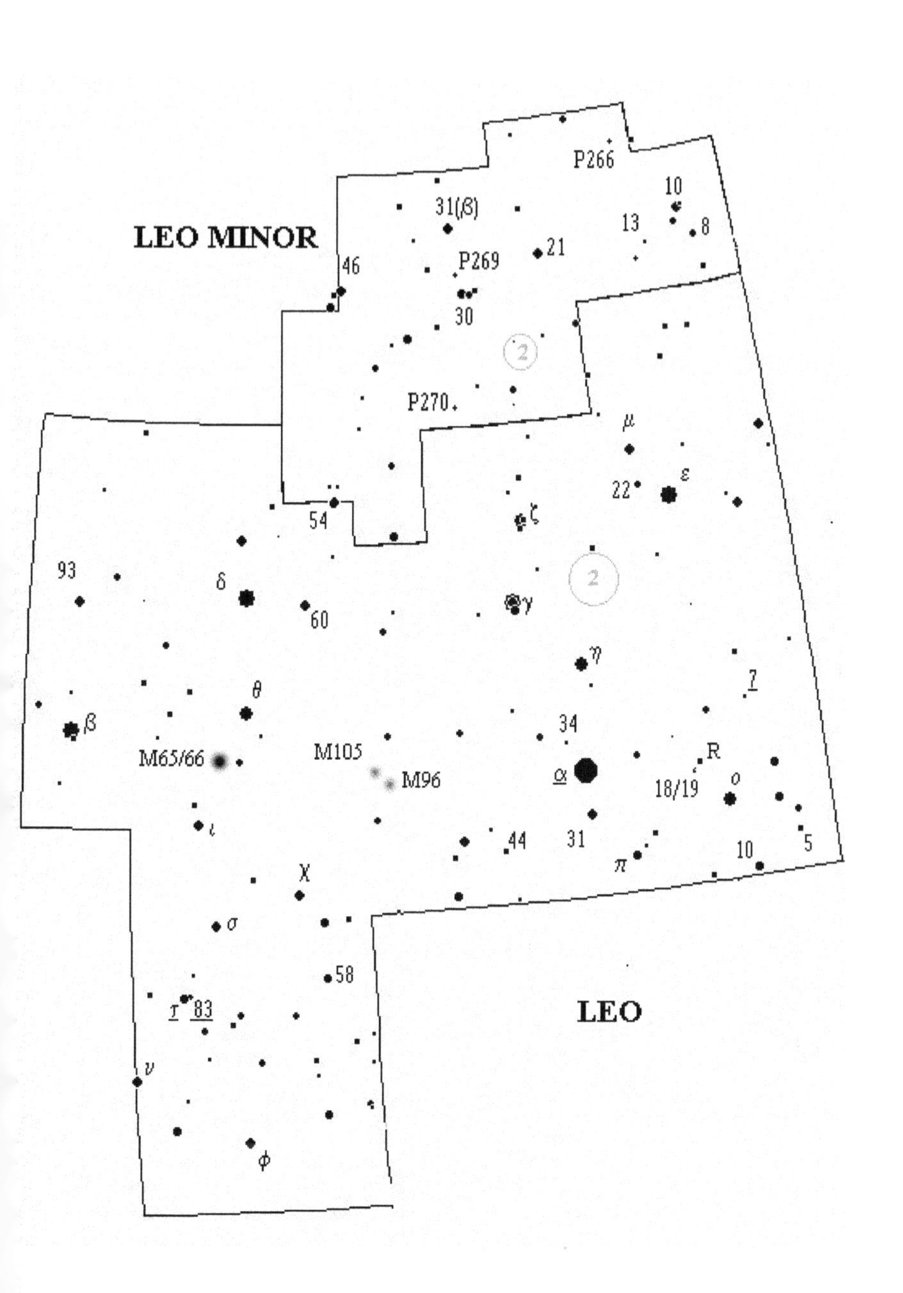
LEO MINOR
46
31(β)
P266
10
13
8
21
P269
30
P270
54
LEO
93
δ
60
ζ
γ
μ
22
ε
η
7
θ
β
M65/66
M105
M96
34
α
R
18/19
o
ι
44
31
π
10
5
χ
σ
58
τ
83
ν
φ

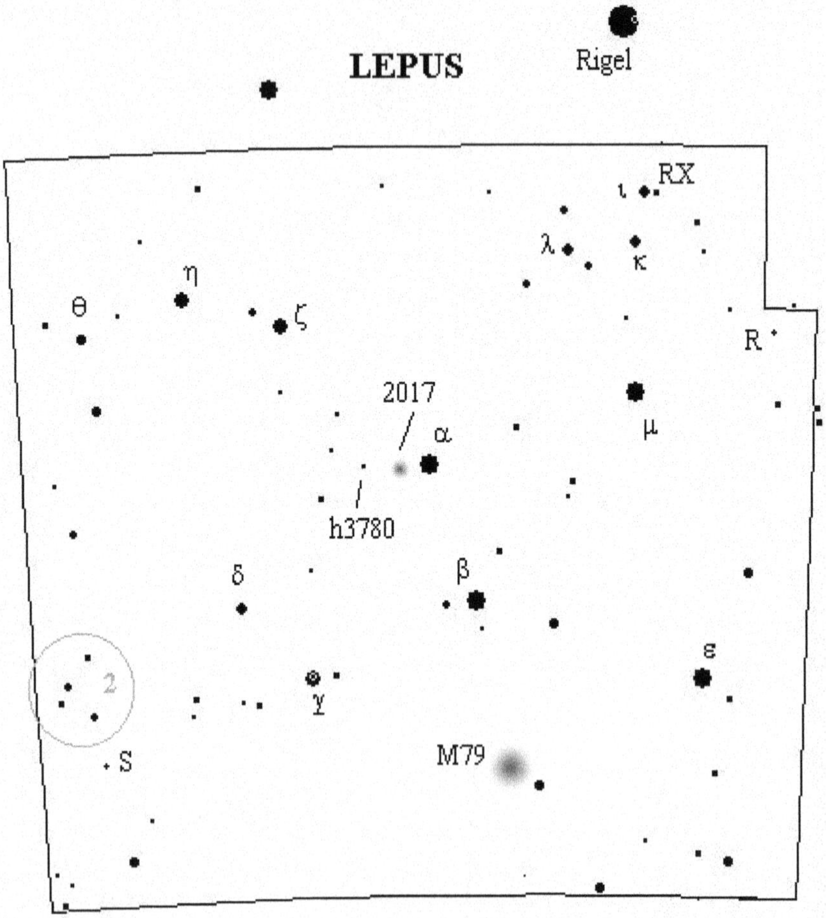

LEPUS

Rigel
RX
ι
λ
κ
η
θ
ζ
R
2017
α
μ
h3780
δ
β
ε
2
γ
S
M79

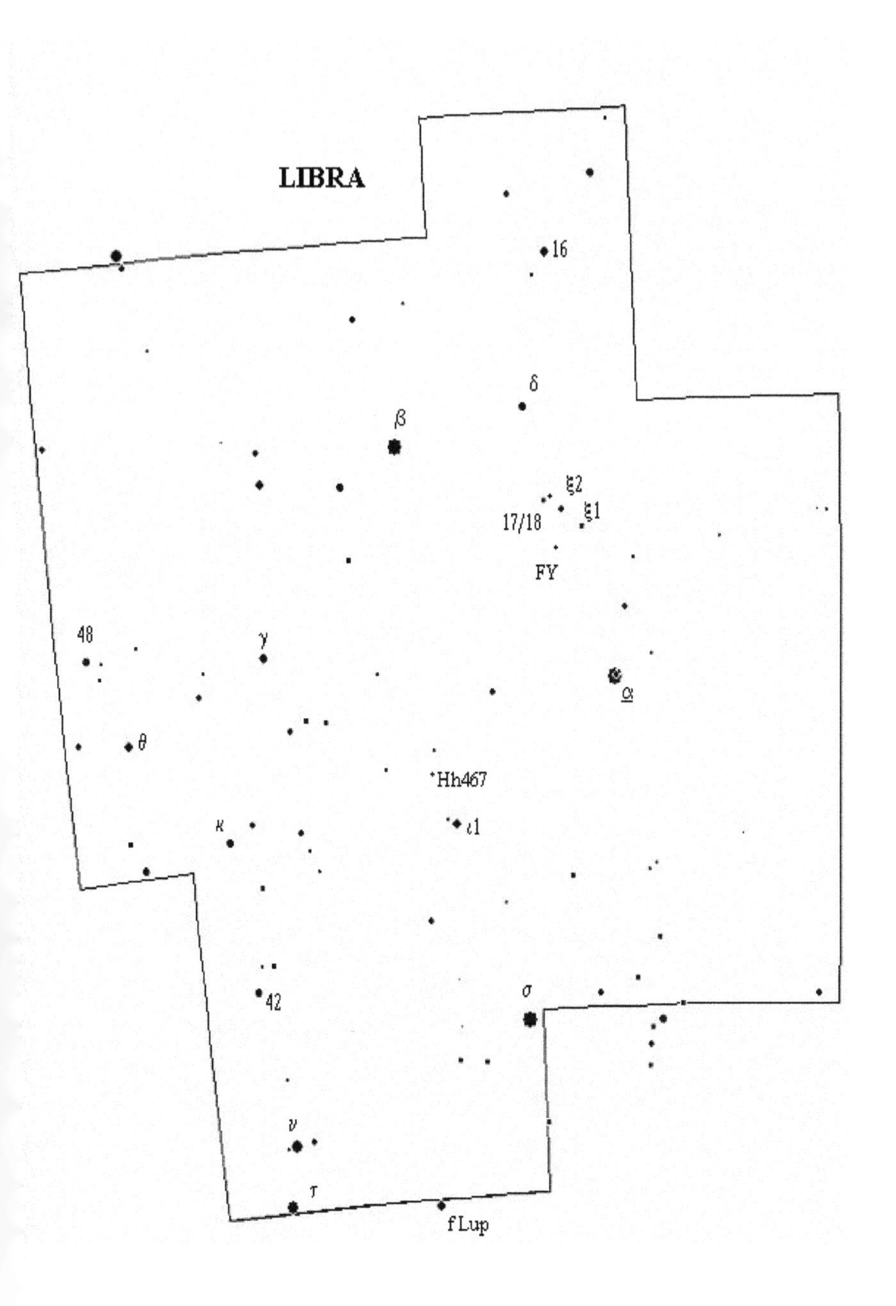
LIBRA
16
δ
β
ξ2
ξ1
17/18
FY
48
γ
α
θ
Hh467
κ
ι1
σ
42
ν
τ
f Lup

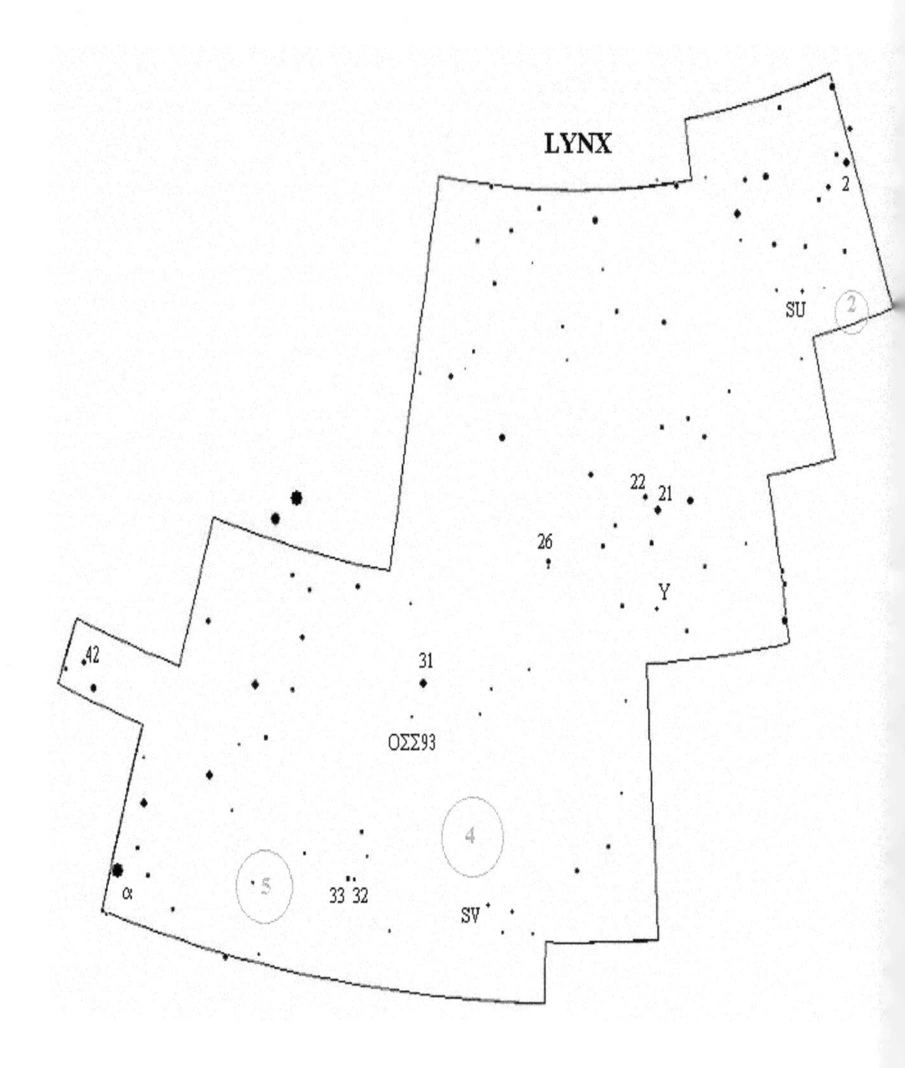
LYNX
2
SU
2
22
21
26
Y
42
31
ΟΣΣ93
4
5
α
33
32
SV

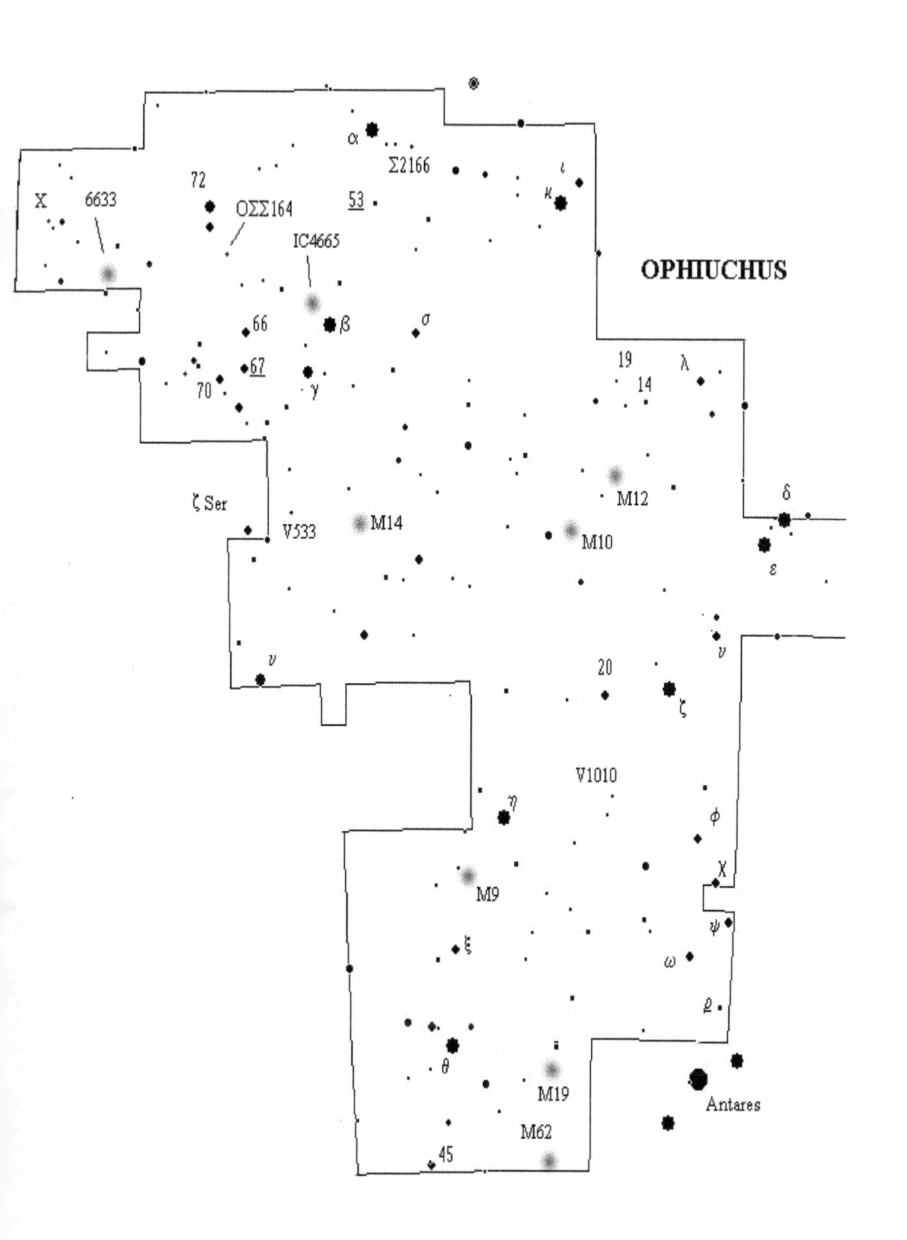
OPHIUCHUS
α
Σ2166
ι
κ
X
6633
72
ΟΣΣ164
53
IC4665
66
β
σ
67
70
γ
19
14
λ
δ
M12
ζ Ser
V533
M14
M10
ε
ν
20
ν
ζ
V1010
η
φ
χ
M9
ψ
ξ
ω
ρ
θ
M19
Antares
M62
45

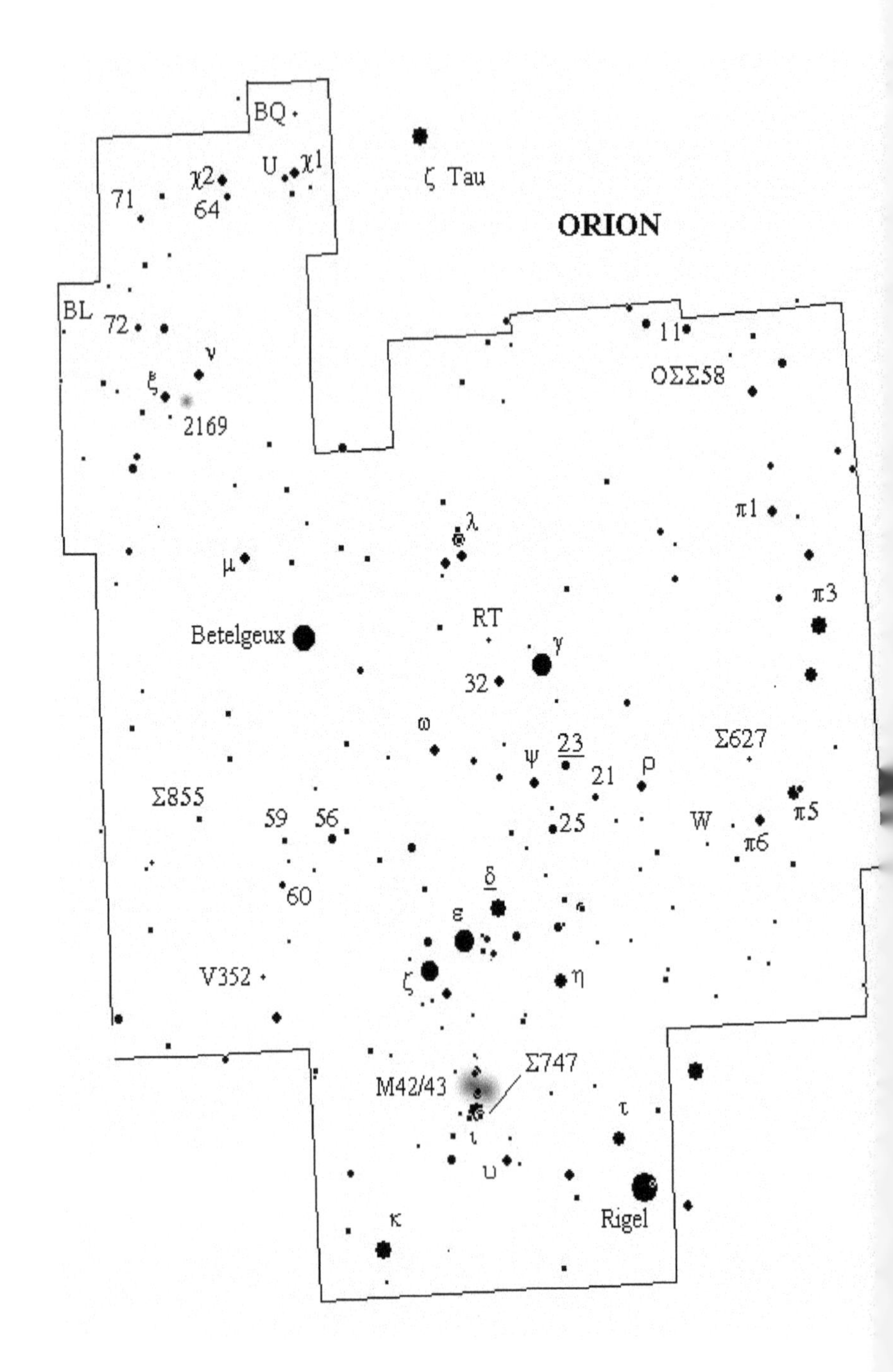
BQ
U
χ1
χ2
64
71
ζ Tau
ORION
BL
72
ν
ξ
2169
11
OΣΣ58
π1
λ
μ
π3
RT
Betelgeux
γ
32
ω
Σ627
ψ
23
21
ρ
Σ855
59
56
25
W
π5
π6
δ
60
ε
V352
ζ
η
Σ747
M42/43
τ
ι
υ
κ
Rigel

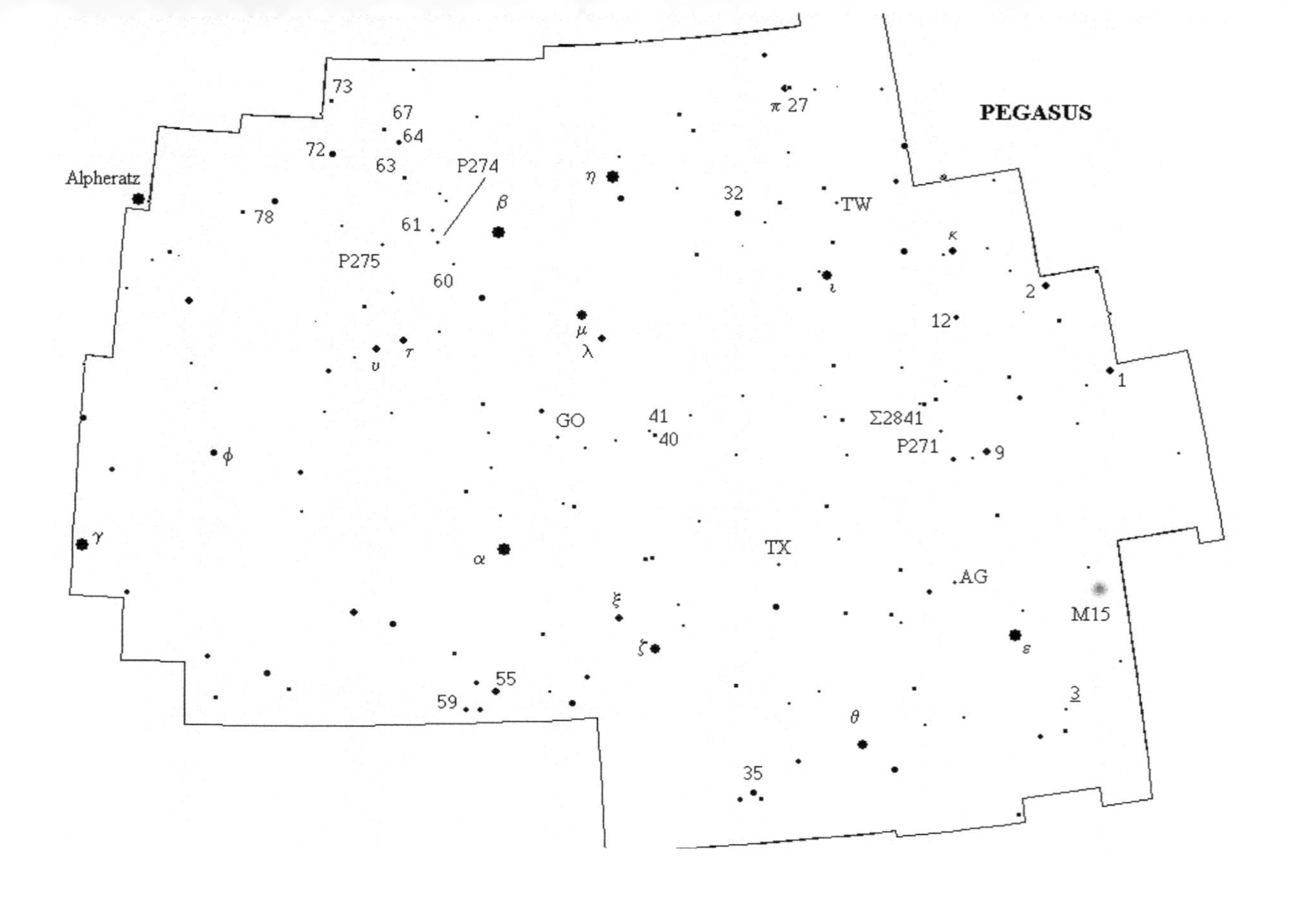
PEGASUS
Alpheratz
73
67
64
72
63
P274
78
61
P275
60
β
η
μ
λ
υ
τ
π 27
32
TW
κ
ι
2
12
1
GO
41
40
Σ2841
P271
9
φ
γ
α
TX
AG
M15
ξ
ζ
ε
55
59
3
θ
35

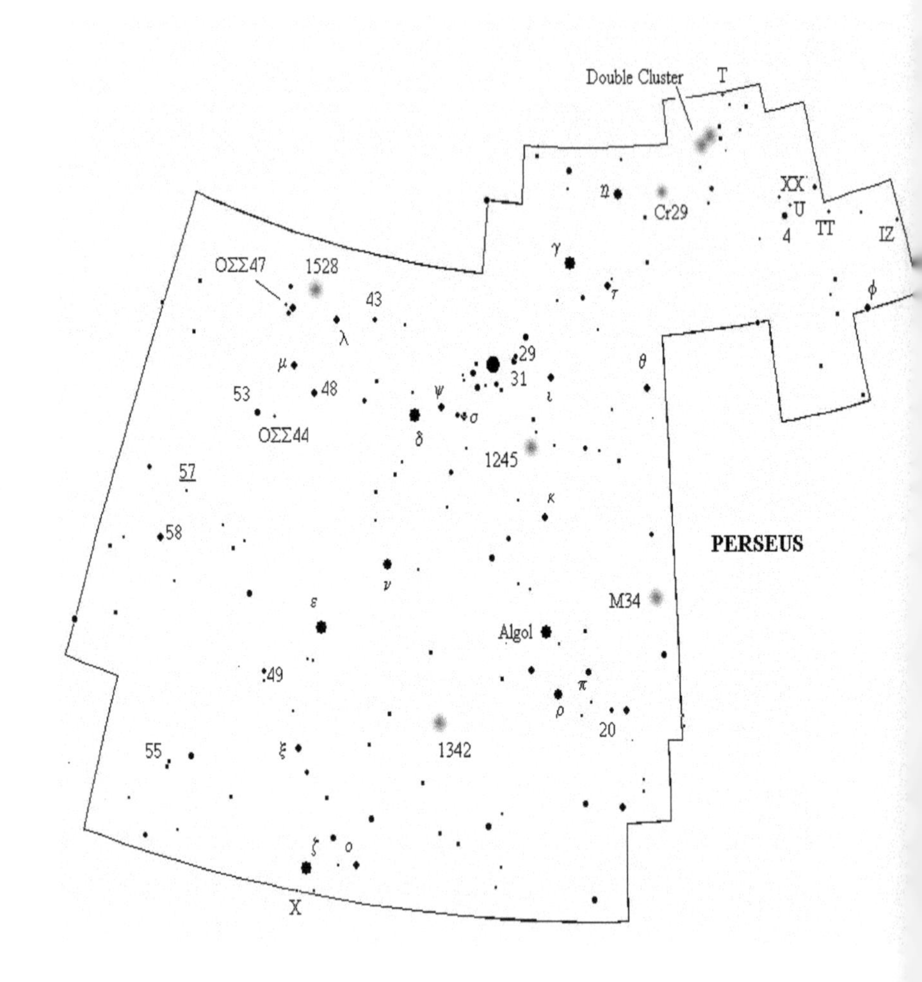

Double Cluster
T
XX
U
4
TT
IZ
η
Cr29
γ
τ
φ
ΟΣΣ47
1528
43
λ
29
31
ι
θ
μ
48
53
ΟΣΣ44
ψ
σ
δ
1245
57
κ
58
PERSEUS
ν
ε
M34
Algol
49
π
ρ
20
55
ξ
1342
ζ
ο
X

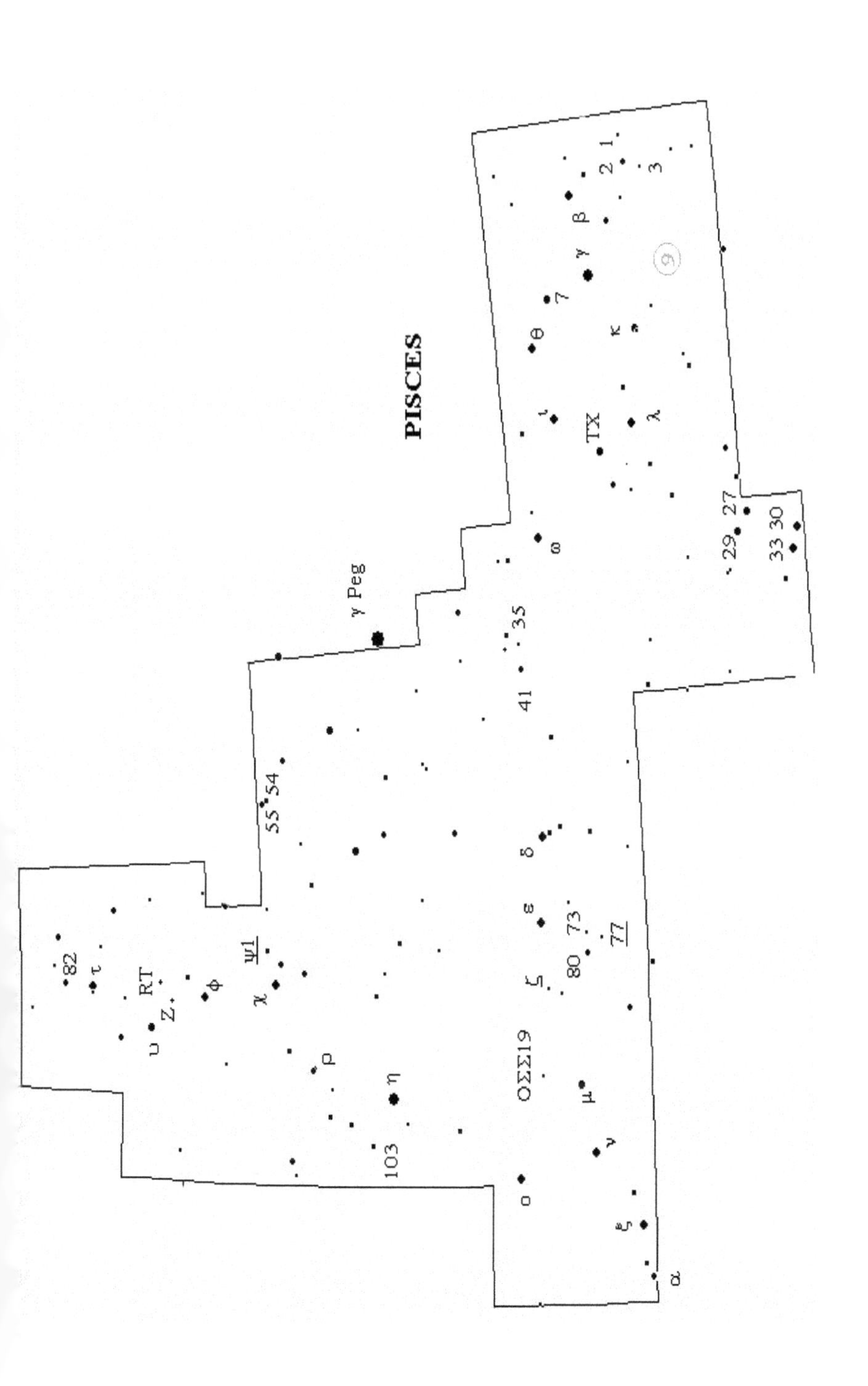

PISCES
γ Peg
TX
RT
Z
ΟΣΣ19

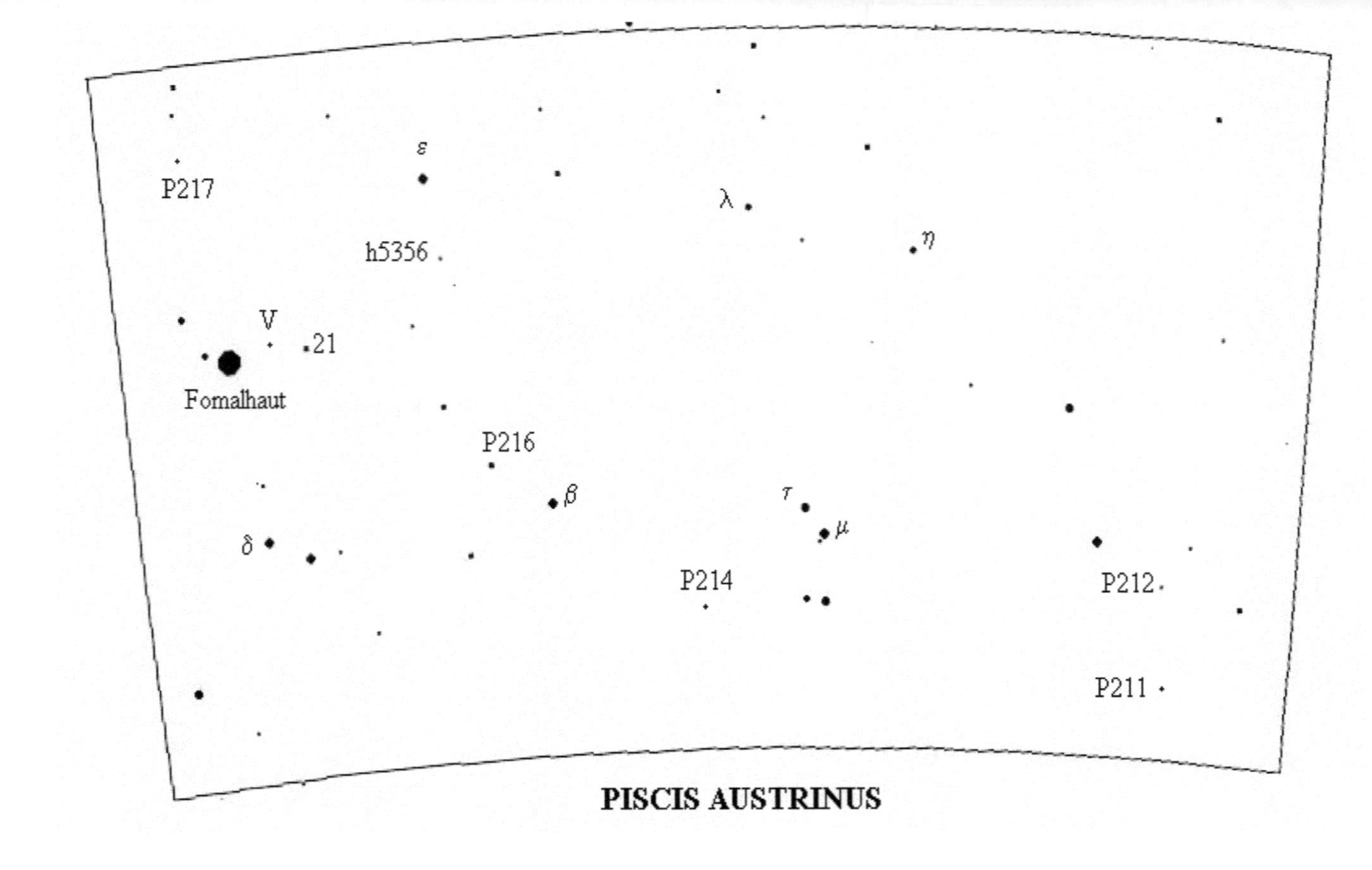

PISCIS AUSTRINUS

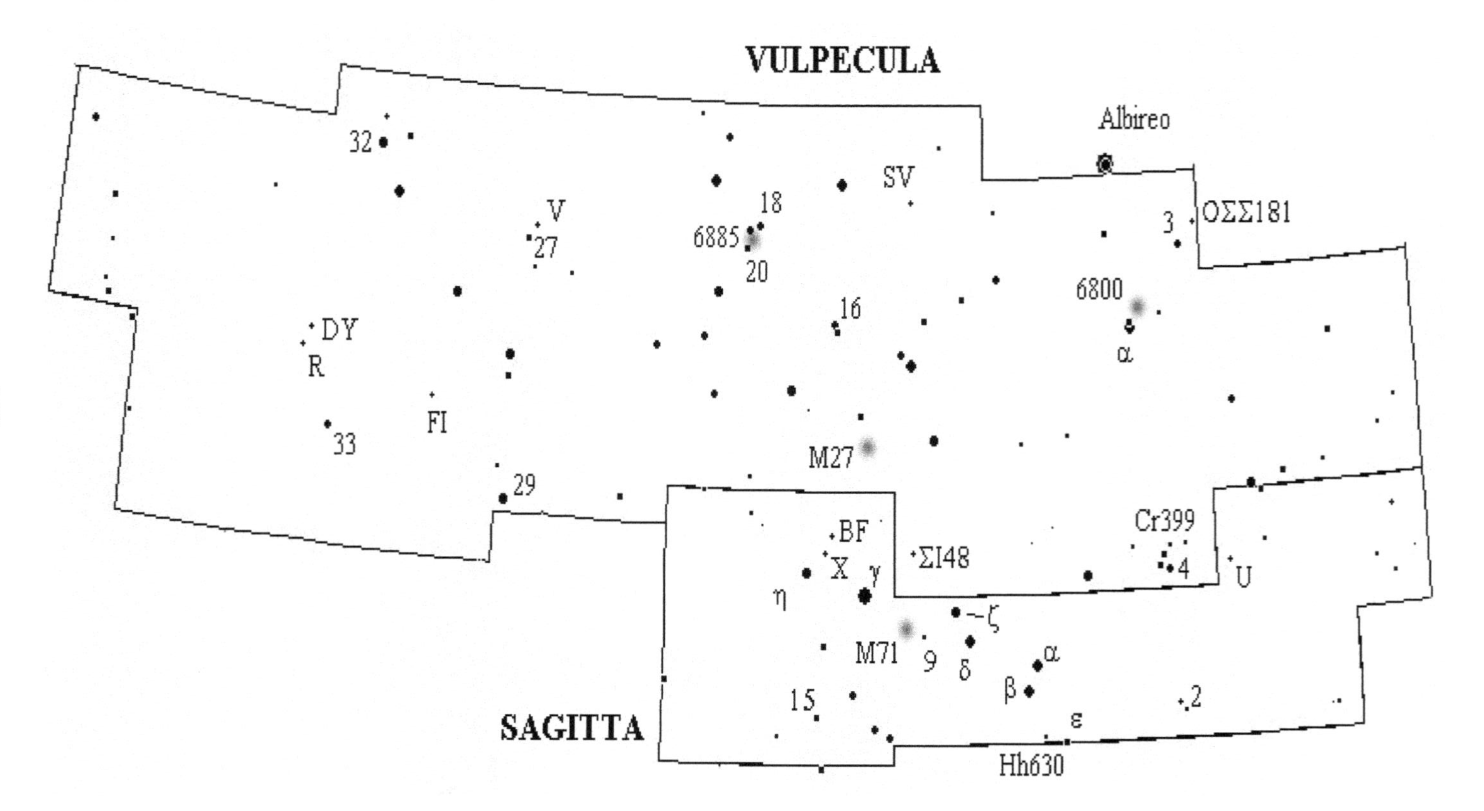
VULPECULA
Albireo
32
SV
V
18
27
6885
20
3
ΟΣΣ181
6800
α
DY
R
16
FI
33
M27
29
BF
ΣI48
Cr399
X
γ
η
4
U
ζ
M71
9
δ
α
β
15
2
ε
SAGITTA
Hh630

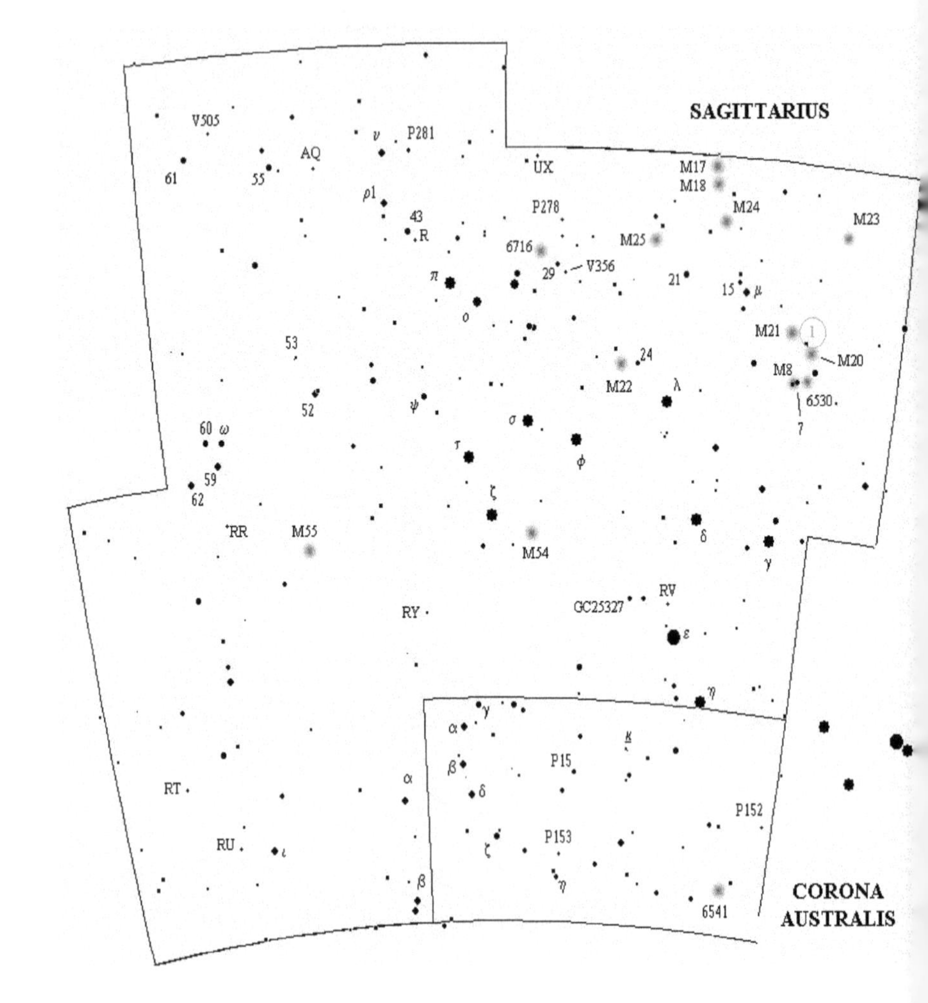
SAGITTARIUS
CORONA AUSTRALIS
V505
AQ
P281
UX
M17
M18
M24
M23
61
55
43
R
P278
M25
6716
29
V356
21
15
M21
M20
53
24
M22
M8
6530
52
60
59
62
RR
M55
M54
RY
GC25327
RV
P15
P152
RT
RU
P153
6541

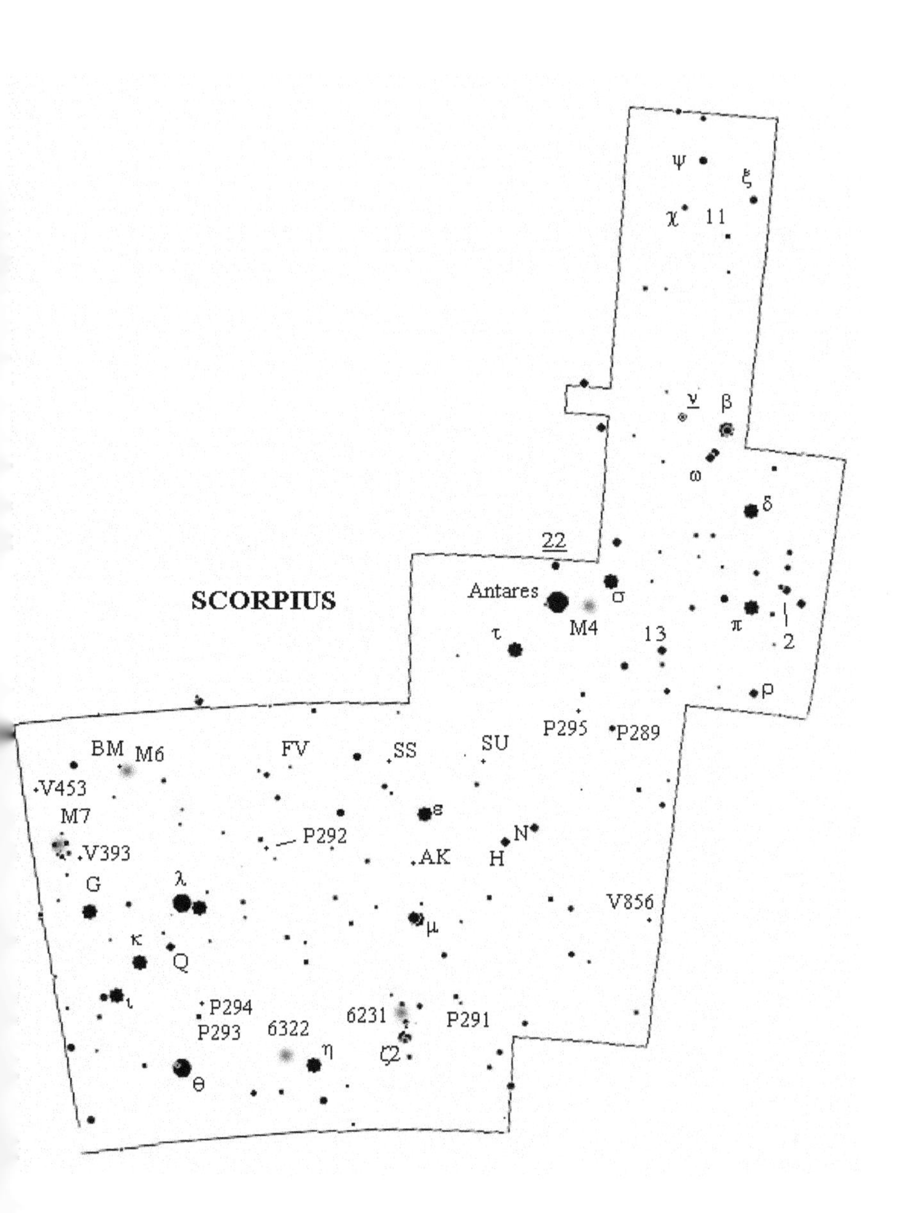
SCORPIUS
ψ
ξ
χ
11
ν
β
ω
δ
22
Antares
σ
M4
τ
13
π
2
ρ
P295
P289
BM
M6
FV
SS
SU
V453
M7
ε
P292
N
V393
AK
H
G
λ
V856
μ
κ
Q
ι
P294
P293
6322
6231
P291
η
ζ2
θ

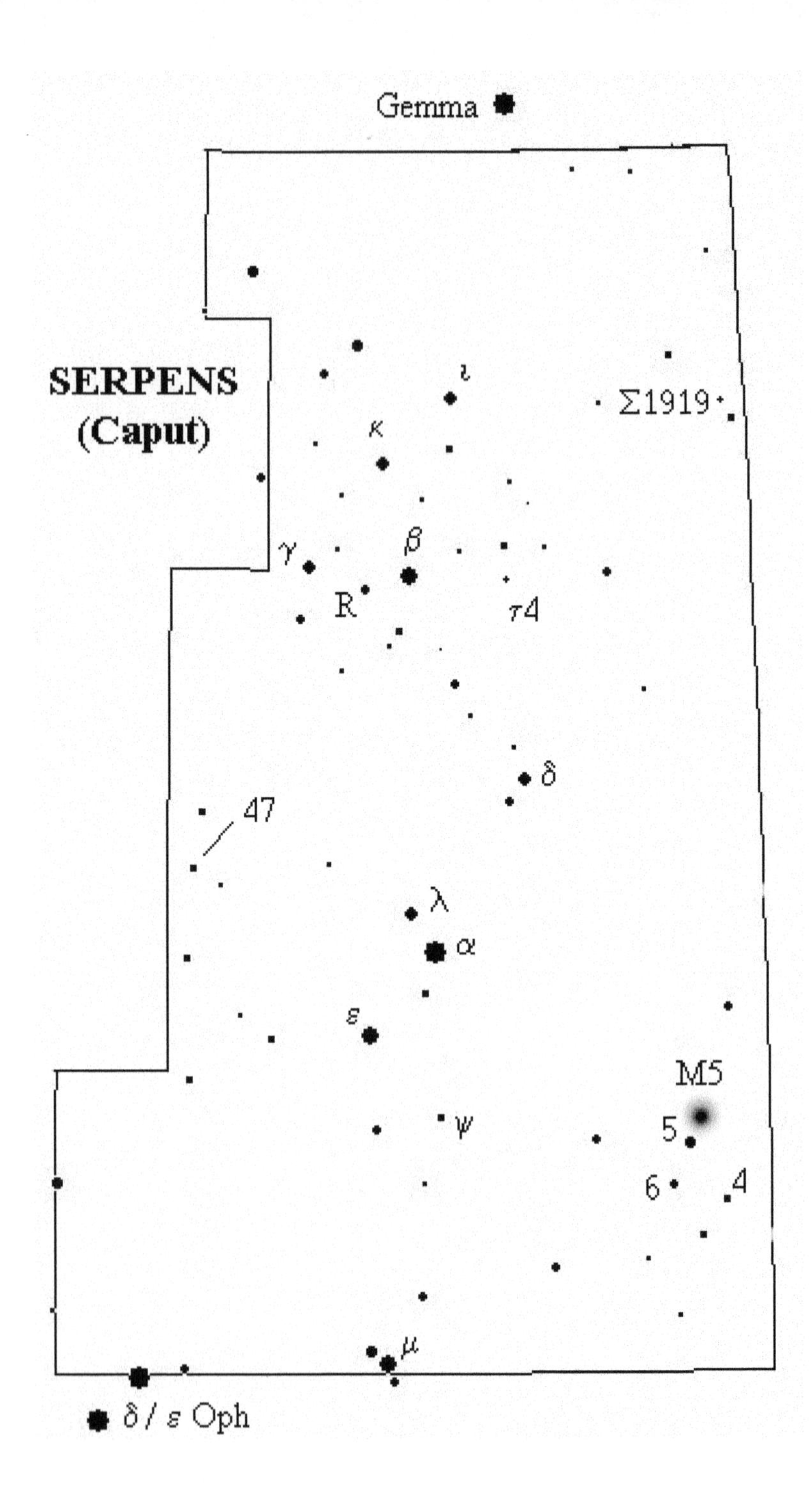
Gemma
SERPENS
(Caput)
ι
Σ1919
κ
γ
β
R
τ4
δ
47
λ
α
ε
M5
ψ
5
6
4
μ
δ / ε Oph

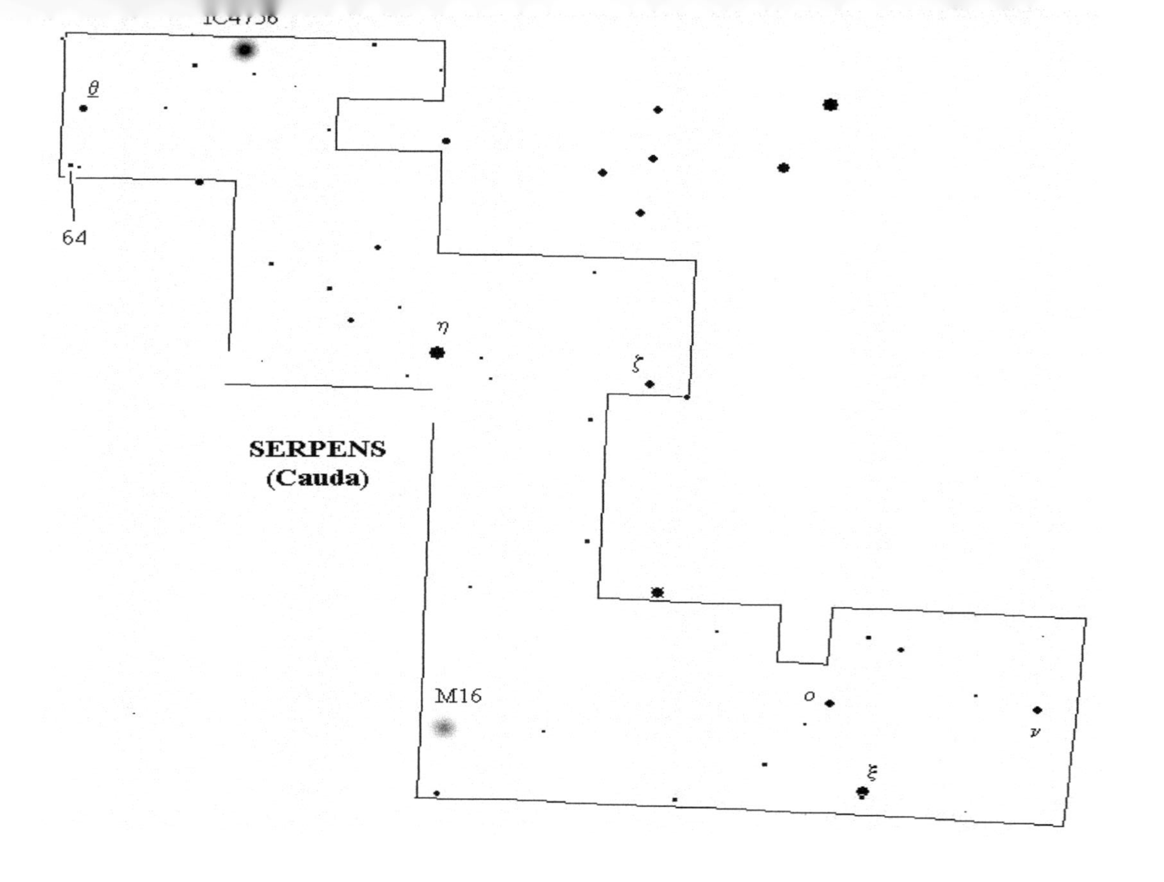
θ
64
η
ζ
SERPENS
(Cauda)
M16
o
ν
ξ

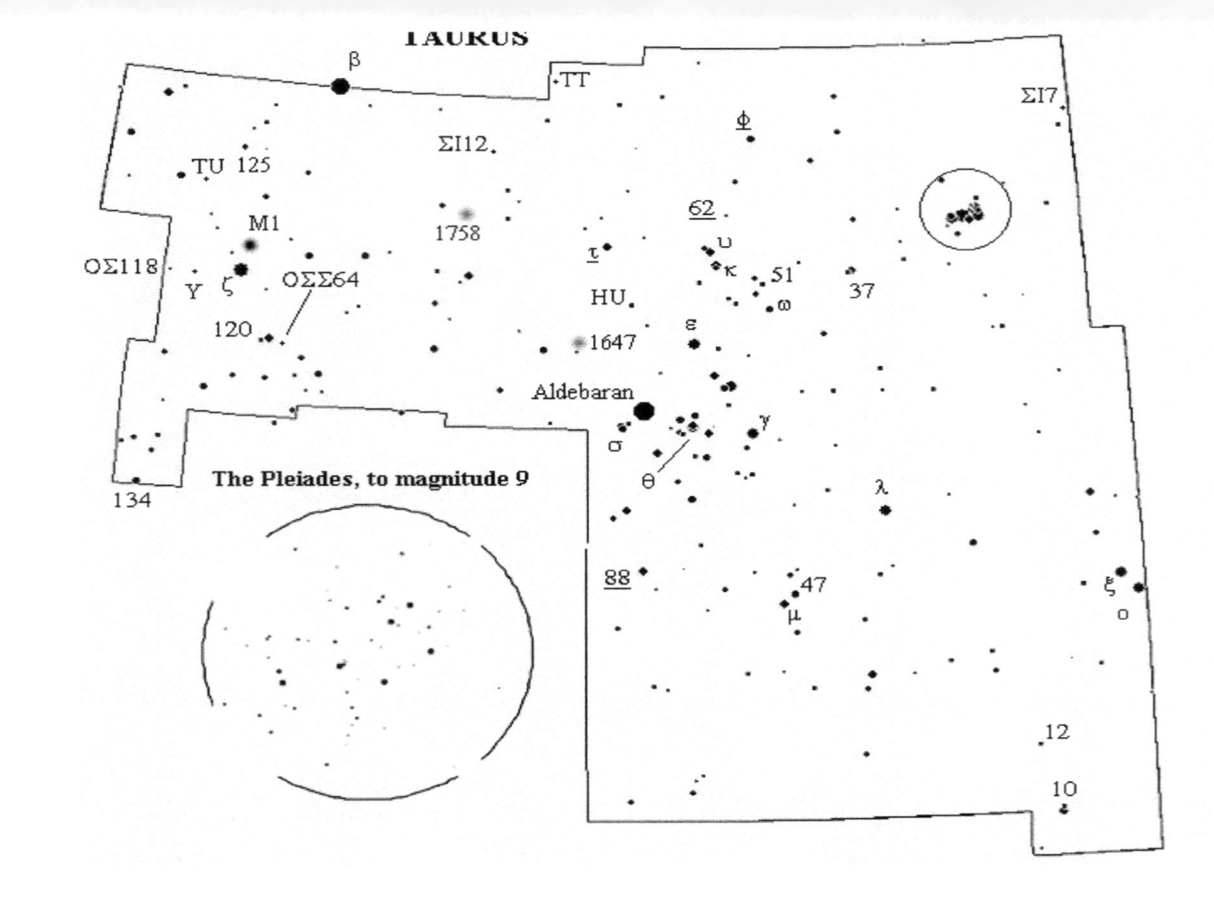

TAURUS
β
TT
ΣI7
Φ
TU 125
ΣI12
62
M1
1758
τ
υ
κ
51
37
ΟΣ118
Y
ζ
ΟΣΣ64
HU
ω
120
1647
ε
Aldebaran
γ
σ
θ
λ
134
The Pleiades, to magnitude 9
88
47
μ
ξ
ο
12
10

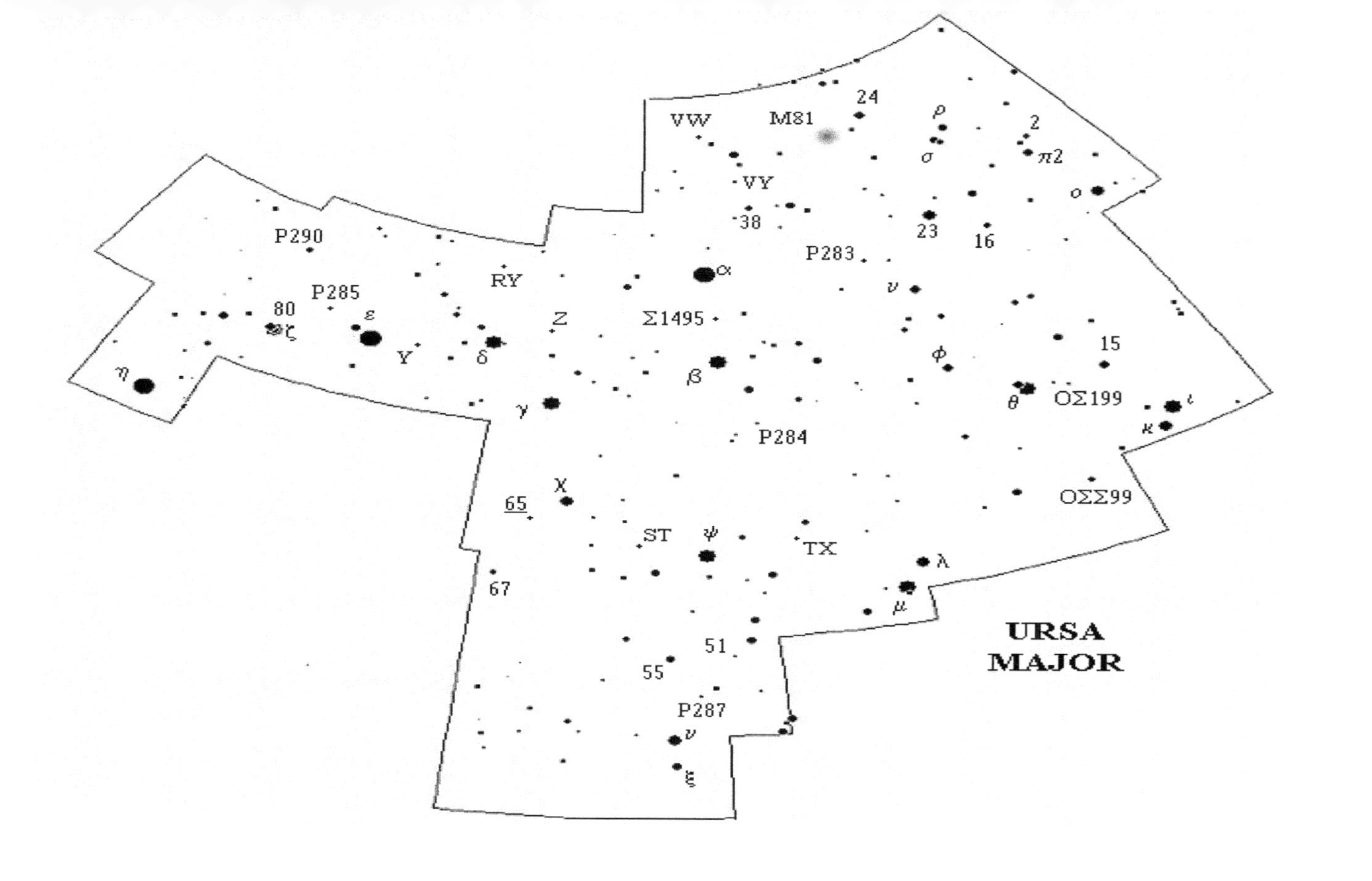
URSA
MAJOR
VW
M81
24
ρ
2
σ
π2
VY
ο
38
23
16
P290
P283
α
RY
ν
P285
80
ε
Z
Σ1495
15
ζ
Y
δ
φ
η
β
θ
ΟΣ199
ι
κ
γ
P284
χ
65
ΟΣΣ99
ST
ψ
TX
λ
67
μ
51
55
P287
ν
ξ

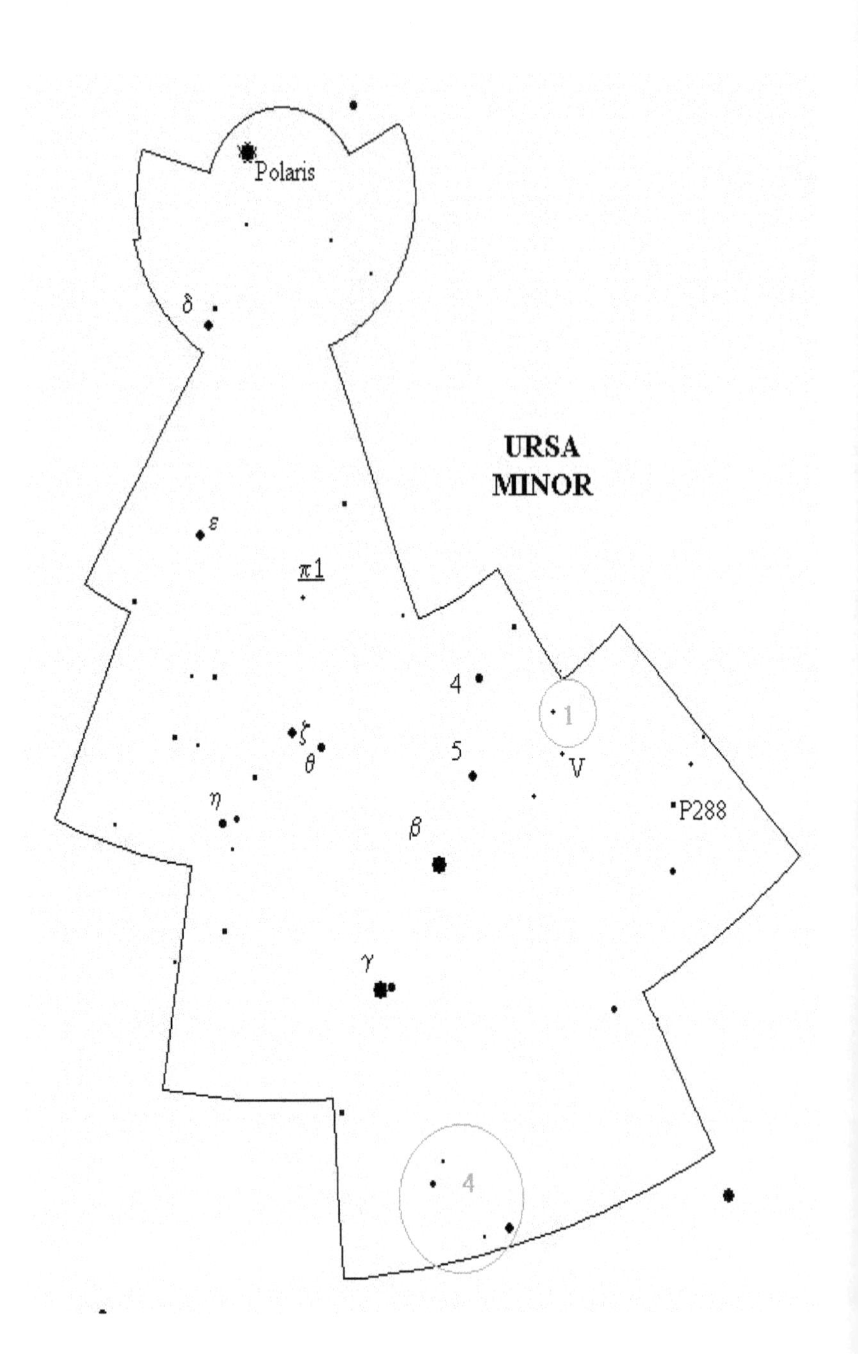
Polaris
δ
URSA
MINOR
ε
π1
4
1
ζ
θ
5
V
η
P288
β
γ
4

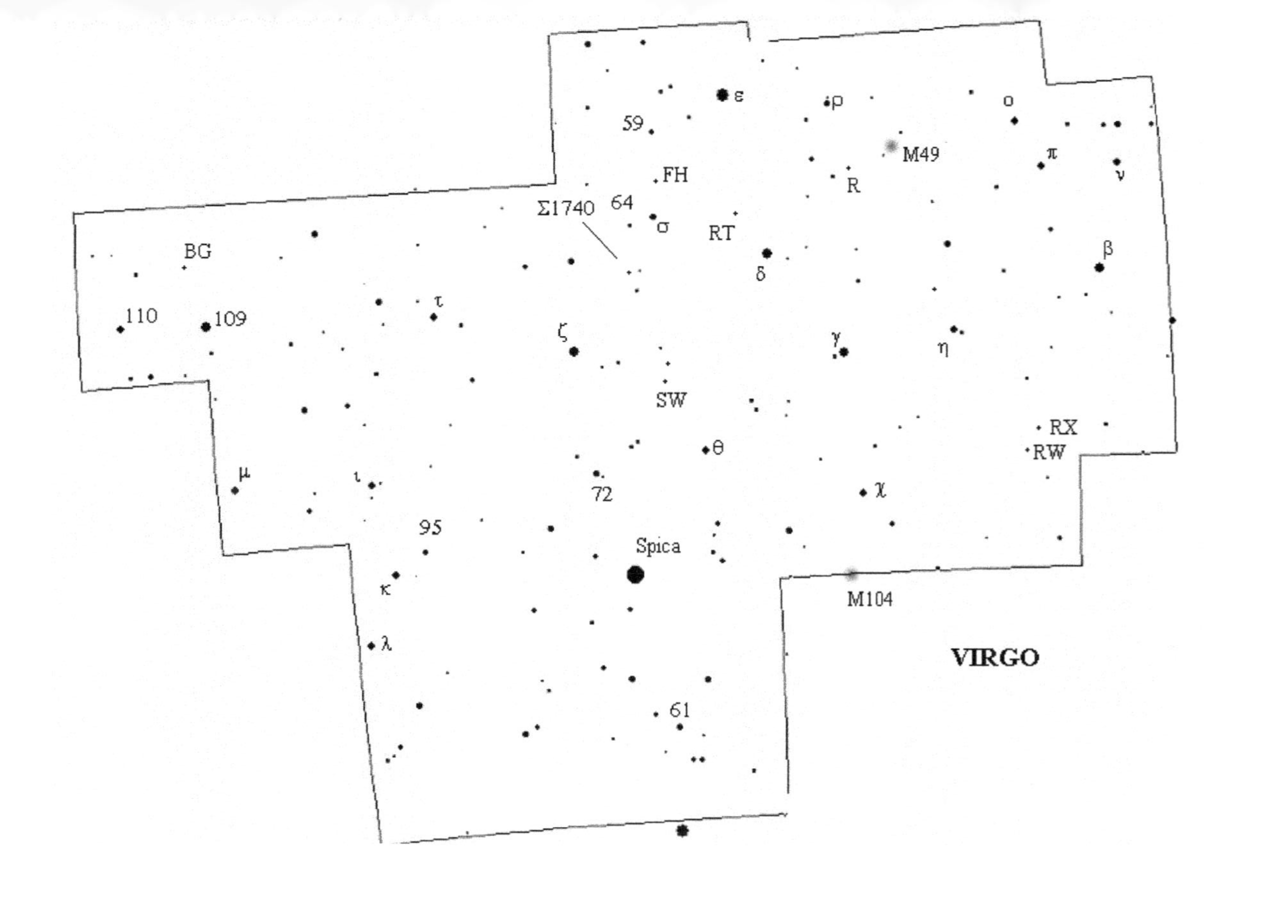
VIRGO
M49
M104
Spica
RX
RW
R
RT
FH
SW
Σ1740
59
64
72
61
95
109
110
BG
ε
ρ
ο
π
ν
β
σ
δ
γ
η
ζ
τ
θ
χ
μ
ι
κ
λ

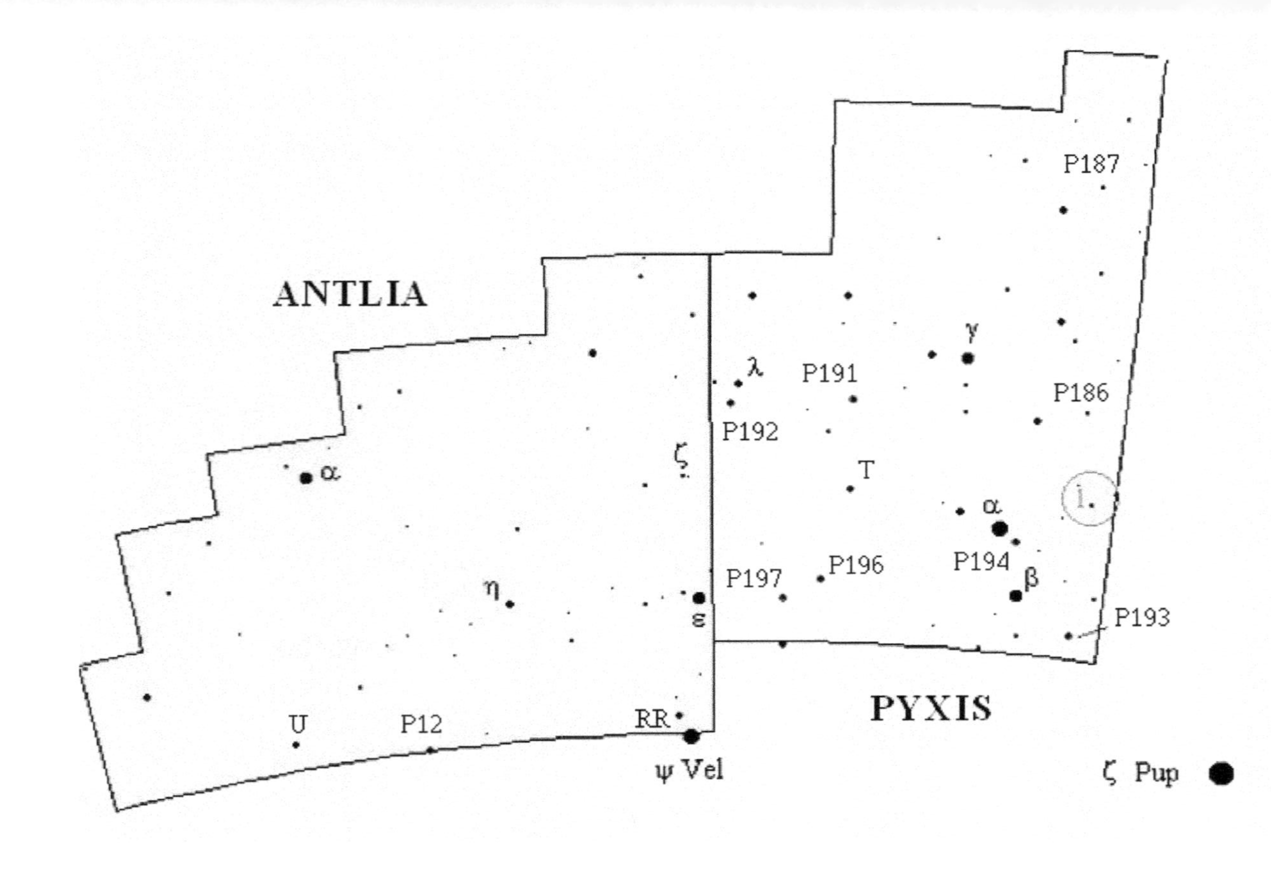
ANTLIA
PYXIS
α
U
P12
η
RR
ψ Vel
ε
ζ
λ
P191
P192
T
P187
γ
P186
α
P194
β
P197
P196
P193
ζ Pup

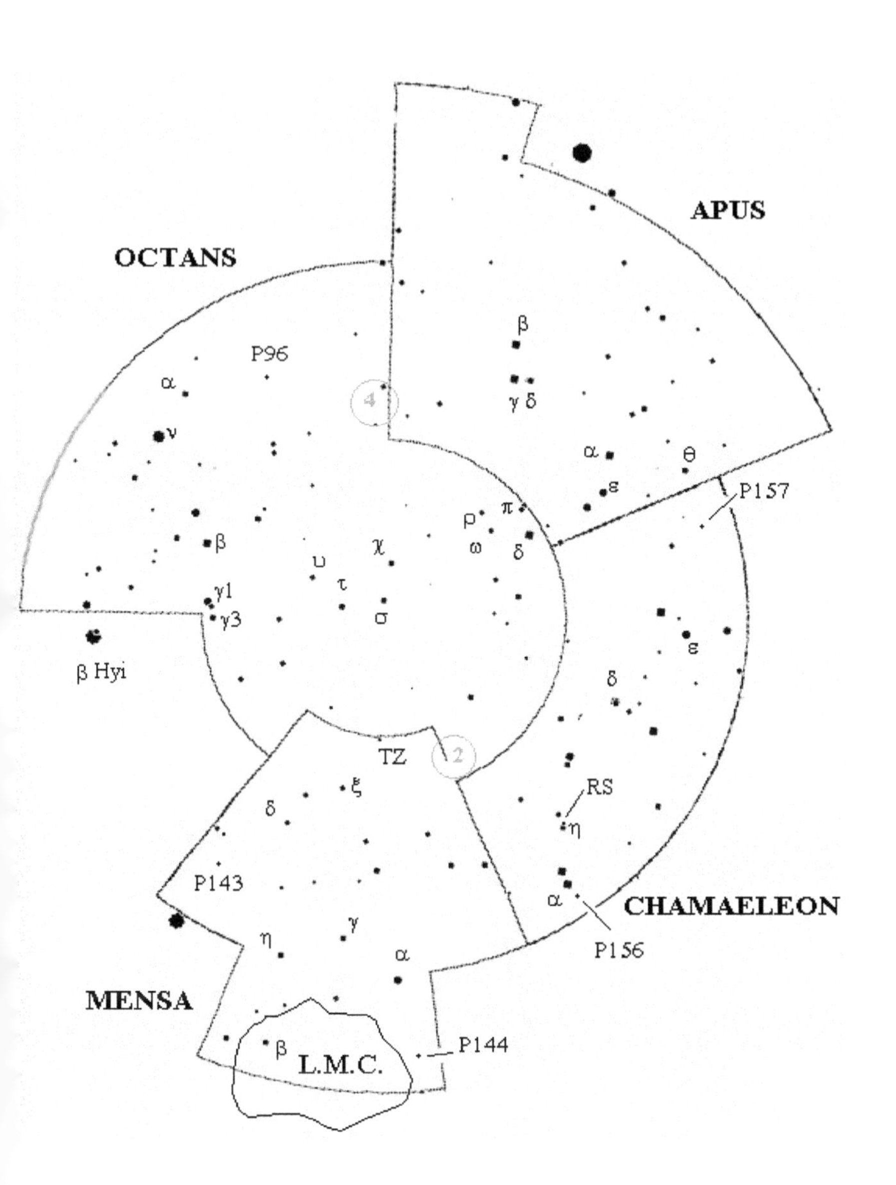
APUS
OCTANS
P96
α
ν
β
γ1
γ3
β Hyi
υ
χ
τ
σ
ρ
π
ω
δ
β
γ δ
α
θ
ε
P157
ε
δ
TZ
ξ
δ
P143
η
γ
α
RS
η
α
CHAMAELEON
P156
MENSA
β
L.M.C.
P144

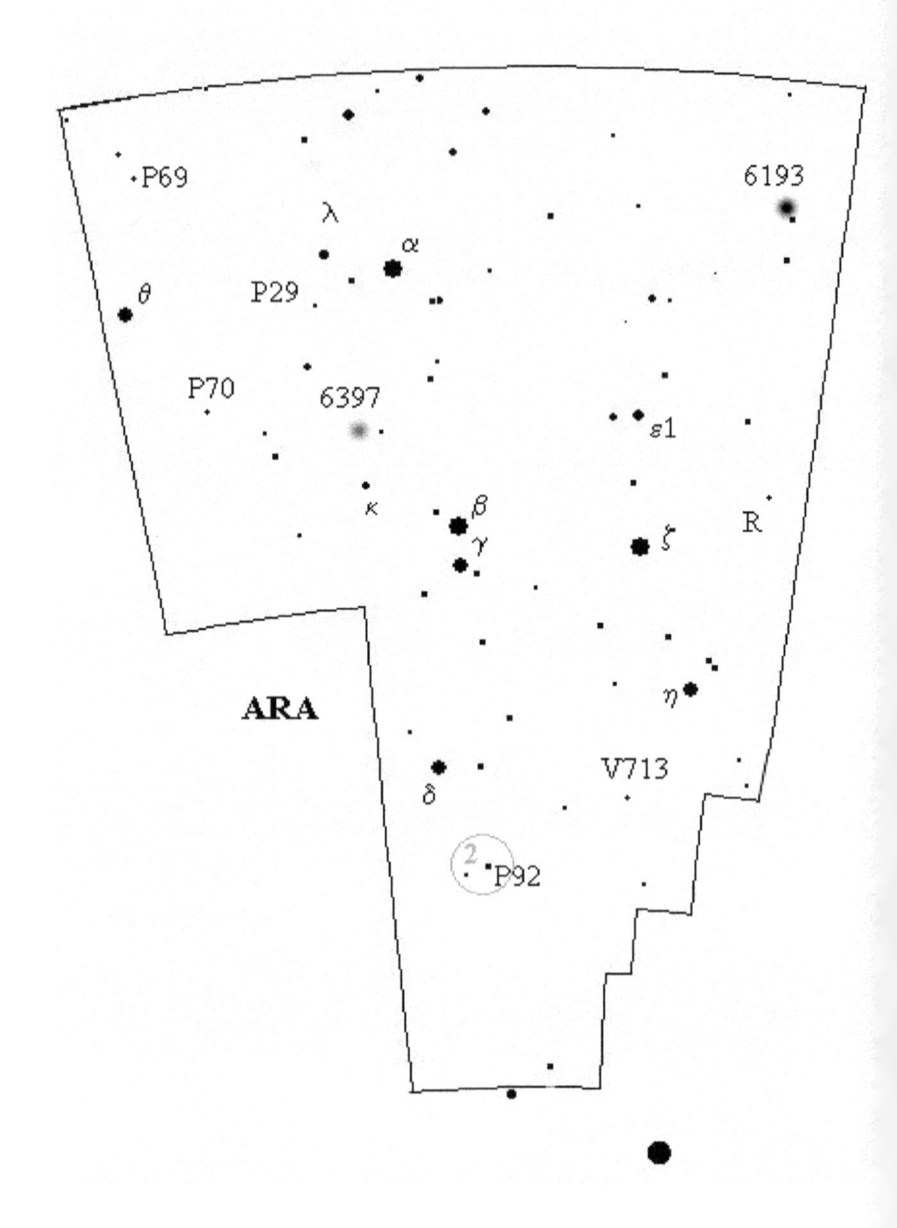
P69
6193
λ
α
θ
P29
P70
6397
ε1
κ
β
R
γ
ζ
ARA
η
V713
δ
2
P92

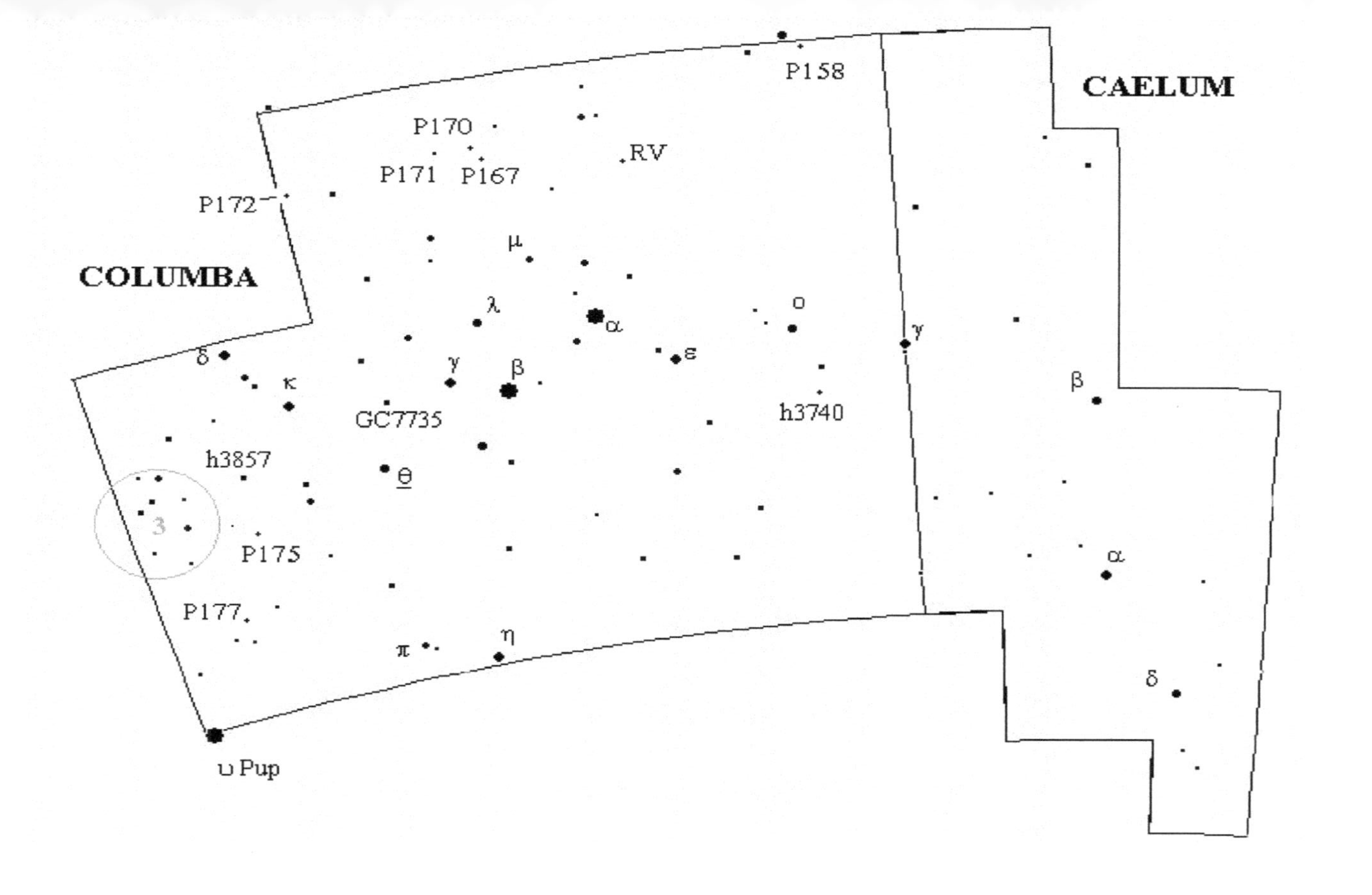

CAELUM
COLUMBA
P158
P170
P171
P167
RV
P172
h3740
GC7735
h3857
P175
P177
υ Pup

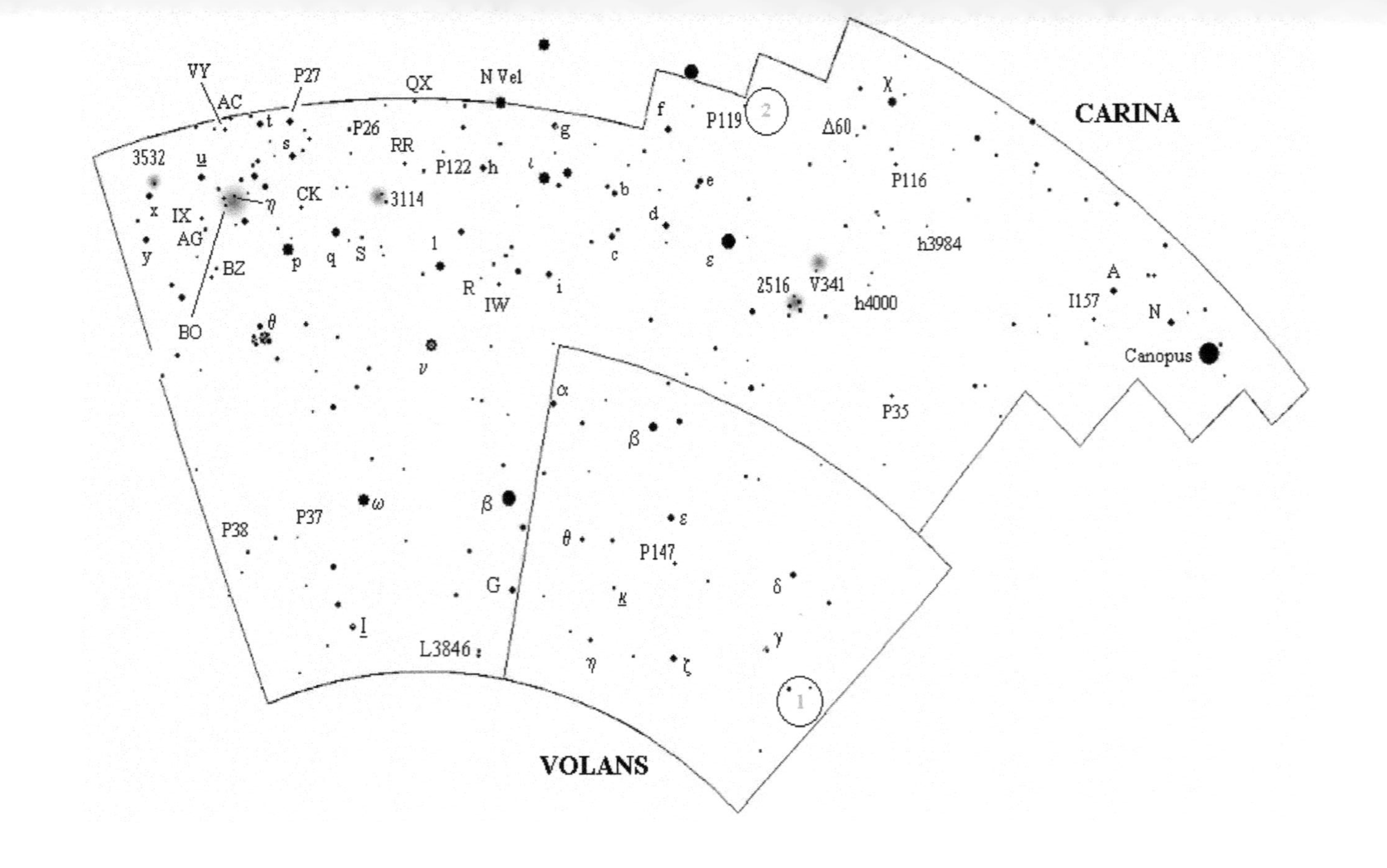
CARINA
VOLANS
VY
AC
P27
QX
N Vel
3532
u
t
s
P26
RR
P122
h
ι
g
f
P119
2
Δ60
χ
x
IX
AG
y
BZ
BO
η
CK
3114
p
q
S
1
R
IW
i
b
c
d
e
ε
2516
V341
h4000
P116
h3984
A
I157
N
Canopus
θ
ν
P35
α
β
ω
P37
P38
ε
θ
P147
G
κ
δ
γ
I
L3846
η
ζ
1

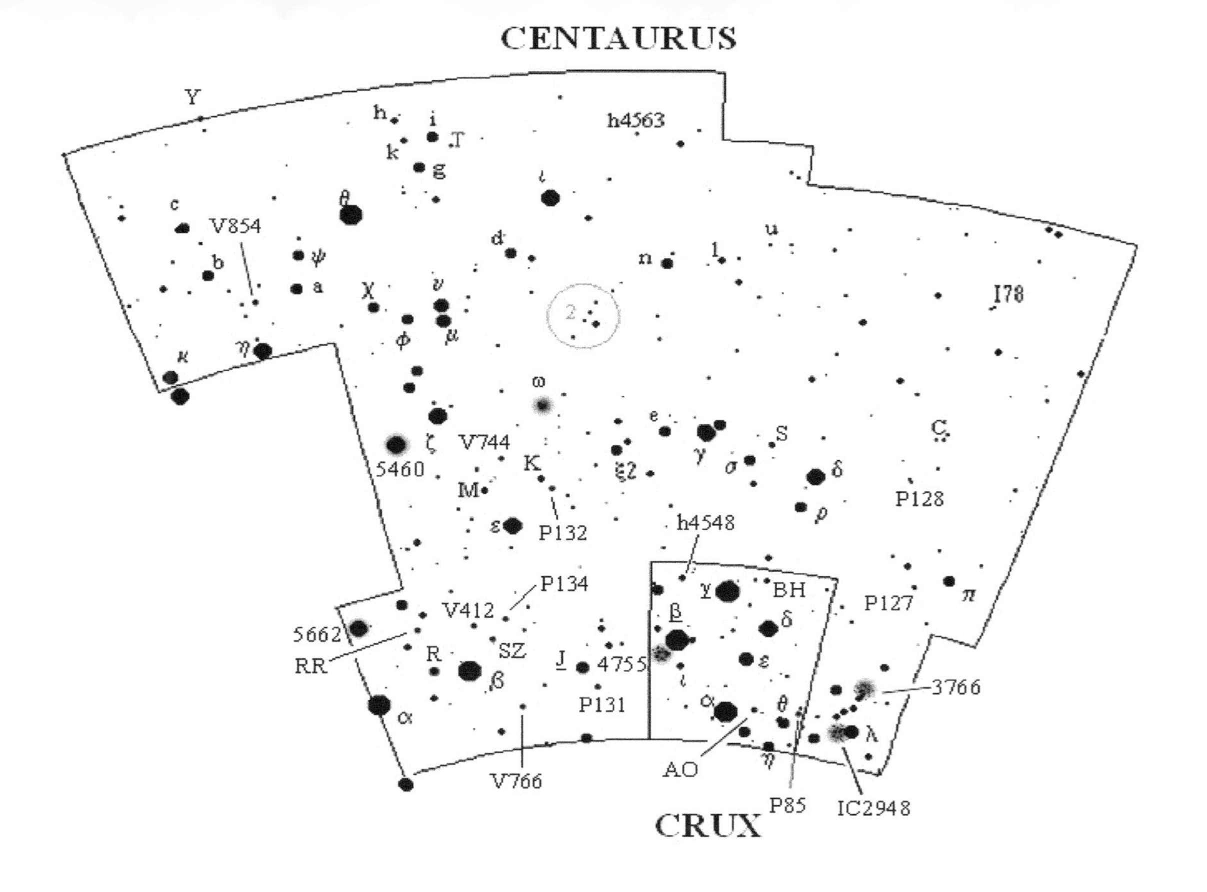
CENTAURUS
CRUX
h4563
I78
P128
P127
3766
IC2948
P85
AO
BH
h4548
4755
P131
P132
P134
V766
SZ
V412
V744
5460
5662
RR
V854

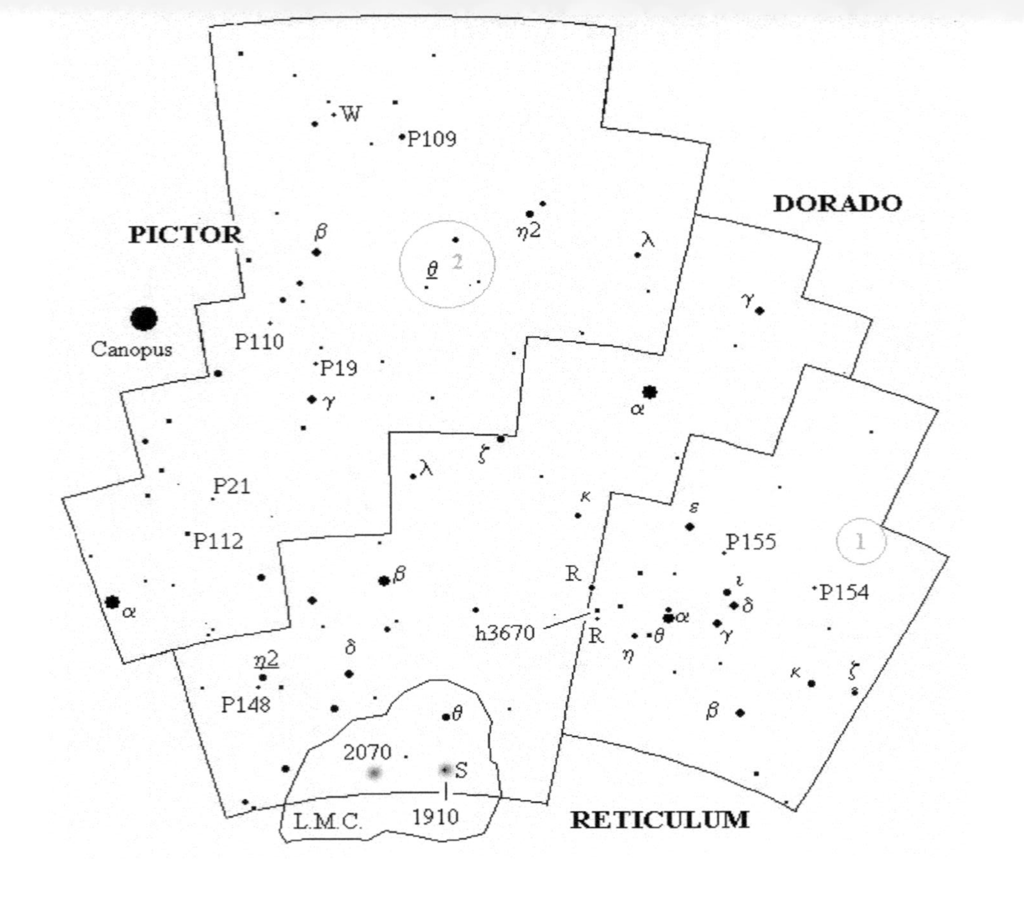
PICTOR
DORADO
RETICULUM
Canopus
W
P109
P110
P19
P21
P112
P148
h3670
R
P155
P154
2070
1910
S
L.M.C.
1
2

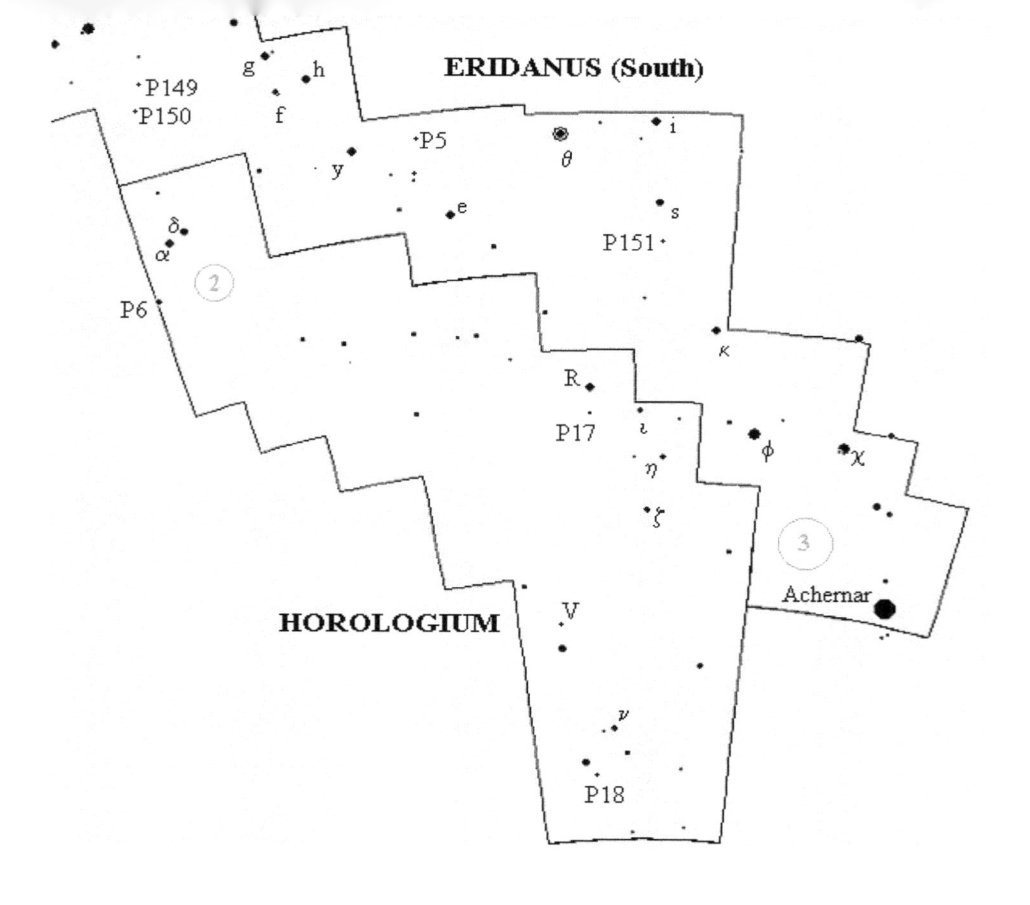
ERIDANUS (South)
HOROLOGIUM
Achernar
P149
P150
P5
P6
P151
P17
P18
R
V

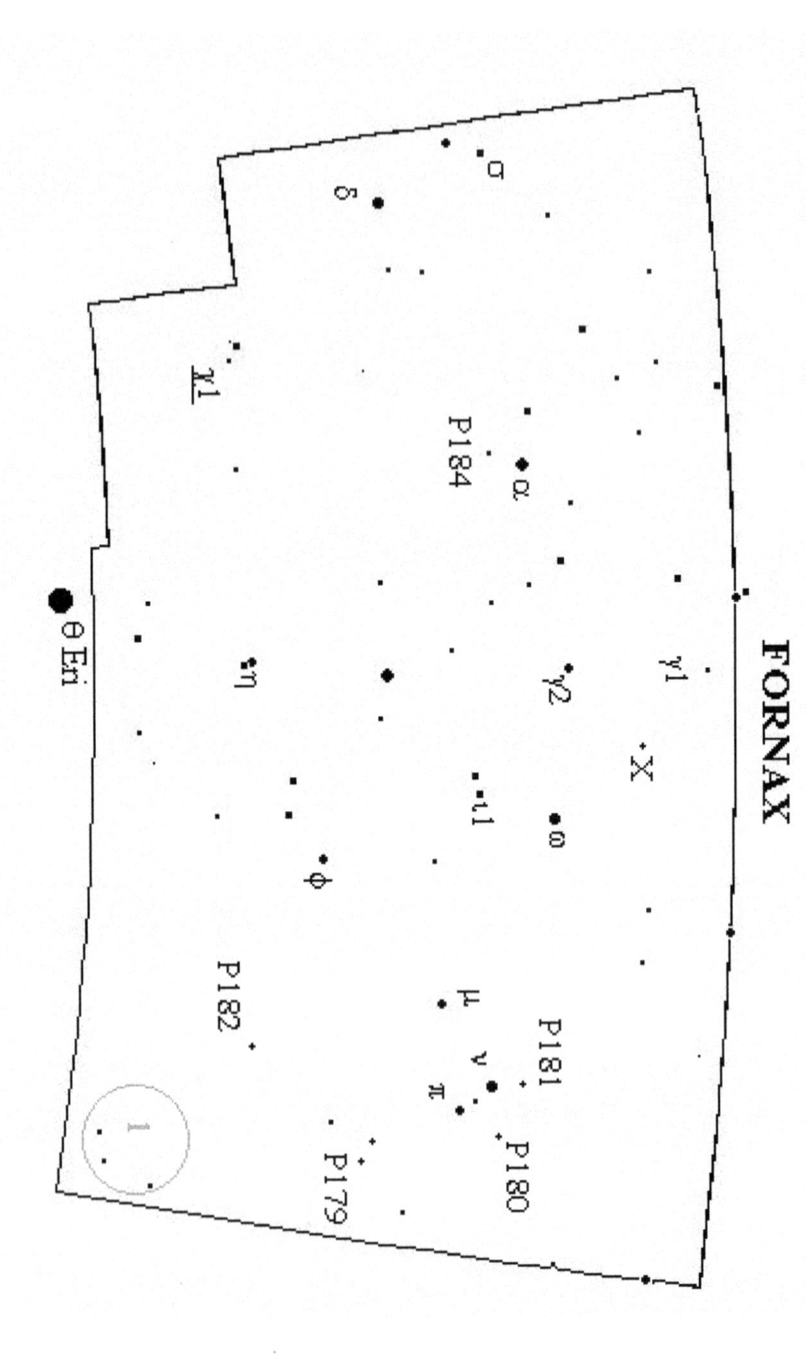

FORNAX
θ Eri
α
P184
χ1
γ1
γ2
η
χ
ι1
ω
φ
P182
μ
P181
ν
π
P180
P179
δ
σ

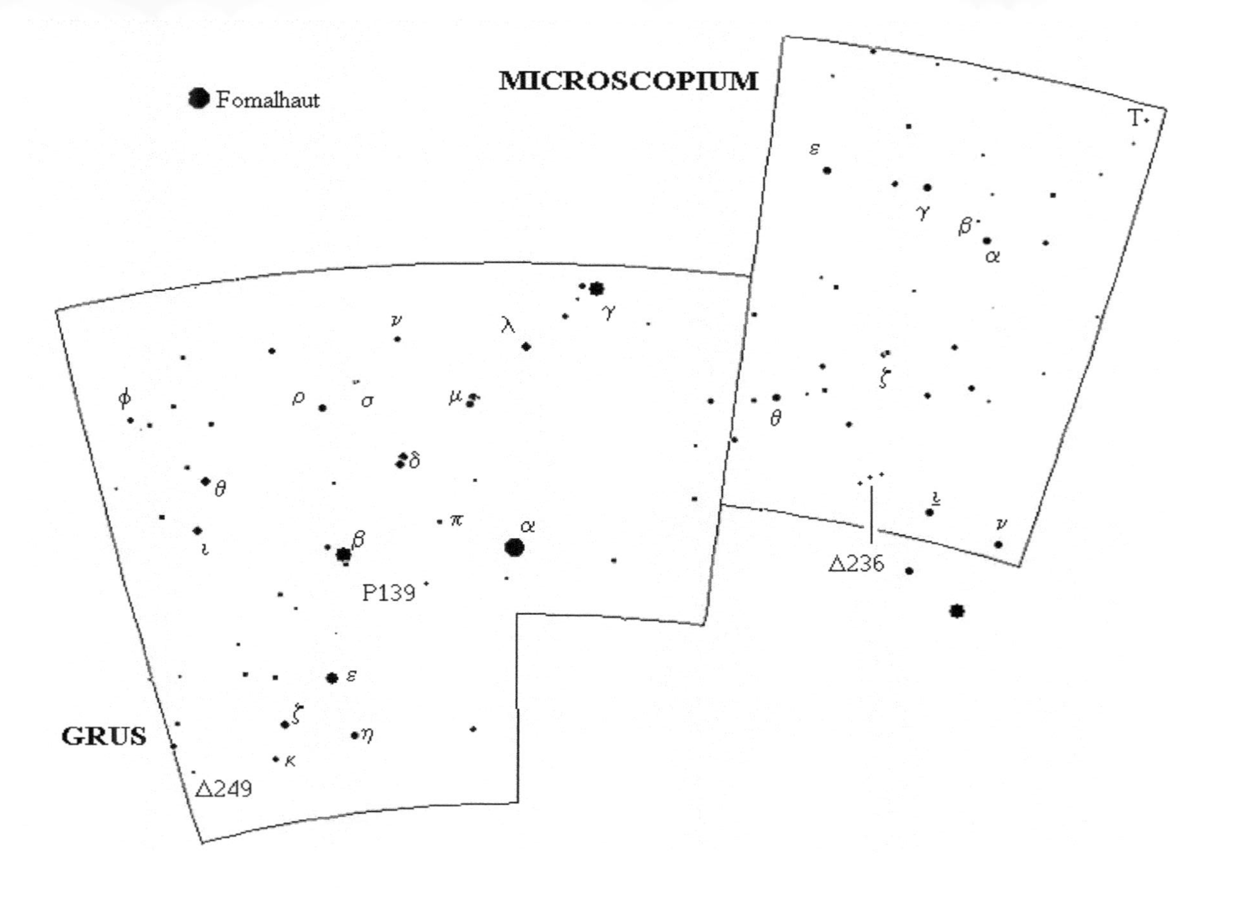
MICROSCOPIUM
Fomalhaut
GRUS
P139
Δ236
Δ249
T

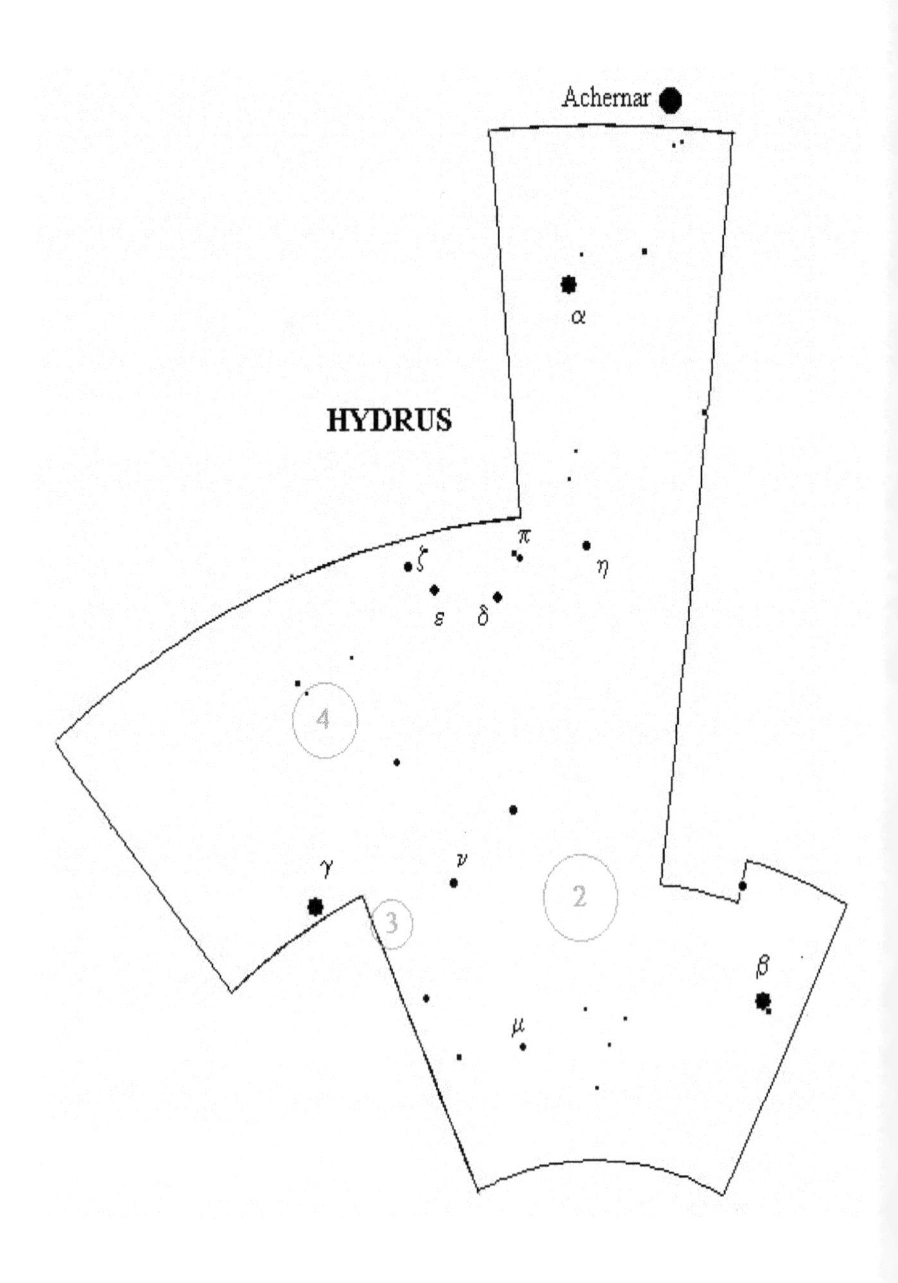
Achernar
α
HYDRUS
π
ζ
η
ε
δ
4
γ
ν
2
3
β
μ

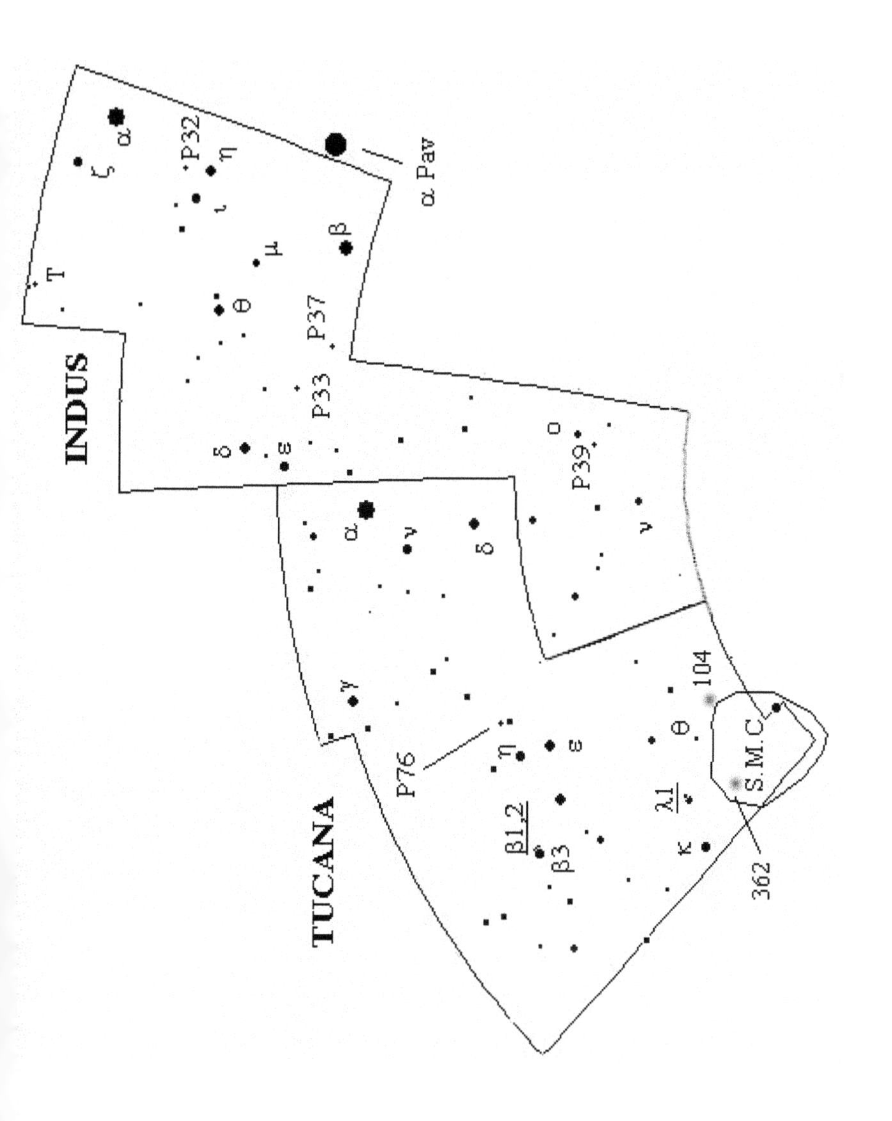
INDUS
TUCANA
α Pav
P32
P37
P33
P39
P76
S.M.C
104
362
β1,2
β3
λ1

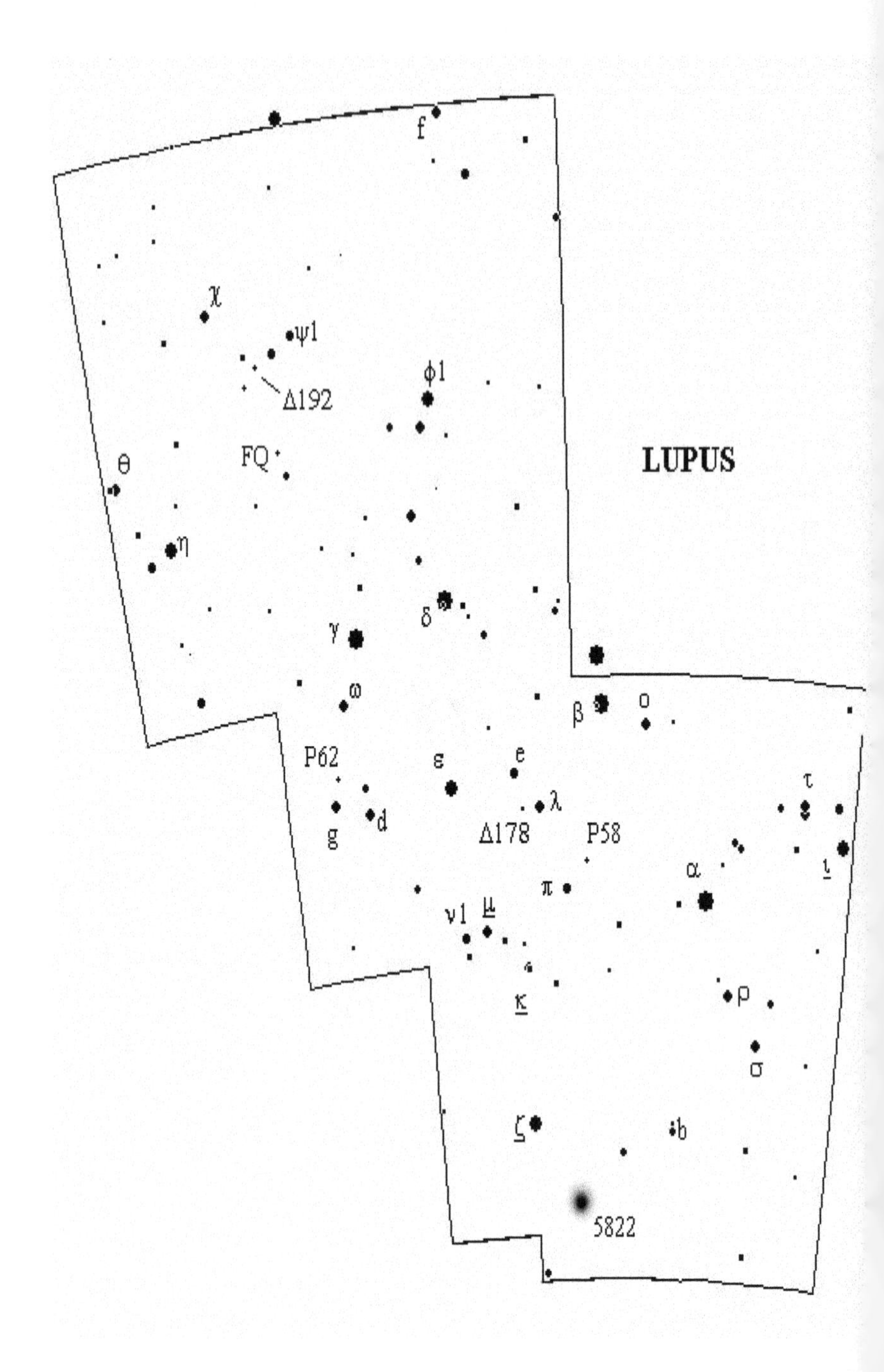

LUPUS
f
χ
ψ1
Δ192
φ1
θ
FQ
η
δ
γ
ω
β
o
P62
ε
e
τ
λ
d
g
Δ178
P58
α
ι
π
ν1
μ
κ
ρ
σ
ζ
b
5822

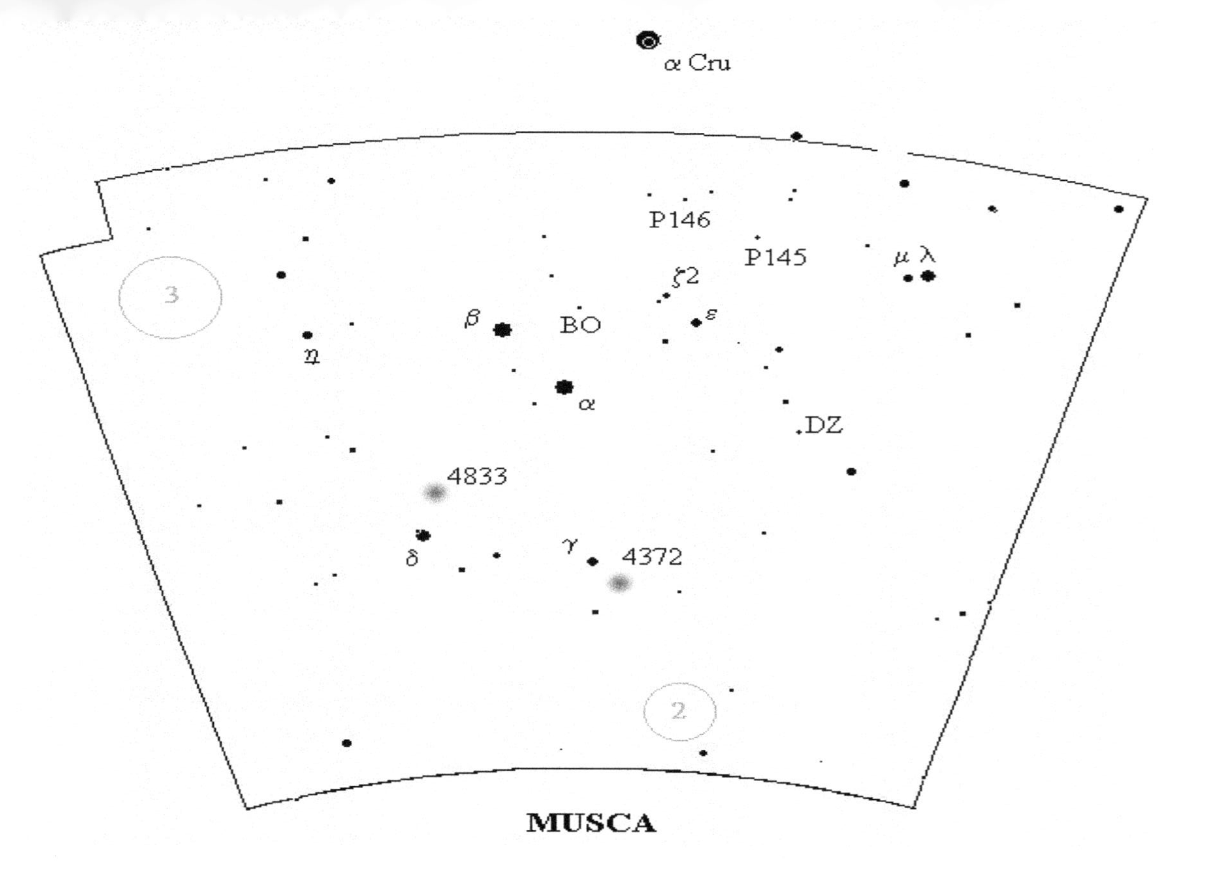

α Cru
P146
P145
μ λ
ζ2
ε
β
BO
α
DZ
4833
γ
δ
4372
3
2
MUSCA

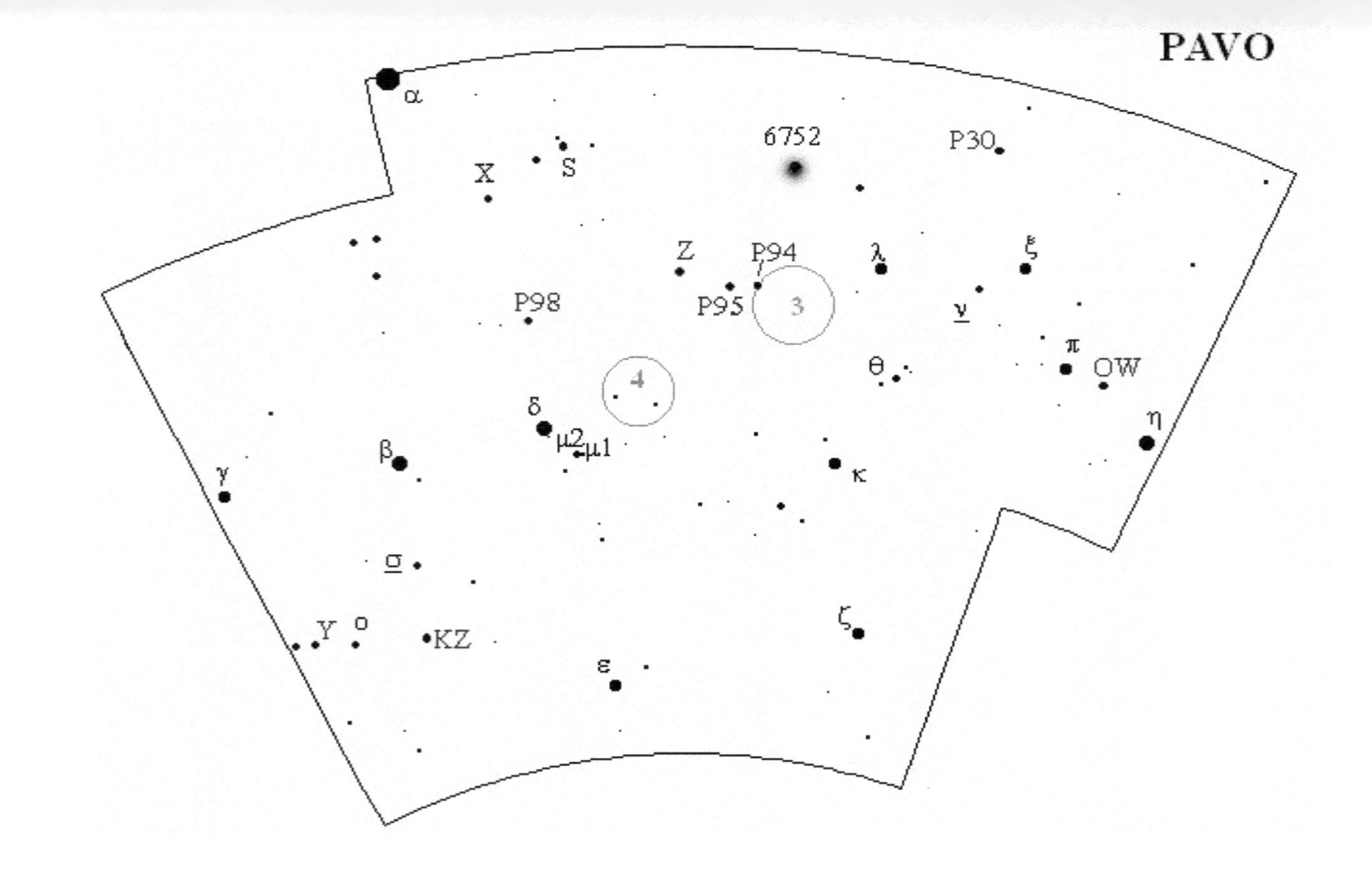
PAVO
α
X
S
6752
P30
Z
P94
λ
ξ
P98
P95
3
ν
π
θ
OW
4
δ
μ2
μ1
η
β
γ
κ
σ
ζ
Y
o
KZ
ε

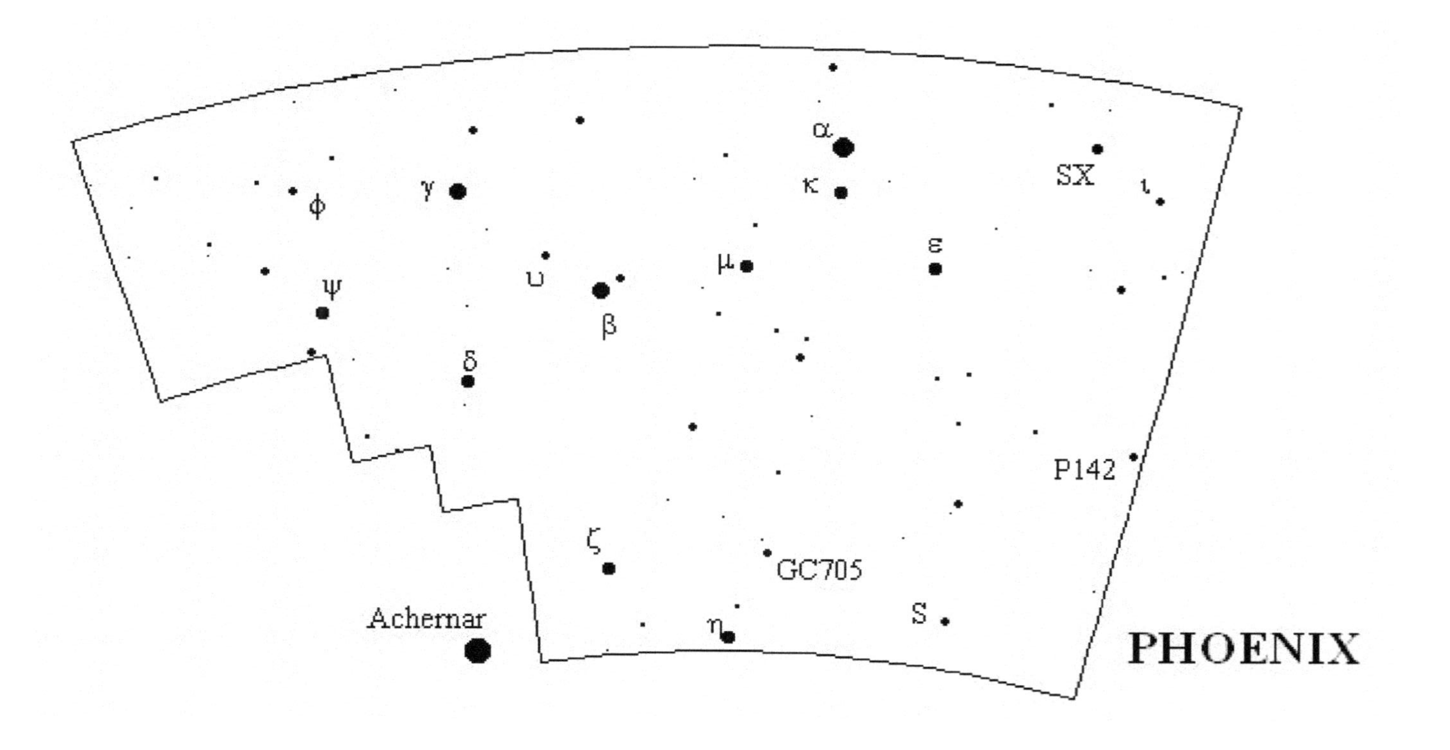
α
SX
ι
γ
κ
ϕ
ε
υ
μ
ψ
β
δ
P142
ζ
GC705
Achernar
S
η
PHOENIX

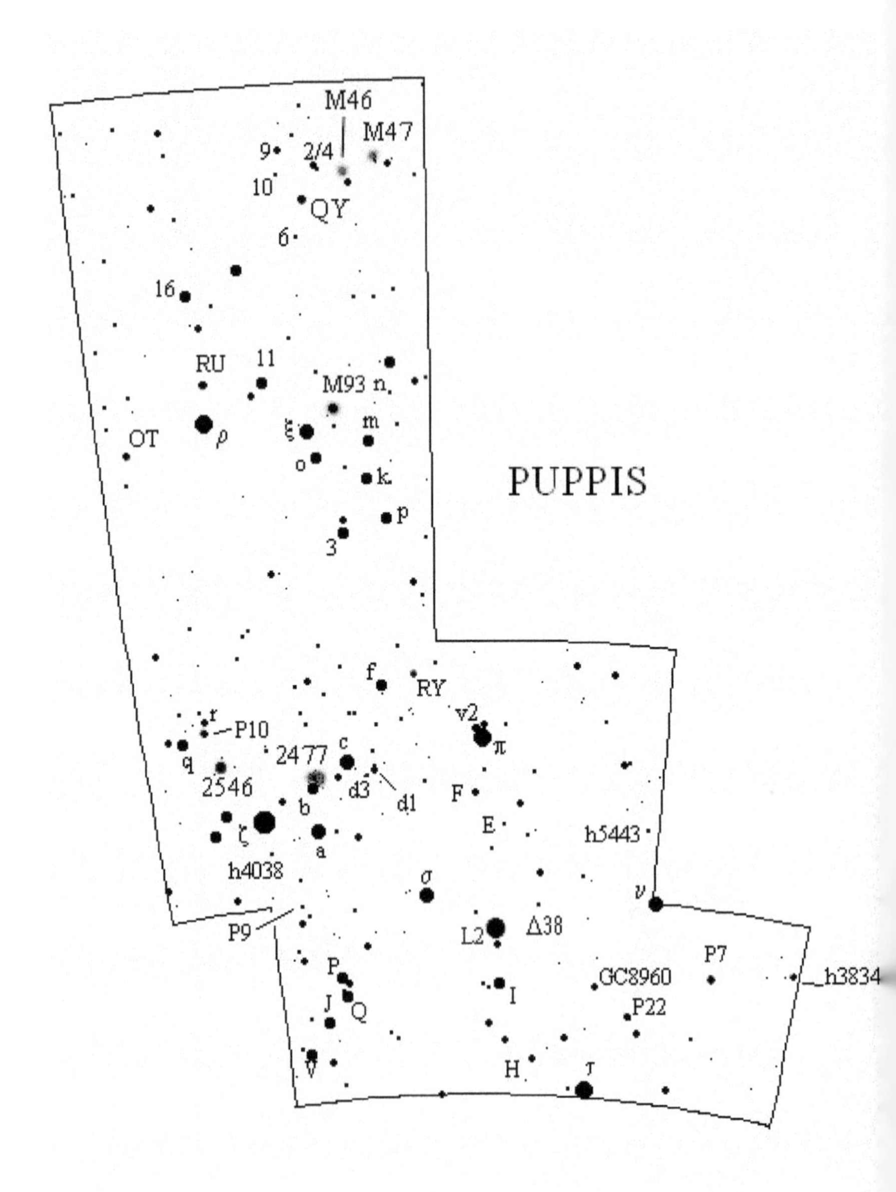
M46
M47
9
2/4
10
QY
6
16
RU
11
M93
n
OT
ρ
ξ
m
o
k
PUPPIS
p
3
f
RY
r
P10
v2
π
q
2477
c
2546
d3
d1
F
b
E
h5443
ζ
a
h4038
σ
ν
P9
L2
Δ38
P7
P
GC8960
h3834
J
Q
I
P22
V
H
τ

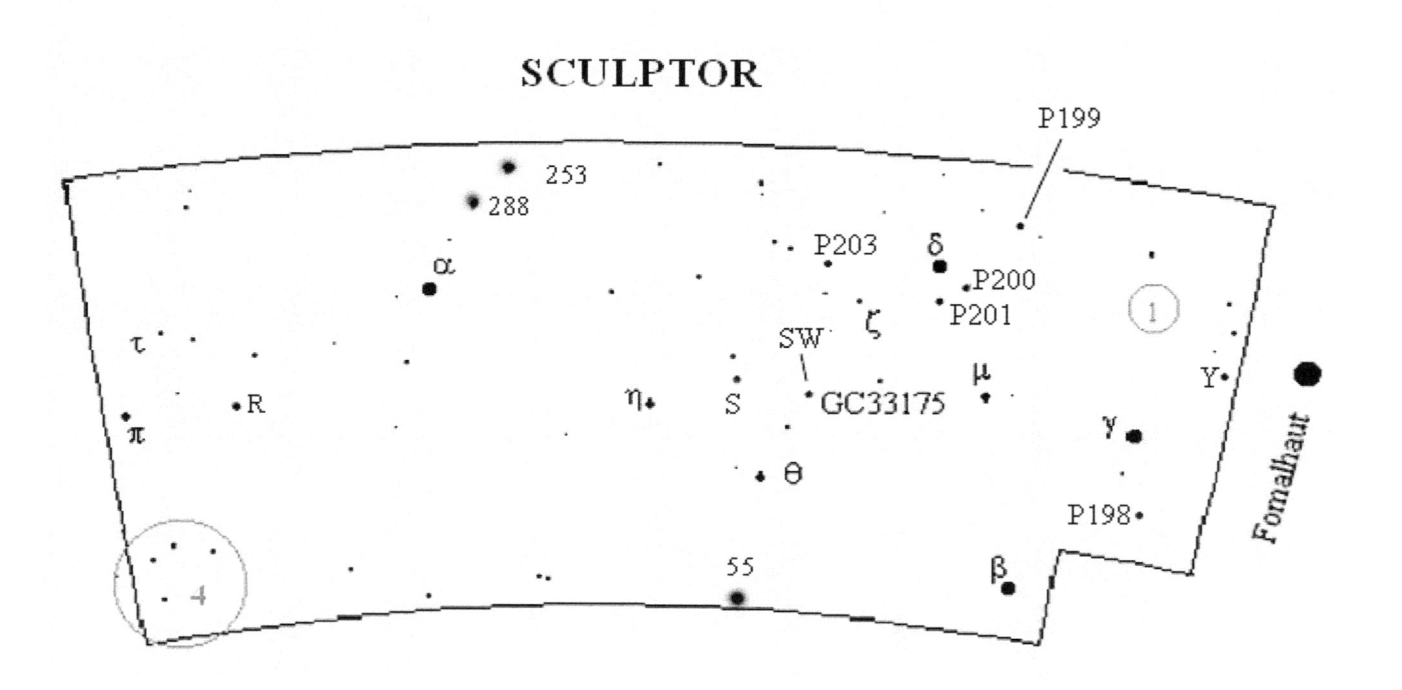
β Cet
SCULPTOR
P199
253
288
P203
δ
P200
P201
α
ζ
SW
τ
μ
Y
R
η
S
GC33175
π
γ
θ
Fomalhaut
P198
55
β

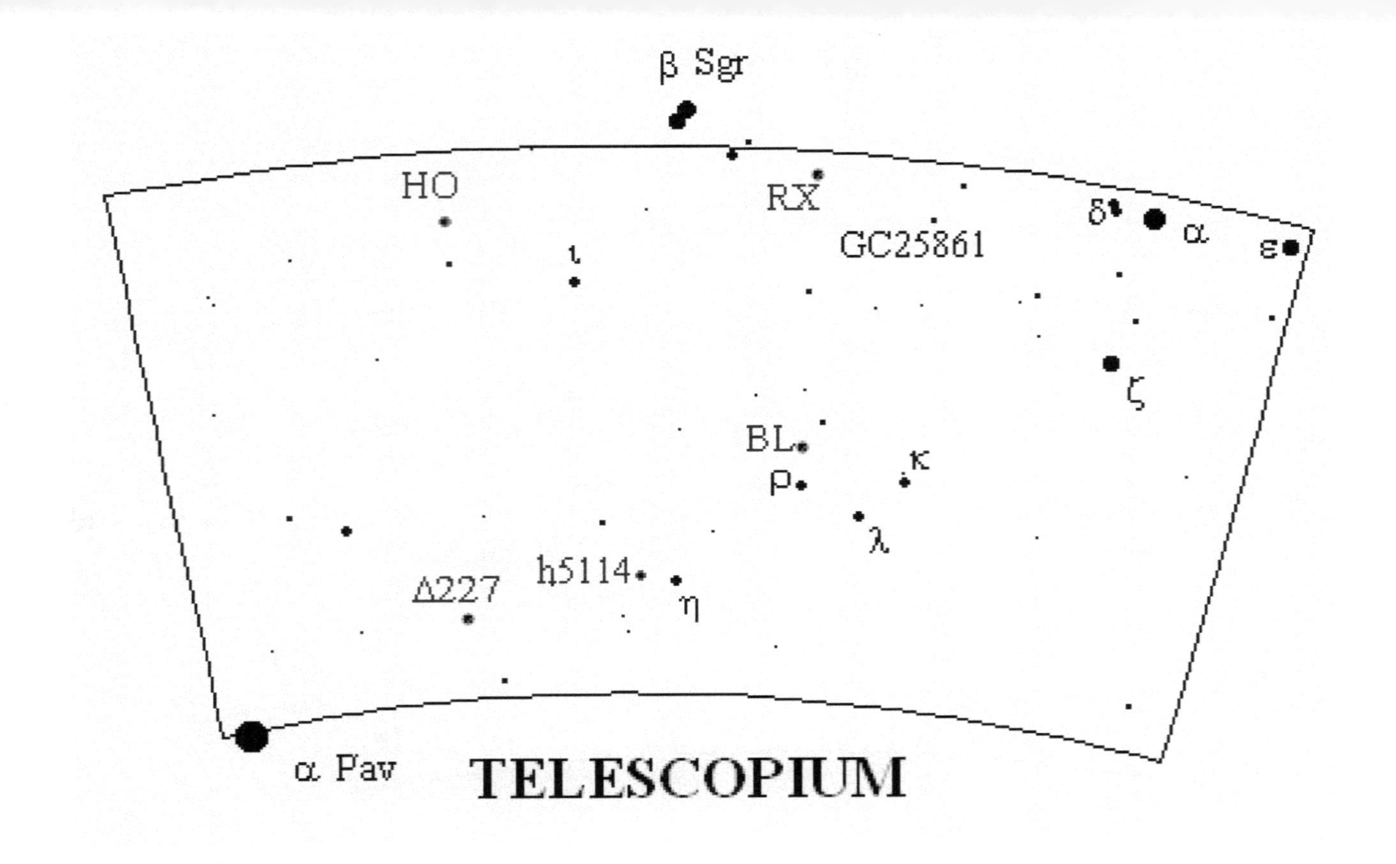

β Sgr
HO
RX
δ1
α
ε
GC25861
ι
ζ
BL
κ
ρ
λ
h5114
η
Δ227
α Pav
TELESCOPIUM

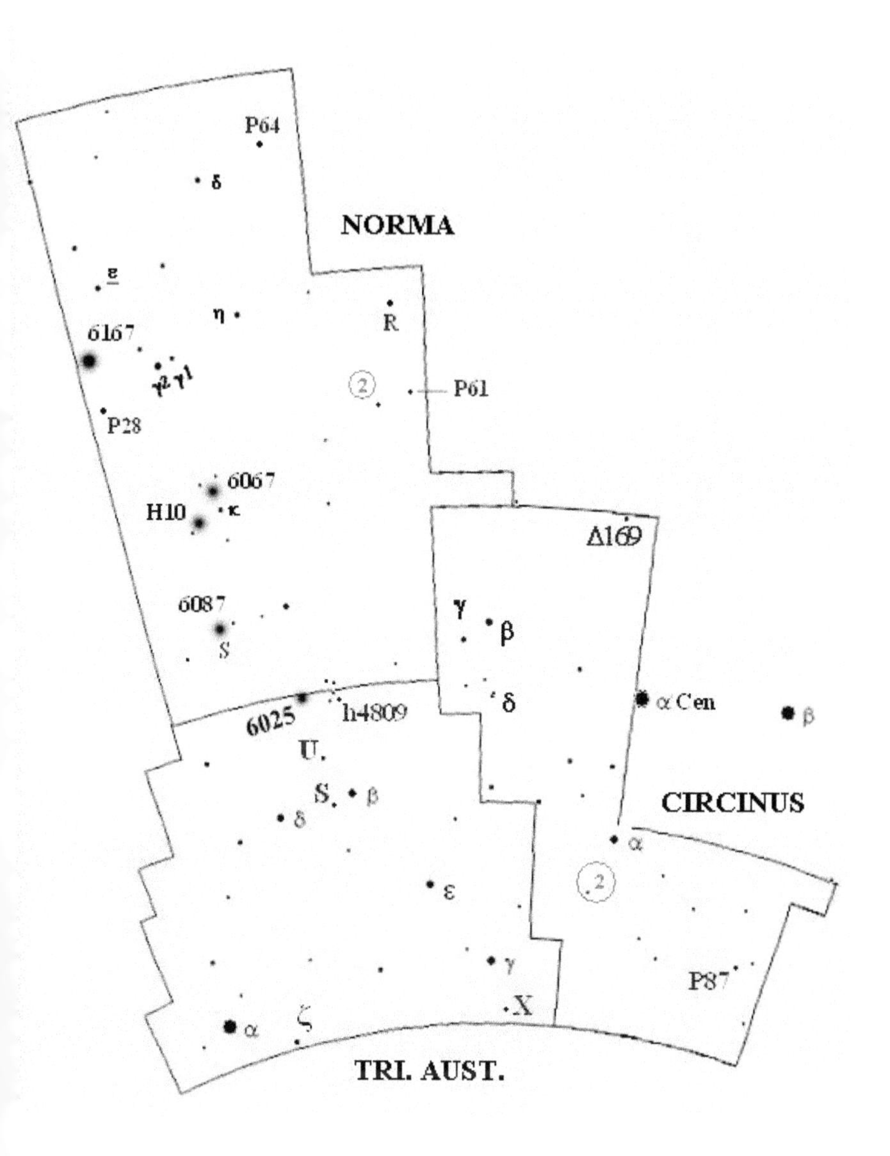
NORMA
P64
δ
ε
η
R
6167
γ2 γ1
P61
P28
6067
H10
κ
Δ169
6087
S
γ
β
δ
6025
h4809
α Cen
β
U.
S.
β
δ
CIRCINUS
α
ε
γ
P87
X
α
ζ
TRI. AUST.

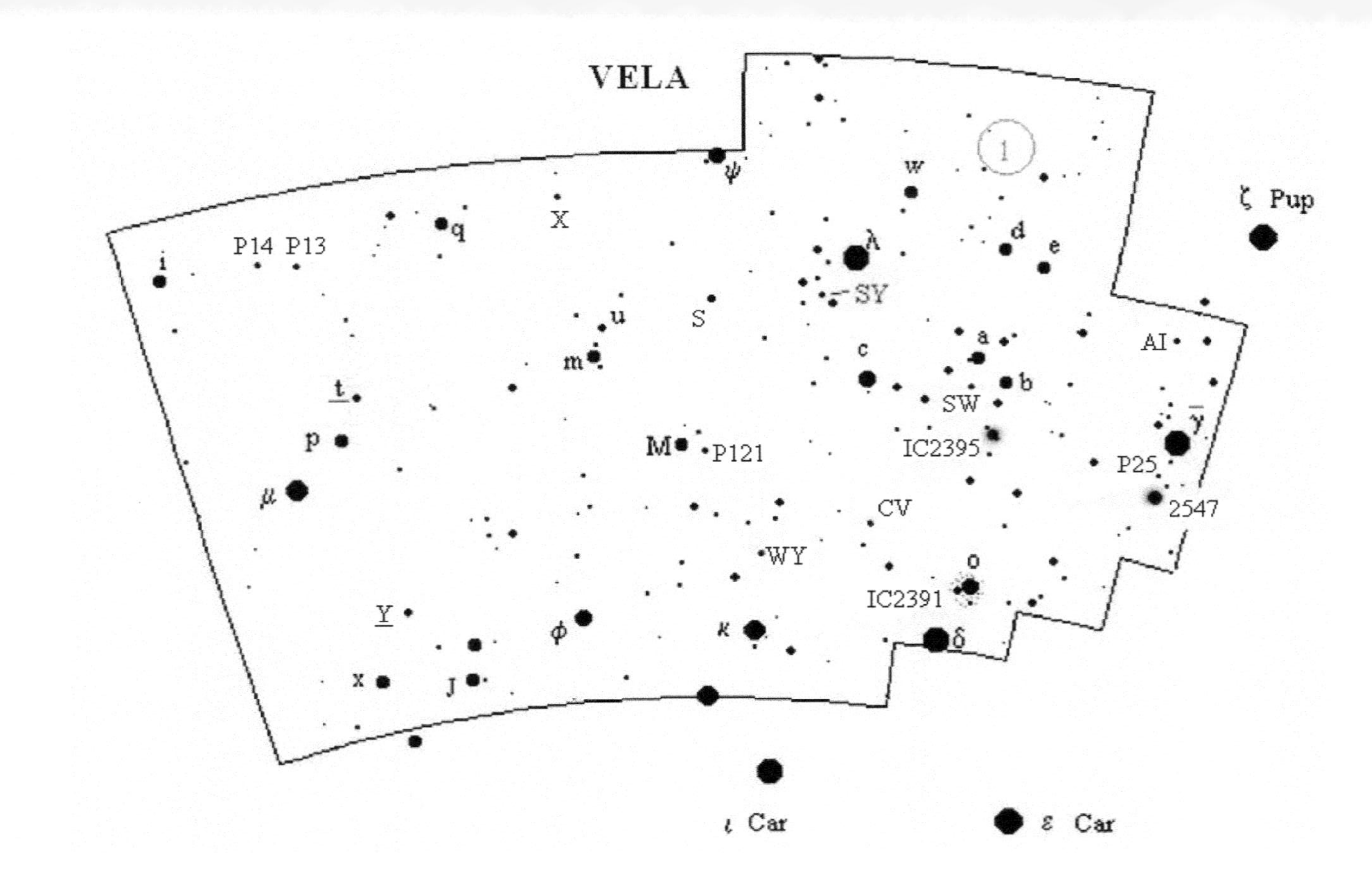
VELA
P14
P13
μ
p
t
x
Y
j
q
X
m
u
φ
M
P121
S
ψ
κ
WY
ι Car
SY
λ
c
CV
IC2391
IC2395
w
SW
a
o
δ
d
b
e
ε Car
AI
P25
γ
2547
ζ Pup

Appendix 1.

Suppliers

Here is a list of worthwhile suppliers of binoculars for astronomy. Since the author is based in the UK (and to keep down the list to a manageable size!) this is where most of the suppliers here are also based, but with the use of a search engine readers all over the world should be able to find a supplier to suit them. Bear in mind that not all models will be available for testing as described in Chapter 1, so a degree of caution might need to be exercised here. Also, bear in mind that web addresses can change!

Pulsar Optical

Large range of all types of astronomy binoculars, from 7 x 50 Porro Prism models through to telescope-sized ones like 25 x 100 with dual (removable) eyepieces. Some world-famous brands such as Bresser also sold. Online purchase option. Run by astronomers.
Pulsar Optical, Astronomy & Nature Centre, 39c Lancaster Way, Witchford, Ely, Cambs. CB6 3NW.
sales@pulsar-optical.co.uk
www.pulsar-optical.co.uk
0845 634 9192

Monk Optics

Specialise in huge binoculars such as 25 x 100 and larger, many mounted. Also make image intensifiers and other specialist equipment, often for bodies such as the government and the RNLI. Sell the

high-quality (and high-price!) Fujinon range but also in more modest sizes like 10 x 50.
Monk Optics Ltd, Wye Valley Observatory, The Old School, Brockweir, Chepstow NP16 7NW
msales@monkoptics.co.uk
www.monkoptics.co.uk
01291 689858

SCS Astro
Sell a good range of 'Orion' binoculars, including binocular telescopes (not actually binoculars, but a matching pair of telescopes arranged in a binocular style). Many of the larger models in this range include attachments for tripod mounting. Also sell Solar Filters suitable for binoculars.
The Astronomy Shop, 1 Tone Hill, Wellington, Somerset TA21 0AU
enquiry@scsastro.co.uk
www.scsastro.co.uk
01823 665510

Optical Vision Ltd
Good range of the reasonably-priced Helios binoculars in both Porro and Roof Prism styles, with a good choice both in the more advanced types as well as 'shop' sized models.
Optical Vision Ltd., Unit 2b, Woolpit Business Park, Woolpit, Bury St Edmunds, Suffolk IP30 9UP
www.opticalvision.co.uk
01359 244200

Strathspey

Online Binocular store, with lots of useful 'before you buy' advice, and also a list of UK Astronomy societies! Pages are well-divided into categories, which include monoculars and tripods.

Cyber Services Ltd., Robertland Villa, Railway Terrace, Aviemore, Scotland PH22 1SA

john@unixnerd.demon.co.uk

www.strathspey.co.uk

01479 812549

Optics Planet.com

US-based website offering good advice on using binoculars for astronomy, with an online shop offering models of all types from such prestigious companies as Celestron, Bushnell and Carl Zeiss - but bear in mind that some firms (often US ones) allow other companies to use their names! Many models offered at sale prices. A wide range of tripods and stands are also available, and the site includes a forum.

www.opticsplanet.net/binastro.html

Appendix 2.
The Greek Alphabet

Used for naming the chief naked-eye stars in a constellation. For this purpose only the small letters are used, though the capital delta (Δ) and sigma (Σ) are used to denote certain double-star catalogue entries, mentioned in Chapter 4.

α	alpha	ν	nu
β	beta	ξ	xi
γ	gamma	ο	omicron
δ	delta	π	pi
ε	epsilon	ρ	rho
ζ	zeta	σ	sigma
η	eta	τ	tau
θ	theta	υ	upsilon
ι	iota	φ	phi
κ	kappa	χ	chi
λ	lambda	ψ	psi
μ	mu	ω	omega

Appendix 3
Positions of Uranus and Neptune
2010 - 2019

This table lists the positions of these planets on January 1st for each year. The magnitude of Uranus will be between about 5.5 to 6.0 while that of Neptune around magnitude 7.5 to 8.0.

URANUS			NEPTUNE		
2010	23h 35m	-03° 28’	2010	21h 48m	-13° 45’
2011	23 49	-01 56 [1]	2011	21 56	-13 06
2012	00 04	-00 23	2012	22 04	-12 25
2013	00 18	+01 10	2013	22 12	-11 43
2014	00 32	+02 44	2014	22 21	-11 00
2015	00 47	+04 18	2015	22 29	-10 17
2016	01 01	+05 51	2016	22 37	-09 33
2017	01 16	+07 23	2017	22 46	-08 47 [2]
2018	01 31	+08 53	2018	22 54	-08 02
2019	01 46	+10 23	2019	23 02	-07 15

1. Only about a degree from Jupiter
2. Very near Mars

Appendix 4
The Constellations

Name	abbrev.	genitive	meaning
Andromeda	And	Andromedae	Andromeda
Antlia	Ant	Antliae	Air Pump
Apus	Aps	Apodis	Bird Of Para dise
Aquarius	Aqr	Aquarii	Water Car rier
Aquila	Aql	Aquilae	Eagle
Ara	Ara	Arae	Altar
Aries	Ari	Arietis	Ram
Auriga	Aur	Aurigae	Charioteer
Bootes	Boo	Bootis	Herdsman
Caelum	Cae	Caeli	Chisel
Camelopardalis	Cam	(Same)	Giraffe
Cancer	Cnc	Cancri	Crab
Canes Venatici	CVn	Canum Venaticorum	Hunting dogs
Canis Major	CMa	Canis Majoris	Greater Dog
Canis Minor	CMi	C. Minoris	Lesser Dog
Capricornus	Cap	Capricorni	Sea-Goat
Carina	Car	Carinae	Keel
Cassiopeia	Cas	Cassiopeiae	Cassiopeia
Centaurus	Cen	Centauri	Centaur

Cepheus	Cep	Cephei	Cepheus
Cetus	Cet	Ceti	Whale
Chamaeleon	Cha	Chamaeleontis	Chameleon
Circinus	Cir	Circini	Compasses
Columba	Col	Columbae	Dove
Coma Berenices	Com	Comae	Berenice's Hair
Corona Australis	CrA	Coronae A.	Southern Crown
Corona Borealis	CrB	Coronae B.	Northern Crown
Corvus	Crv	Corvi	Crow
Crater	Crt	Crateris	Cup
Crux	Cru	Crucis	Cross
Cygnus	Cyg	Cygni	Swan
Delphinus	Del	Delphini	Dolphin
Dorado	Dor	Doradus	Swordfish
Draco	Dra	Draconis	Dragon
Equuleus	Equ	Equulei	Foal
Eridanus	Eri	Eridani	(River) Eri danus
Fornax	For	Fornacis	Furnace
Gemini	Gem	Geminorum	Twins
Grus	Gru	Gruis	Crane
Hercules	Her	Herculis	Hercules
Horologium	Hor	Horologii	Clock
Hydra	Hya	Hydrae	Snake

Hydrus	Hyi	Hydri	Water Snake
Indus	Ind	Indi	Indian
Leo	Leo	Leonis	Lion
Leo Minor	LMi	Leonis Minoris	Little Lion
Lepus	Lep	Leporis	Hare
Libra	Lib	Librae	Scales
Lupus	Lup	Lupi	Wolf
Lynx	Lyn	Lyncis	Lynx
Lyra	Lyr	Lyrae	Lyre
Mensa	Men	Mensae	Table Moun tain
Microscopium	Mic	Microscopii	Microscope
Monoceros	Mon	Monocerotis	Unicorn
Musca	Mus	Muscae	Fly
Norma	Nor	Normae	Rule
Octans	Oct	Octantis	Octant
Ophiuchus	Oph	Ophiuchi	Serpent Holder
Orion	Ori	Orionis	Orion
Pavo	Pav	Pavonis	Peacock
Pegasus	Peg	Pegasi	Pegasus
Perseus	Per	Persei	Perseus
Phoenix	Phe	Phoenicis	Phoenix
Pictor	Pic	Pictoris	Painter
Pisces	Psc	Piscium	Fishes
Piscis Austrinus	PsA	Piscis Austrini	Southern Fish
Puppis	Pup	Puppis	Poop Deck

Pyxis	Pyx	Pyxidis	Compass
Reticulum	Ret	Reticuli	Net
Sagitta	Sge	Sagittae	Arrow
Sagittarius	Sgr	Sagittarii	Archer
Scorpius	Sco	Scorpii	Scorpion
Sculptor	Scl	Sculptoris	Scupltor
Scutum	Sct	Scuti	Shield
Serpens	Ser	Serpentis	Serpent
Sextans	Sex	Sextantis	Sextant
Taurus	Tau	Tauri	Bull
Telescopium	Tel	Telescopii	Telescope
Triangulum	Tri	Trianguli	Triangle
Triangulum australe	TrA	Trianguli australis	S. Triangle
Tucana	Tuc	Tucanae	Toucan
Ursa Major	UMa	Ursae Majoris	Great Bear
Ursa Minor	UMi	Ursae Minoris	Little Bear
Vela	Vel	Velorum	Sails
Virgo	Vir	Virginis	Virgin
Volans	Vol	Volantis	Flying Fish
Vulpecula	Vul	Vulpeculae	Fox

Appendix 5

Variable Star charts

This appendix is divided into, firstly, variables suitable for Northern hemisphere observers and secondly, those suitable for those in the South. In all cases the variable is shown as the usual 'dot and circle'. I have tried to mix chart and comparison star identification types, so that there should be something for everyone.

These charts are provided for interest and experience' sake. It is important to note that if you (like me) become what has been described as "a hopeless variable star junkie" and decide to observe these fascinating stars with a degree of commitment, then you should use the charts issued by one of the internationally known specialist groups, such as some of the following:

AAVSO (www.aavso.org) - US based, but international in scope.
BAA VS section (www.britastro.org/vss) in the UK, or
AFOEV (http://cdsweb.u-strasbg.fr/afoev/) in France.

Many astro societies throughout the world have their own variable star groups. The internet is a good source of these. All of the above groups have active members outside their home countries, for instance.

VX Andromedae

A deep red star that varies between magnitudes 8 and 9 in a period of about a year. Field size is about 5 degrees.

Comparison Stars:

A = 7.2 B = 7.6 C = 7.9 D = 8.5 E = 9.1 F = 9.4

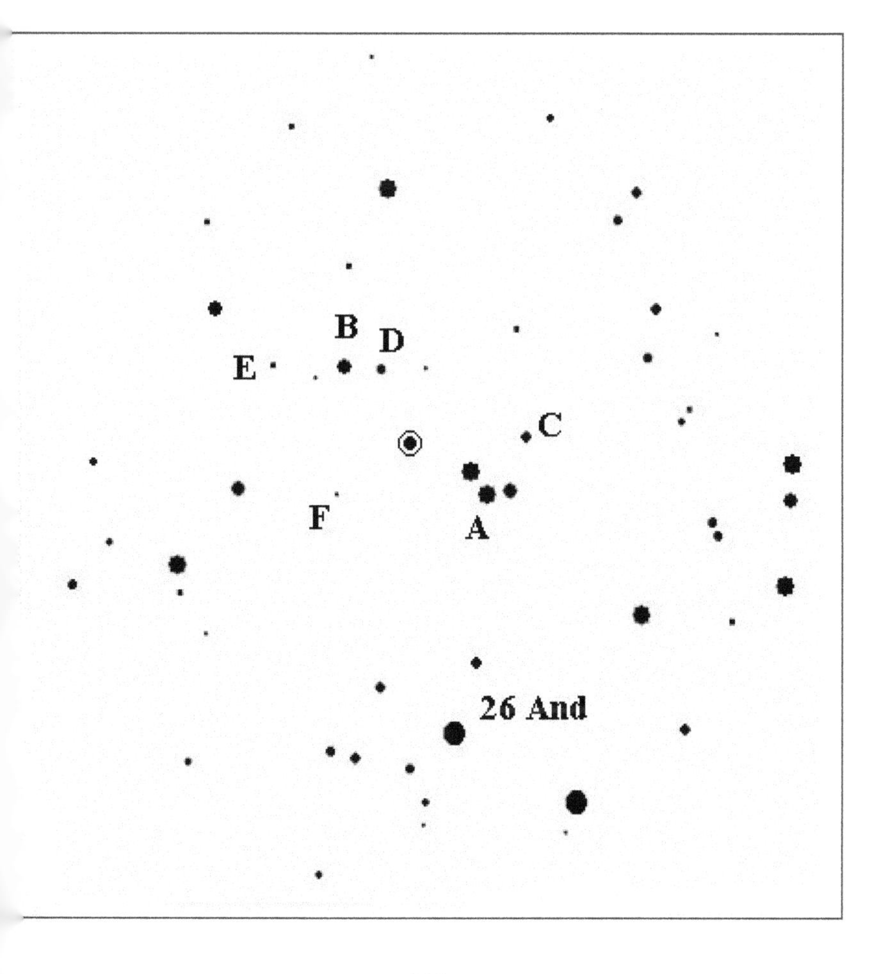

TU Aurigae

A fairly obscure and faint star in a rather out-of-the-way region. Its changes are quite slow, so observe it once a month. Field size 6 degrees.

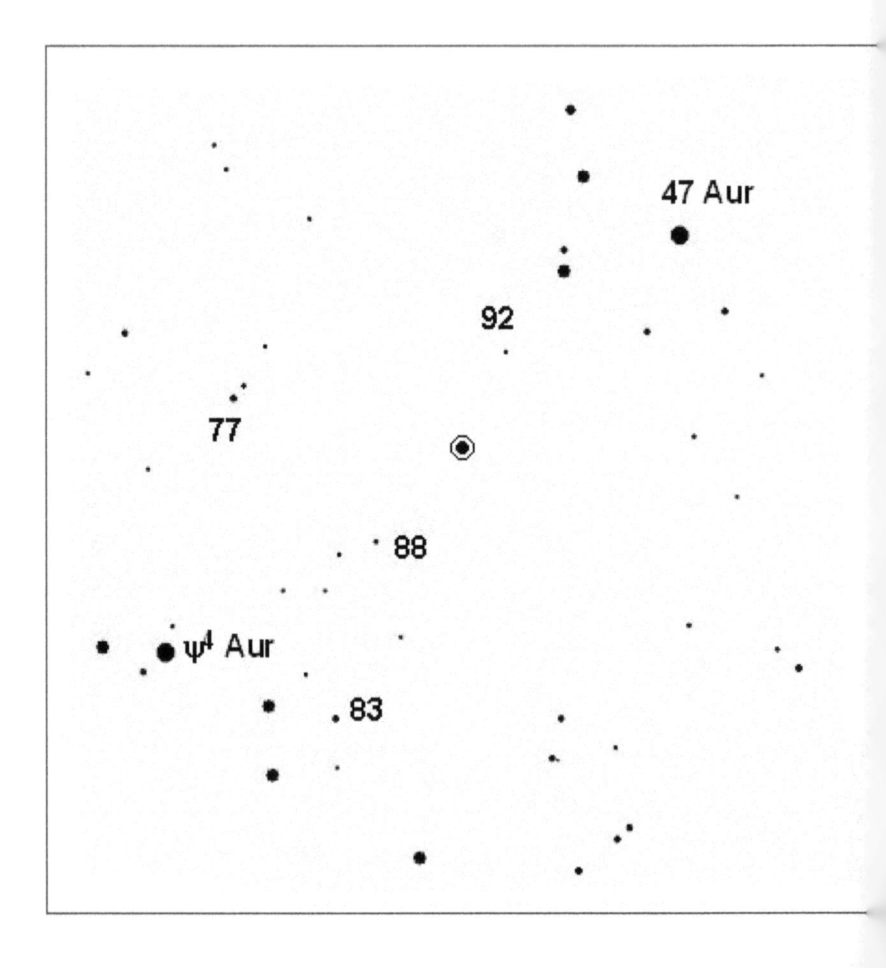

RW and RV Bootis

Two semi-regular variables easily found just North of “Bootes’ Belt” whose chief star is the second-magnitude ε Boo, or Izar. Field size here is about 6°, and observe these stars together once every 3 weeks.

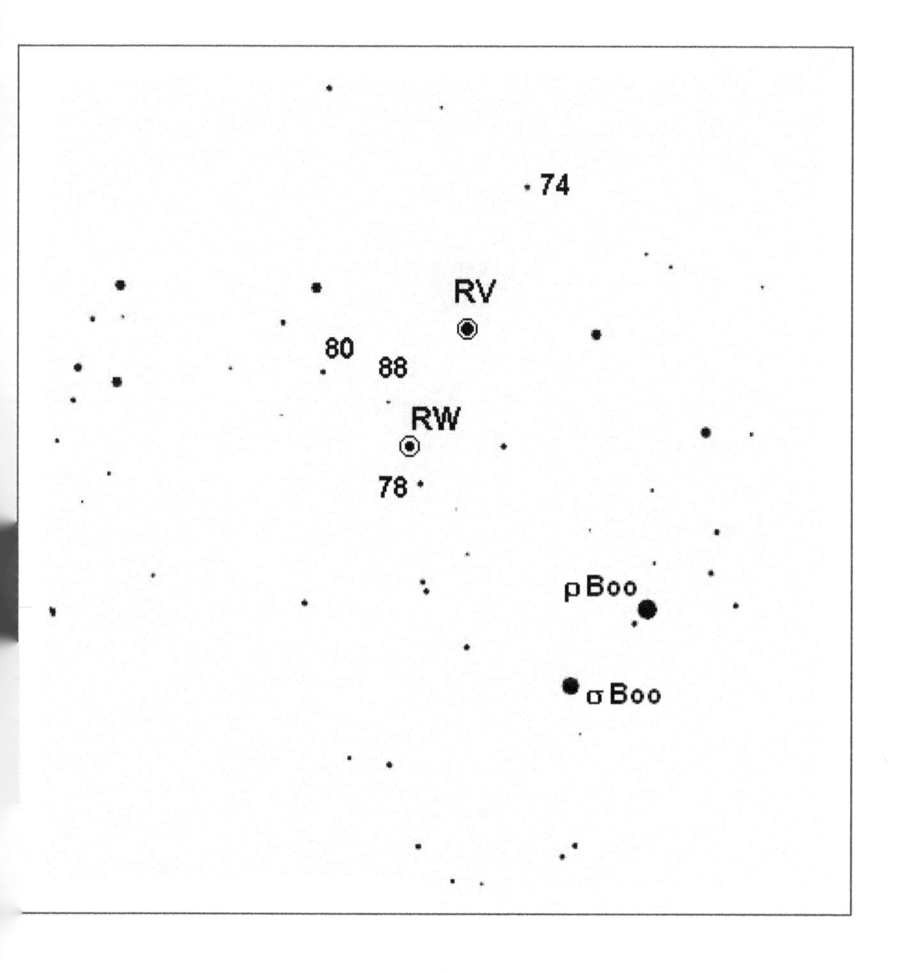

RX Cancri

A red variable in the same field as another star of the same type, BL Cnc, but which has too small a range to be of interest.

Comparison Stars:

A = 7.5 B = 8.1 C = 8.4 D = 8.7 E = 9.1

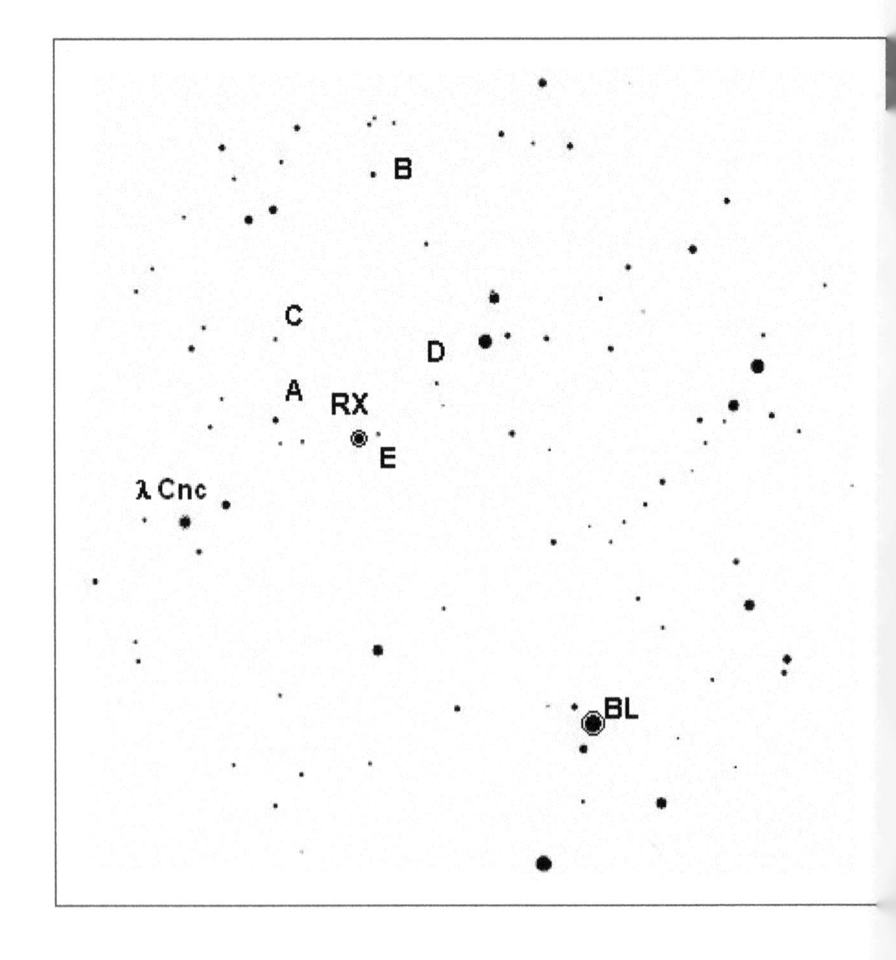

RT Capricorni

A deep red star with a period of 343 days and quite a good range in light. Estimate it once every 3 weeks.

Comparison Stars:

A (4 Cap) = 5.9 B = 6.7 C = 7.4 D = 7.8 E = 8.5

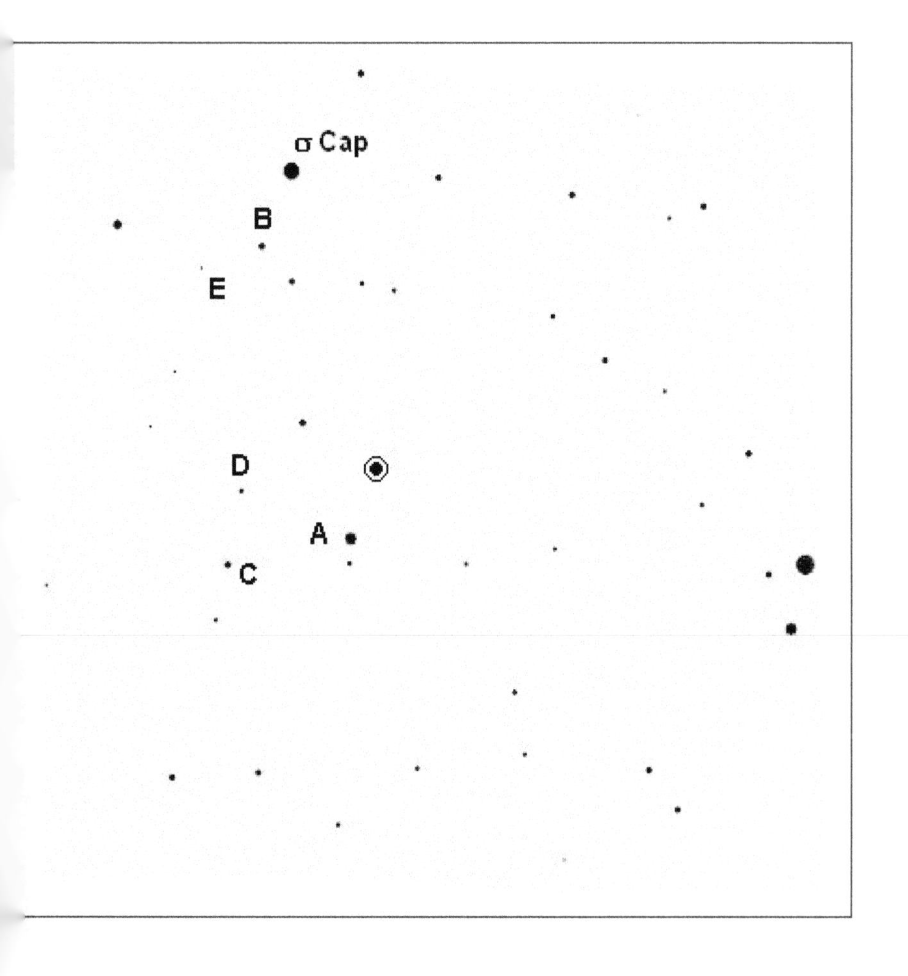

IM Cassiopeiae

This deep red star gives you a great view of several clusters in this rich region of the sky. Field diameter is about 5 degrees and the variable lies midway between δ and ε Cas.

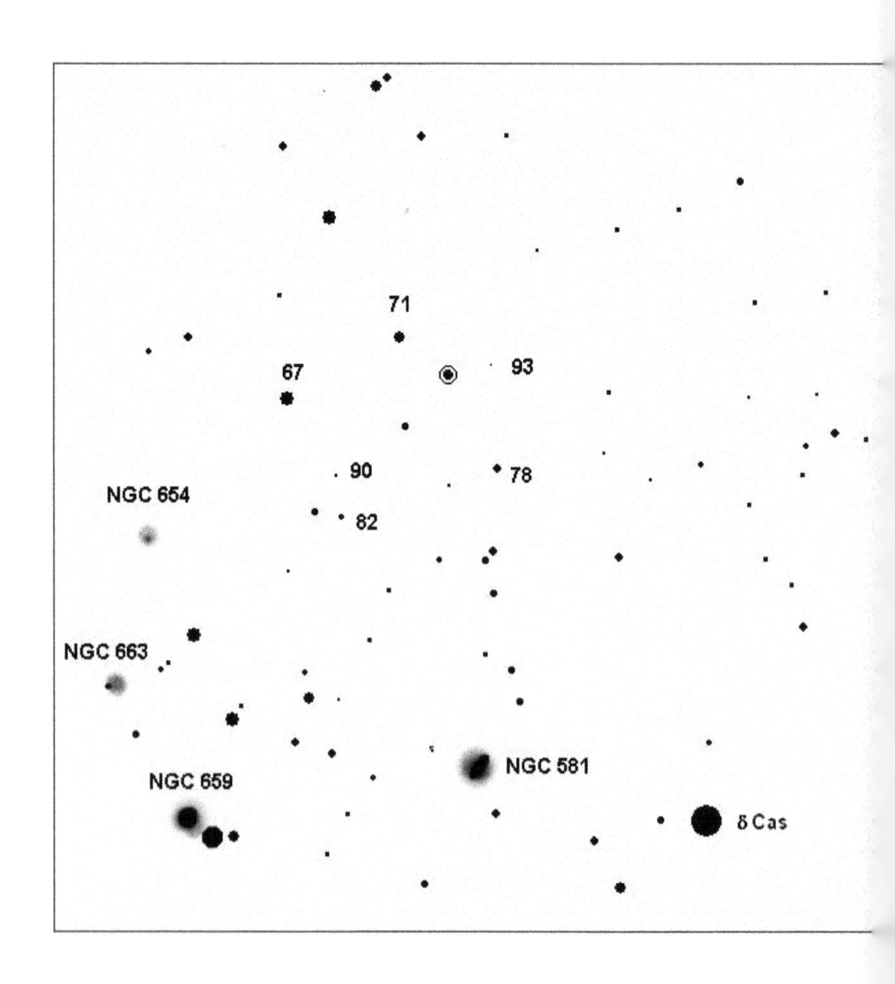

T Ceti

A good star for the smaller glasses, and with quite a good range of variation. This is a wide angle view of about 10 degrees. Star A should be visible with the naked eye.

Comparison Stars:

A = 4.4 B = 5.2 C = 5.8 D = 6.2 E = 6.8 F = 7.4

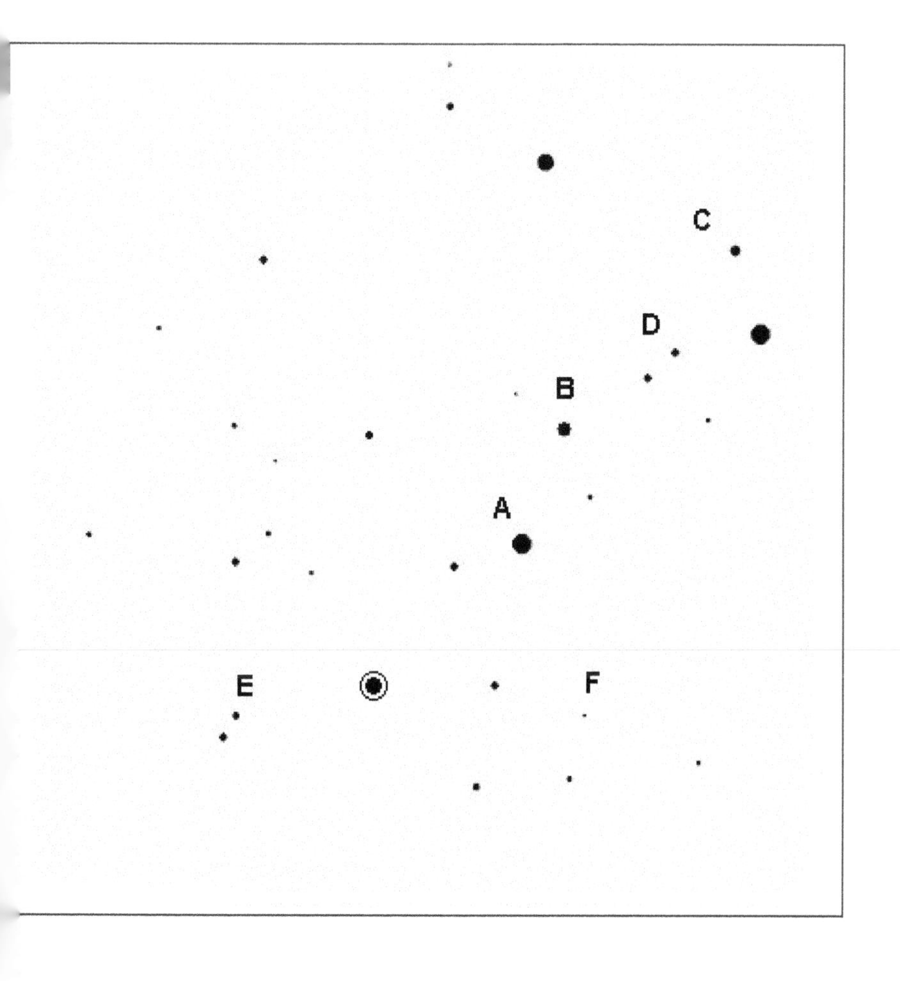

T Coronae Borealis

Known as the "Blaze Star" this is what is known as a recurrent Nova, having exploded in 1866 and 1946. It still shows small variations but should be looked at once a week just to see if anything untoward is happening! A close-up view of about 3 degrees.

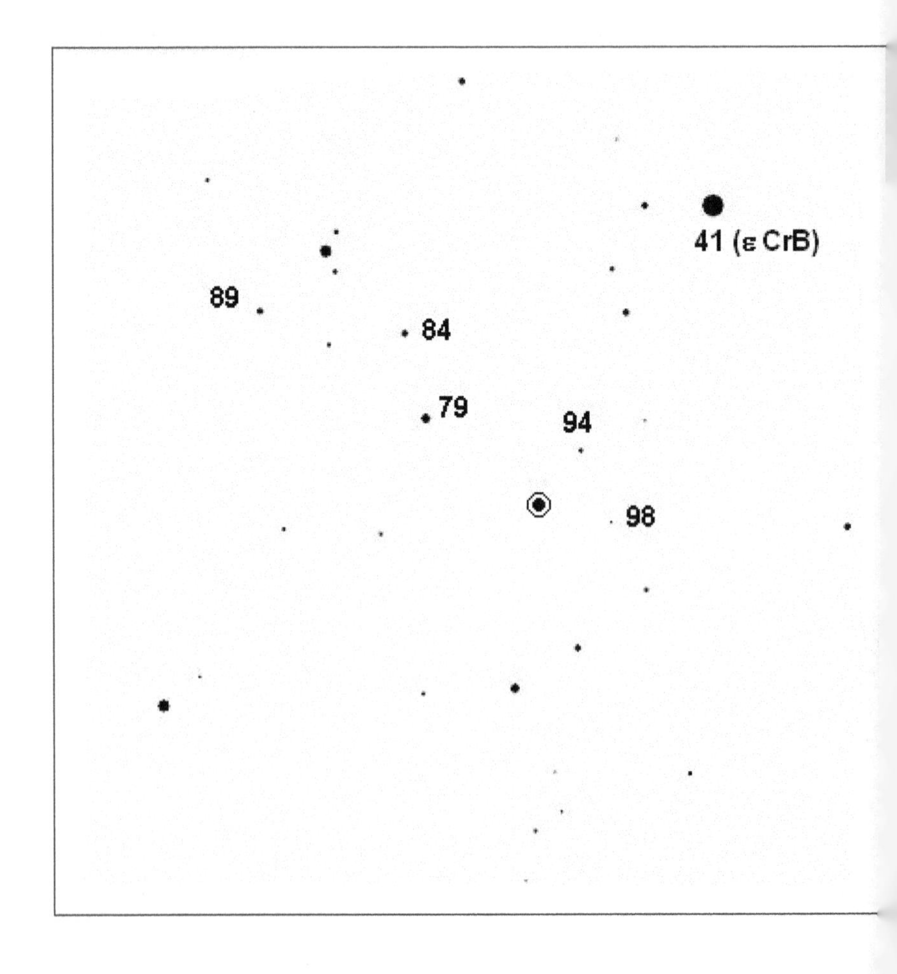

RW Cygni

A faint variable in the 'red region' of Cygnus with the long period of 587 days, so only one observation a month is necessary. Again, a close-up view of 3°.

Comparison Stars:

A = 7.7 B = 8.1 C = 8.7 D = 9.1 E = 9.4

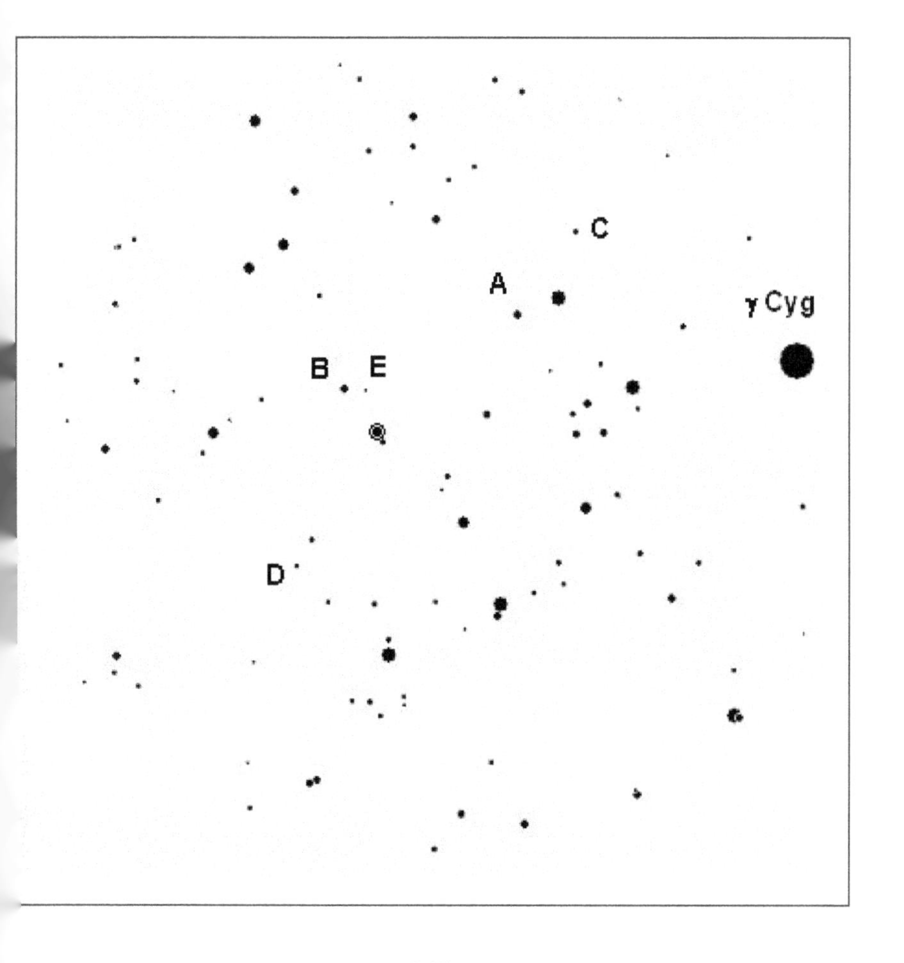

SS Geminorum

Many other variables mentioned in the text can be observed together with this RV Tauri star. SS Gem, however, needs weekly estimates whereas the others only need one observation a month or possibly one per 3 week period. Field size 4 degrees.

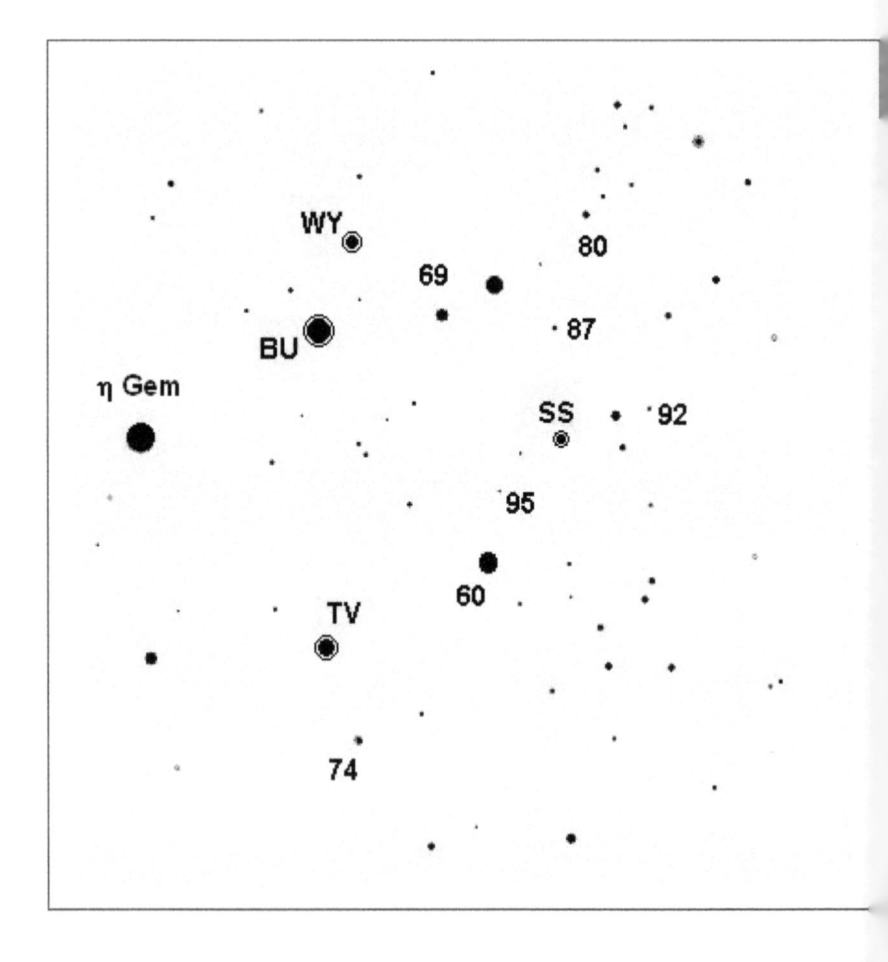

KN Hydrae

A fairly recent discovery, this is a red Mira-type star readily found from the bright stars γ and ψ Hydrae, with which star 65 on the chart forms the South point of an isosceles triangle. Observe it every 2 to 3 weeks. Field size is about 5 degrees.

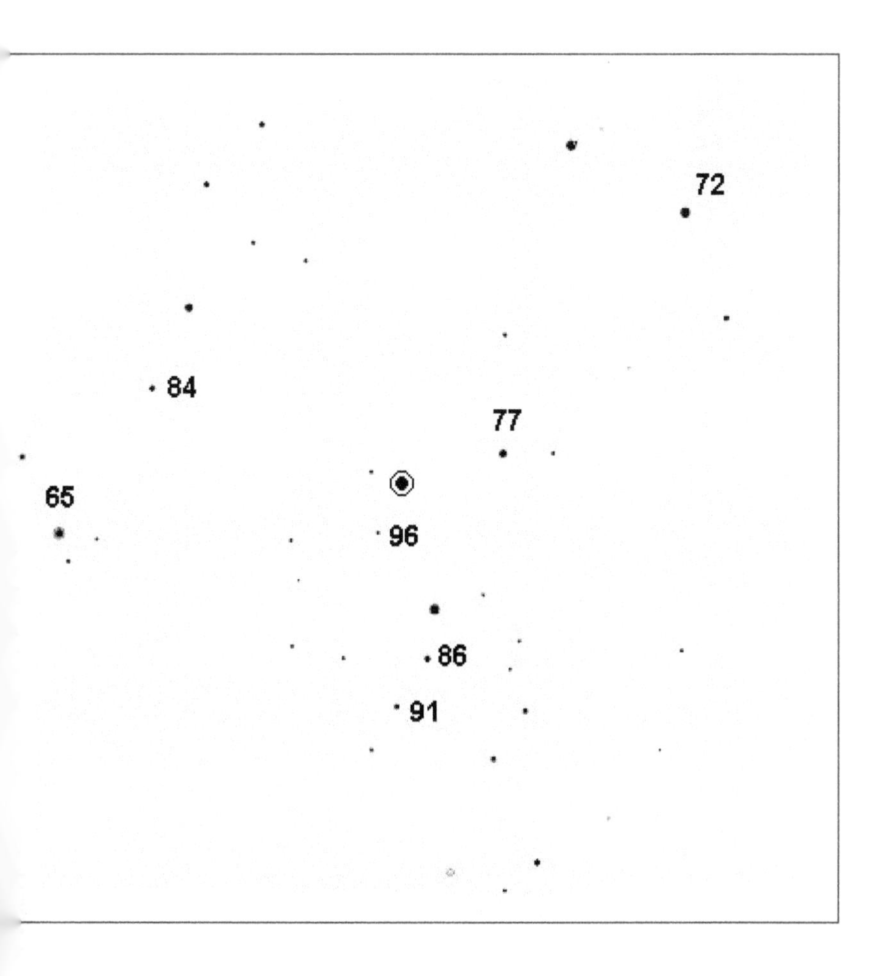

S Leporis

The smallest binoculars can observe this red star which is easily found by way of the 10° chart below.

Comparison Stars:

A = 5.5 B = 6.0 C = 6.4 D = 7.0 E = 7.5

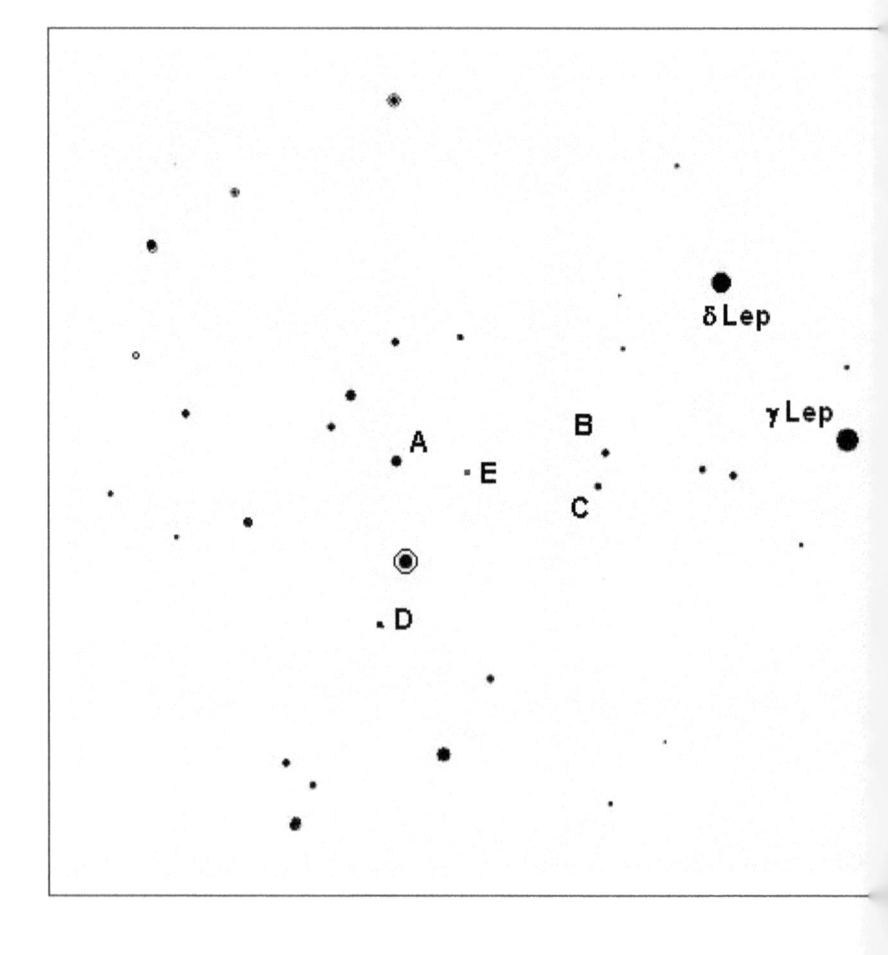

X Monocerotis

A good star to observe, in a crowded field which may be hard to find initially, about 3° North of θ Canis Majoris. Also shown is another binocular variable, RY Mon, that varies between 7th and 9th magnitudes. Observe these stars together every 3 weeks.

Comparison Stars (for X Mon): A = 6.0 B = 6.3 C = 6.9

D = 7.6 E = 7.9 F = 8.5 H = 8.9 K = 9.3

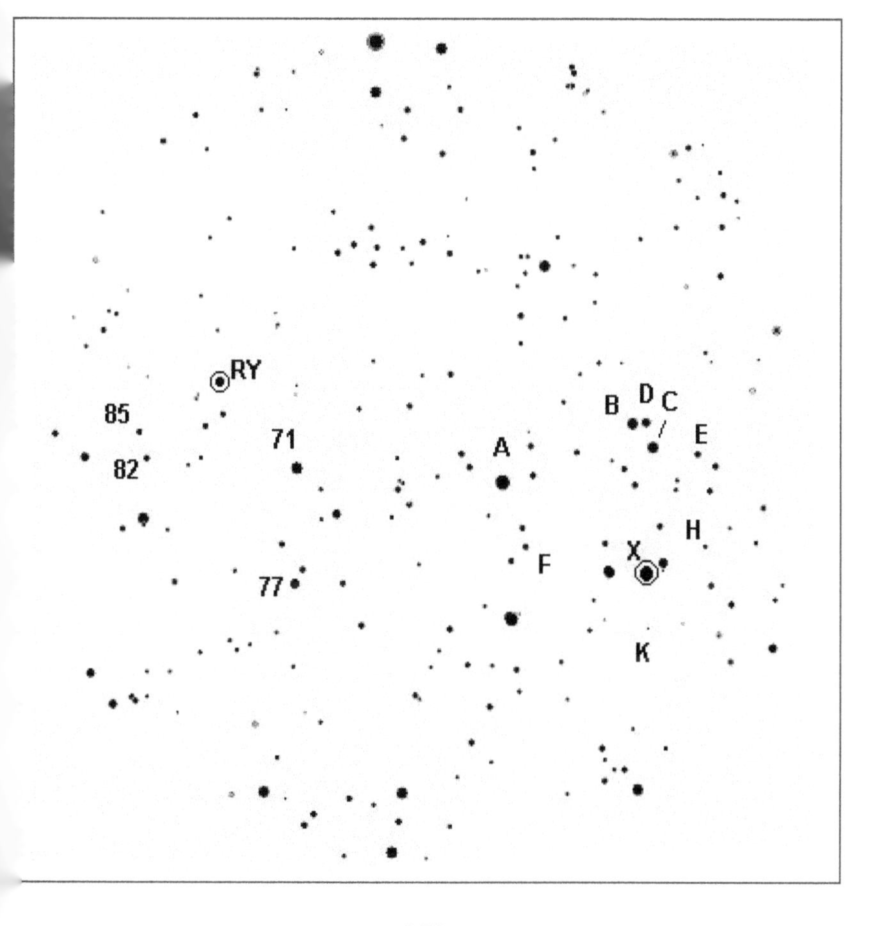

V1010 Ophiuchi

An eclipsing binary, easy with the slightest optical aid. The field size here is about 10 degrees and includes the bright naked eye star ζ Ophiuchi.

Comparison Stars:

A = 6.0 B = 6.4 C = 7.1 D = 7.4

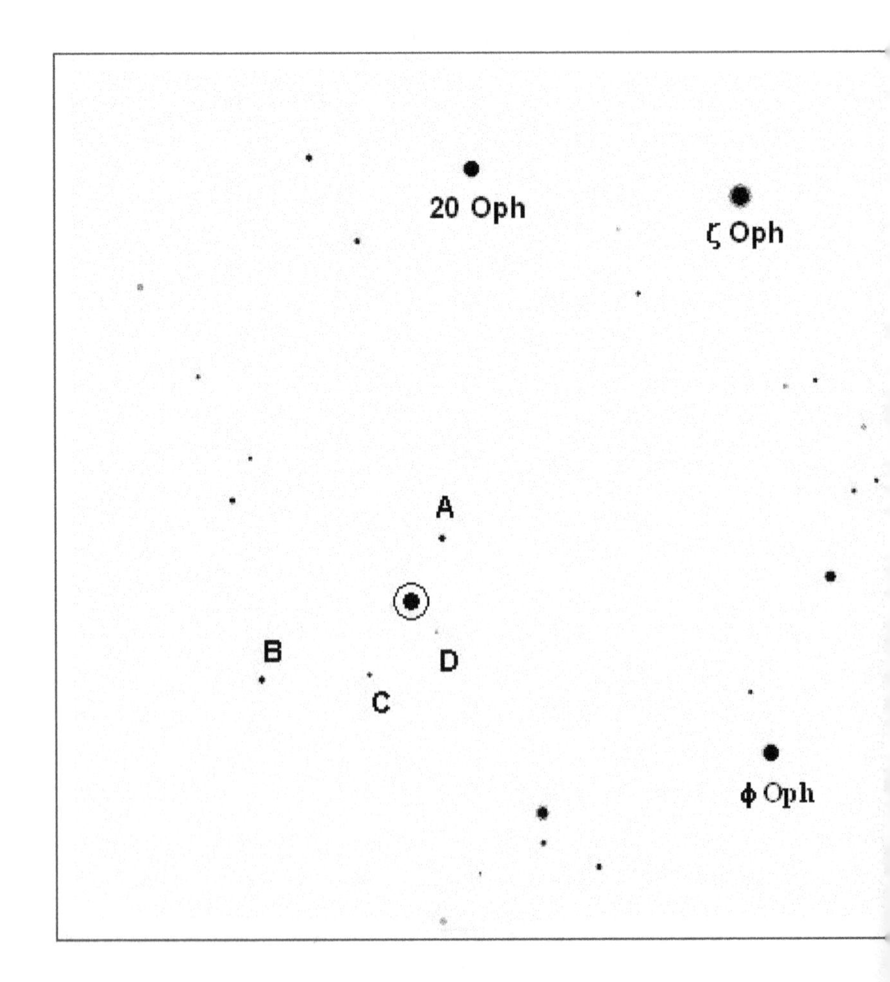

RT Orionis

Observe this deep red star once a fortnight. The chart also includes two interesting variables - BK Ori, a Mira star which can reach binocular visibility, and BN Ori, an eruptive, newly-formed star that spends most of its time near maximum of around magnitude 9, so I have included some comparison stars near it. Observe this one every night!

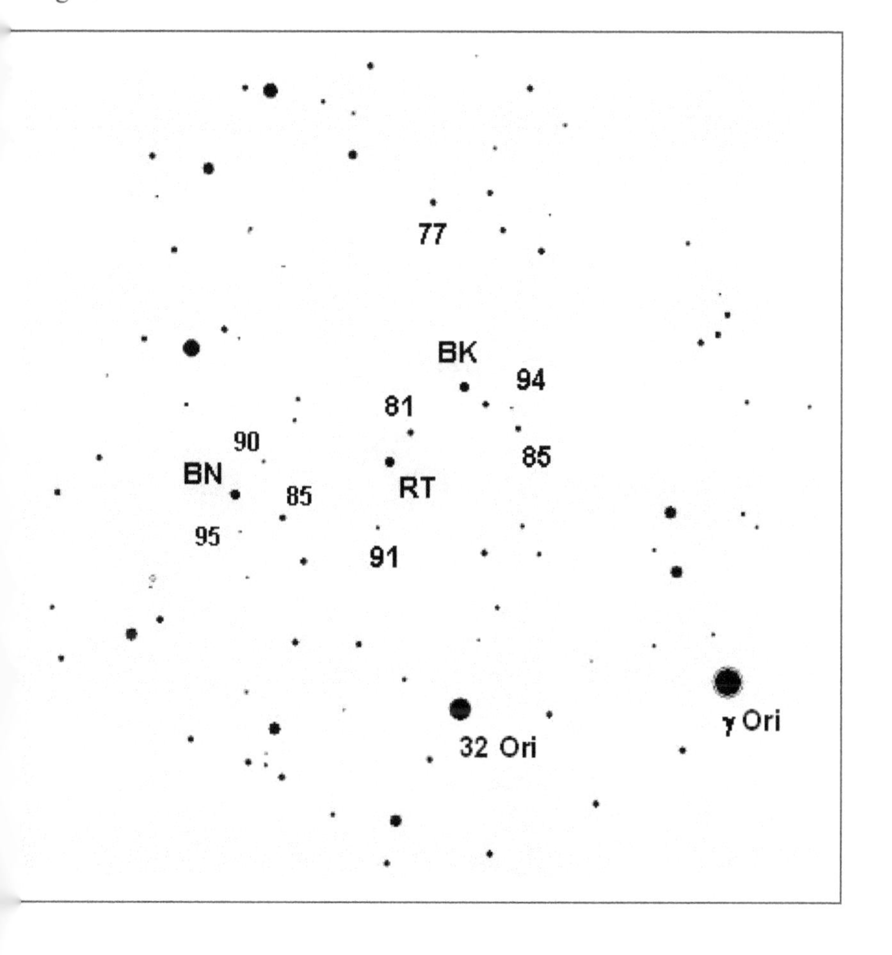

TW Pegasi

Easily found from the star iota Pegasi, look at this variable every month. The size of the field is 5°.

Comparison Stars:

A = 6.4 B = 7.0 C = 7.5 D = 7.7 E = 8.2 F = 8.5 H = 8.9

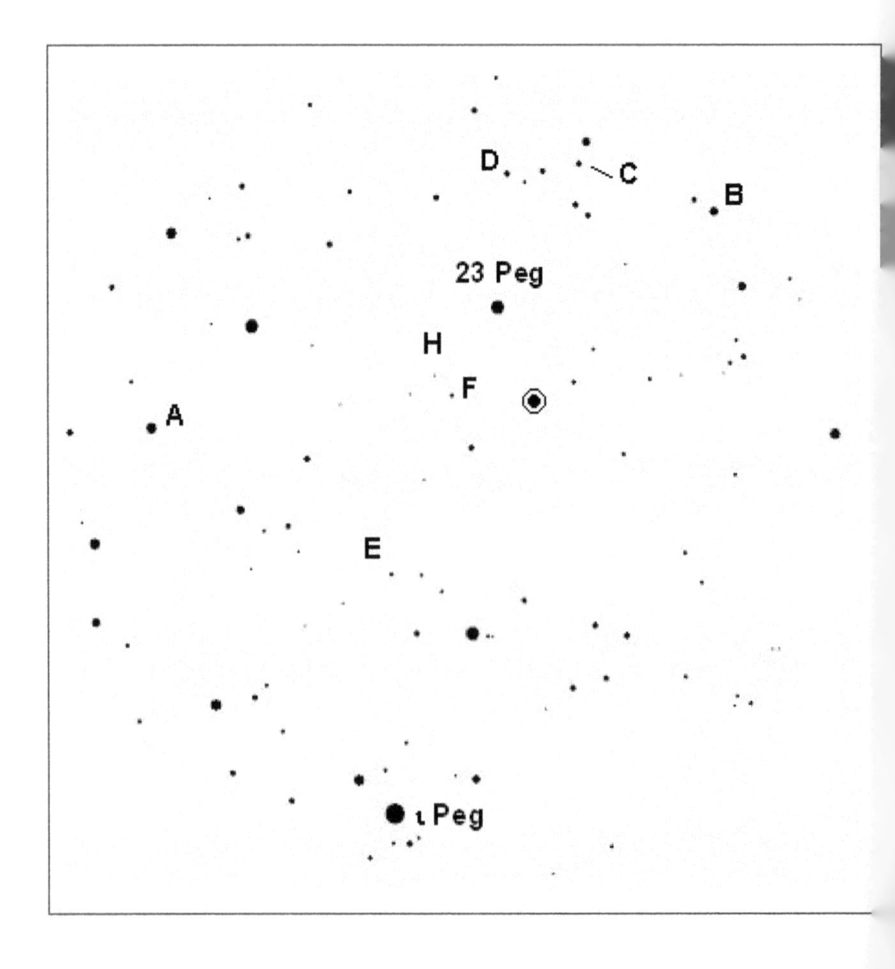

AQ Sagittarii

An easy, deep red variable located in the North of the constellation.

The size of the field here is 8 degrees.

Comparison Stars:

A = 6.2 B = 6.8 C = 7.4 D = 7.9

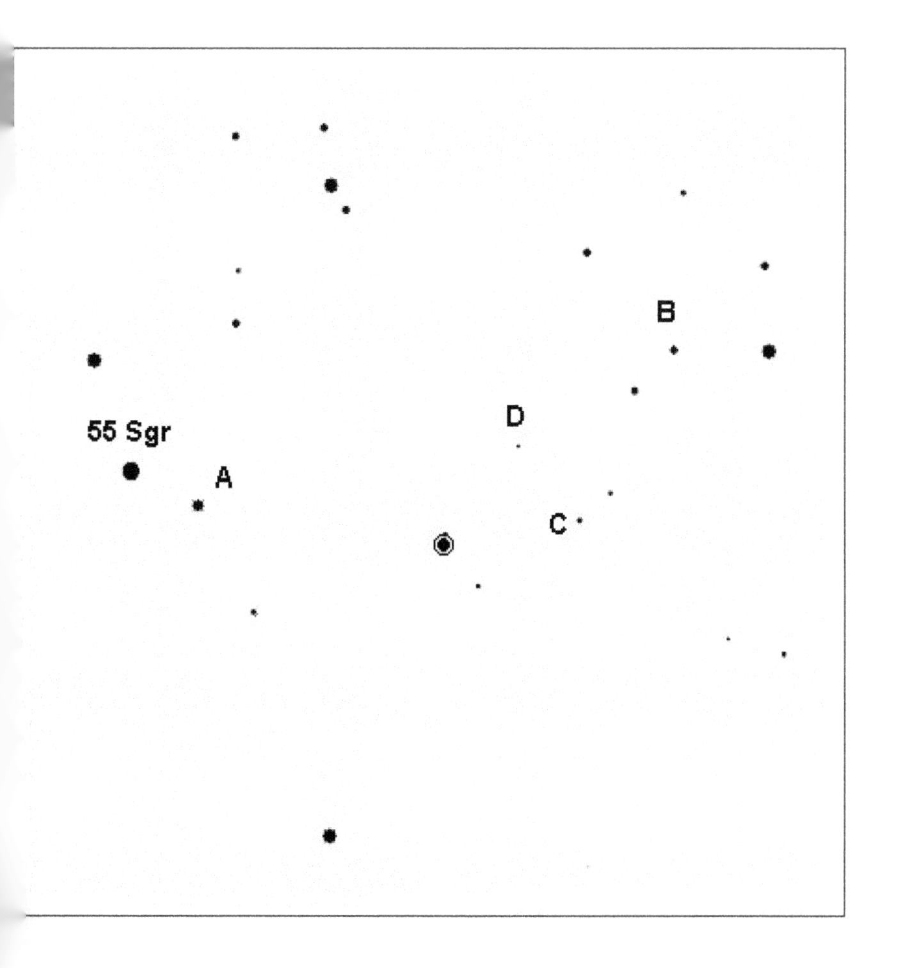

BF Sagittae

This and its neighbour X Sagittae are quite faint objects, so this is a close-up view of about 3 degrees. HU Sagittae nearby has too small a range for effective visual observing, but *does* form a useful triangle with comp star 73 by which both BF and X can be found. Be careful not to confuse X with its 8.7m companion!

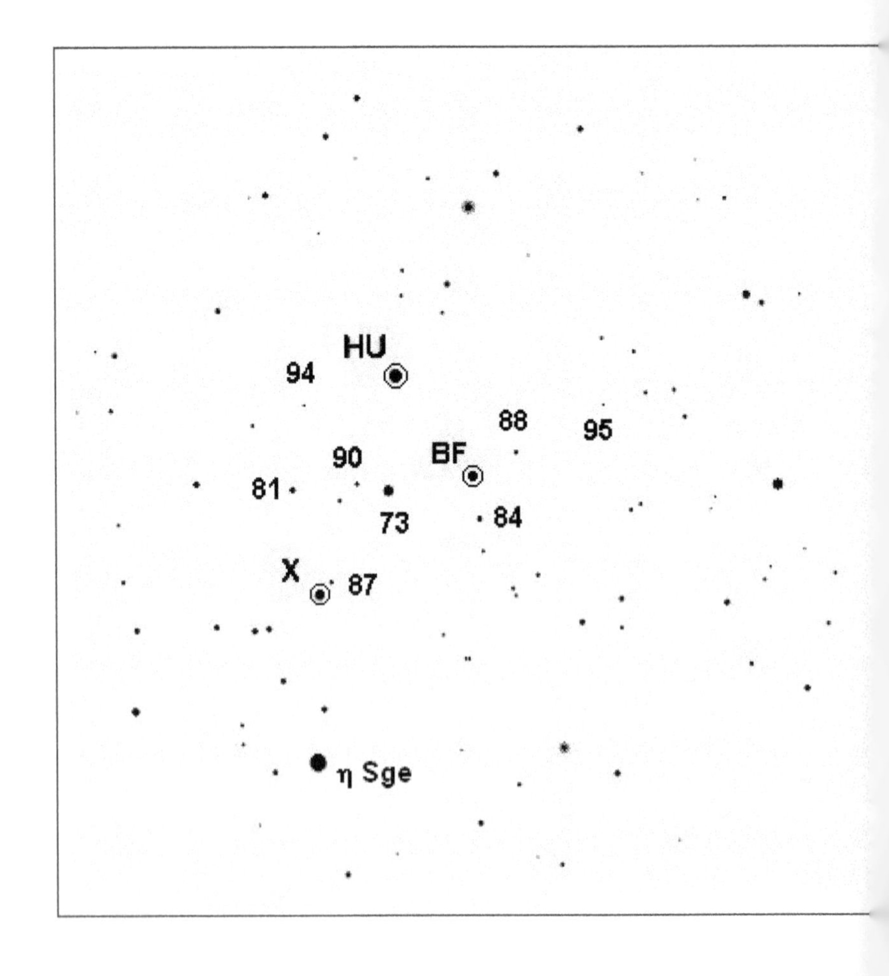

SS Scorpii

Easily found near the bright star Wei (ε Sco) this orange variable needs observing once a fortnight. Field size is 5°.

Comparison Stars:

A = 7.2 B = 7.7 C = 8.1 D = 8.4 E = 8.7 F = 9.2

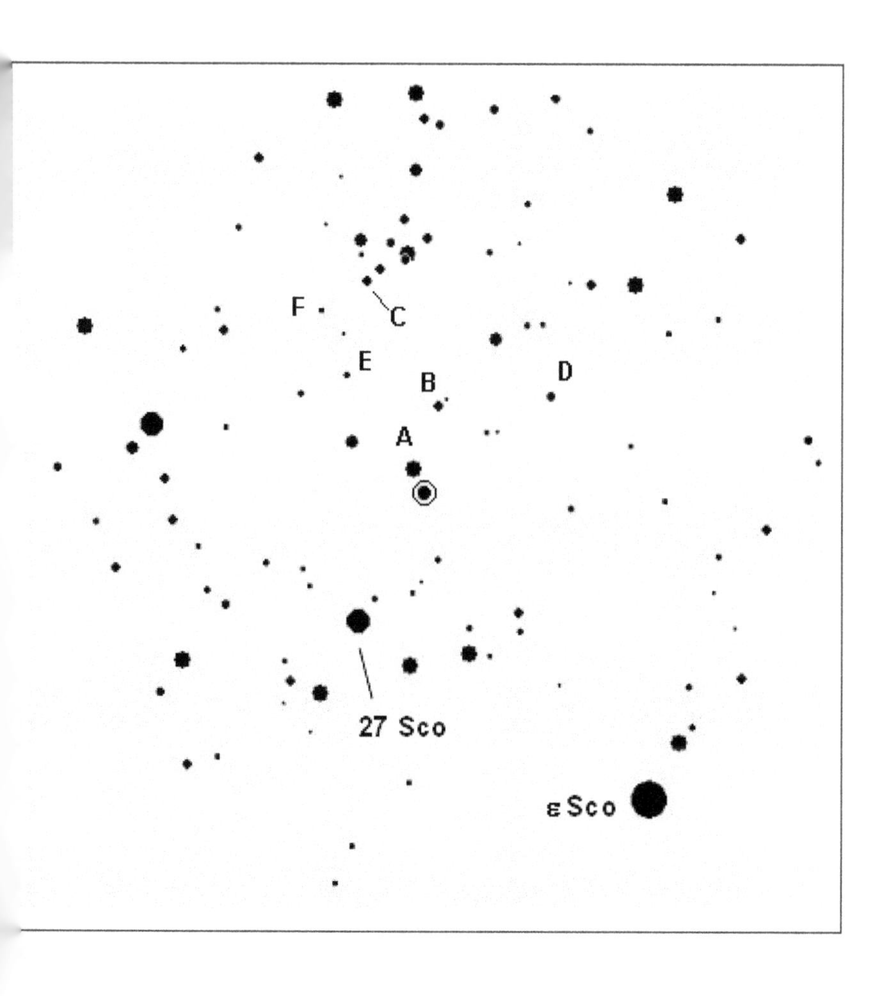

AK Scorpii

Near the beautiful wide double μ Scorpii, this is a faint nebular type variable, so a closeup (3°) view is shown here. For some of the time it may, indeed, be invisible, and on these occasions make a note of the faintest comparison star you can see.

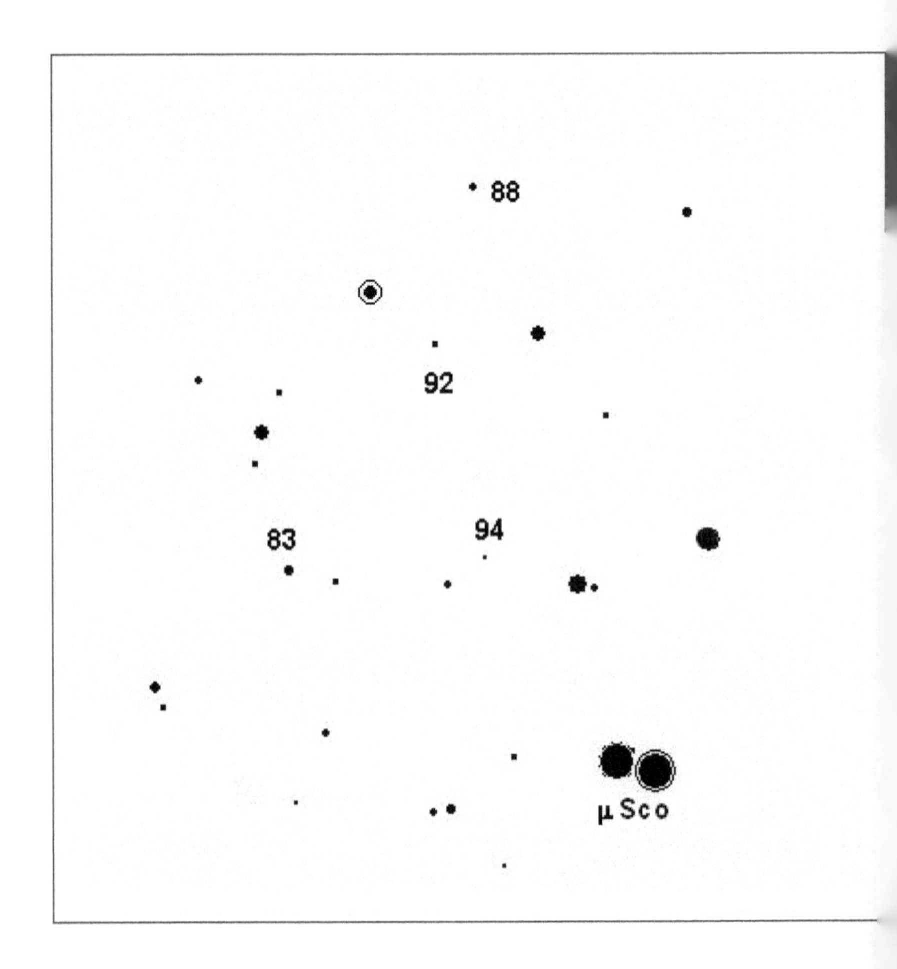

RZ Scuti

An Algol-type eclipsing star with a good range of variation, with dips to minimum every 16 days. Locate it from ζ Sct, which is halfway between the naked-eye stars γ Sct and η Serpentis.

Comparison Stars:

A = 7.3 B = 8.0 C = 8.5 D = 9.2

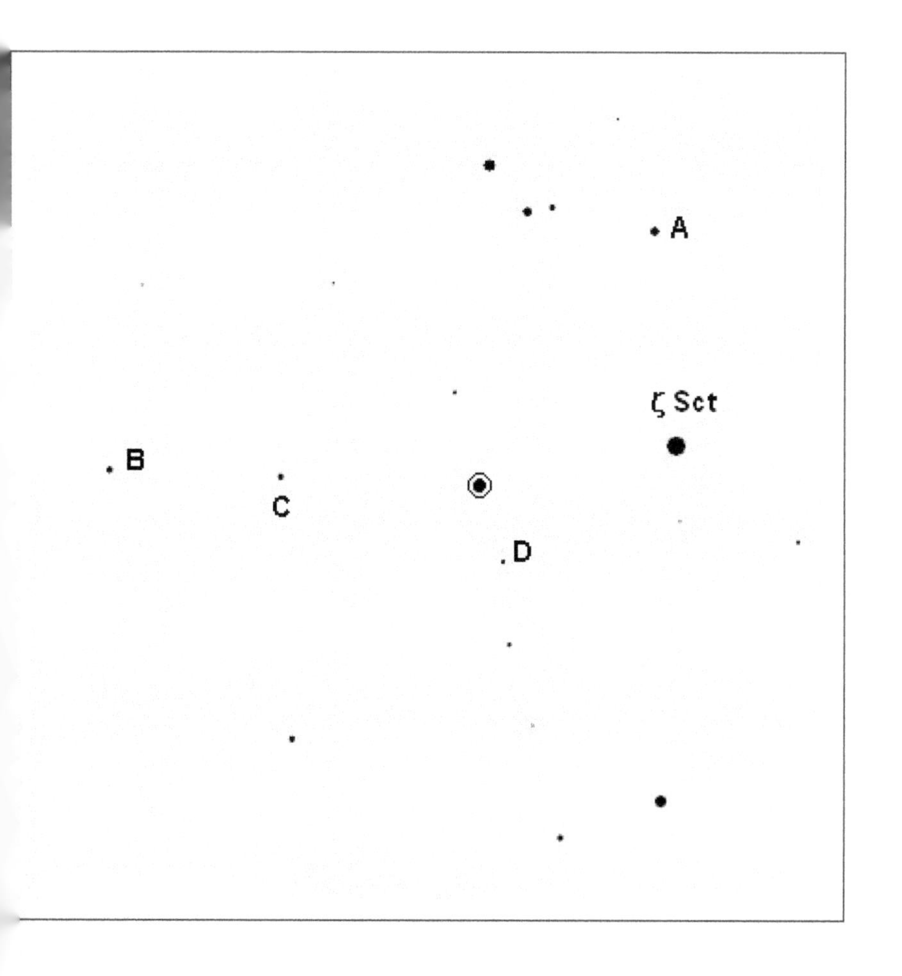

TU Tauri

A deep red variable in a beautiful field of bright stars that can make identification confusing. 132 Tauri is the southernmost member of a bright quadrilateral East of β Tauri and North of ζ. Note the binocular double star near 132; its separation is 94 arc-seconds. Field size 5°.

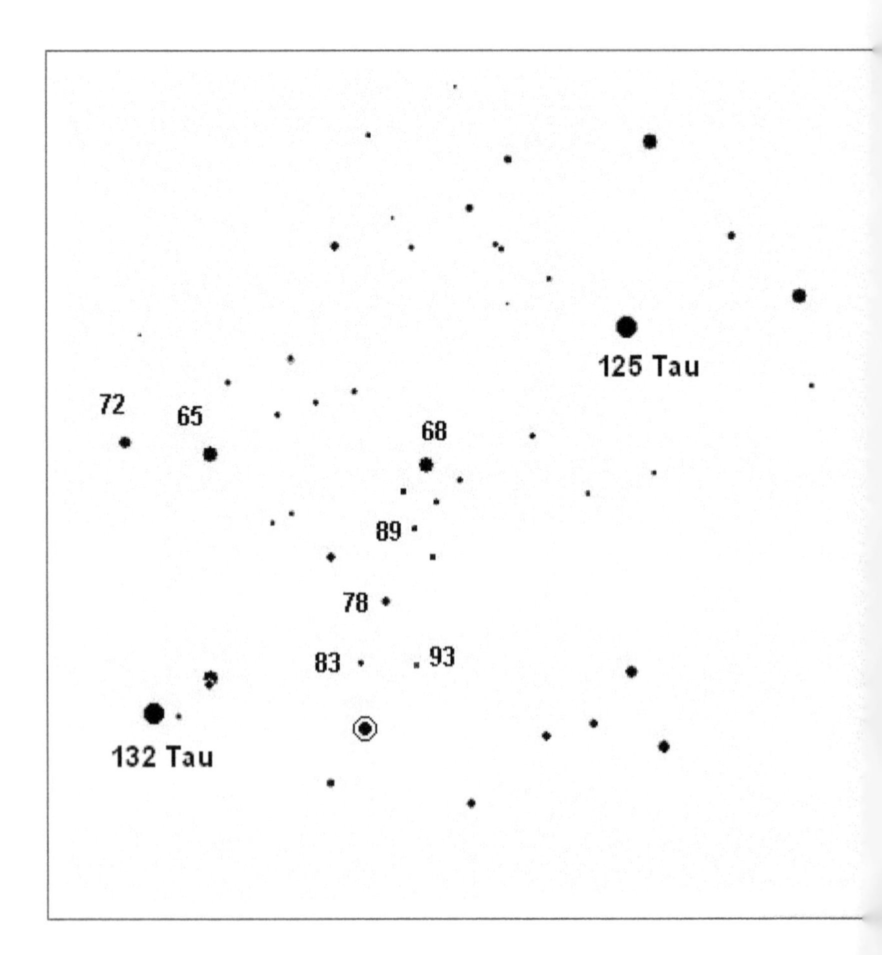

SOUTHERN VARIABLES

U Antliae

A bright red variable with a range of magnitudes 6 to 7. A wide angle field.

Comparison Stars:

A = 4.8 B = 5.3 C = 6.0 D = 6.7 E = 7.2

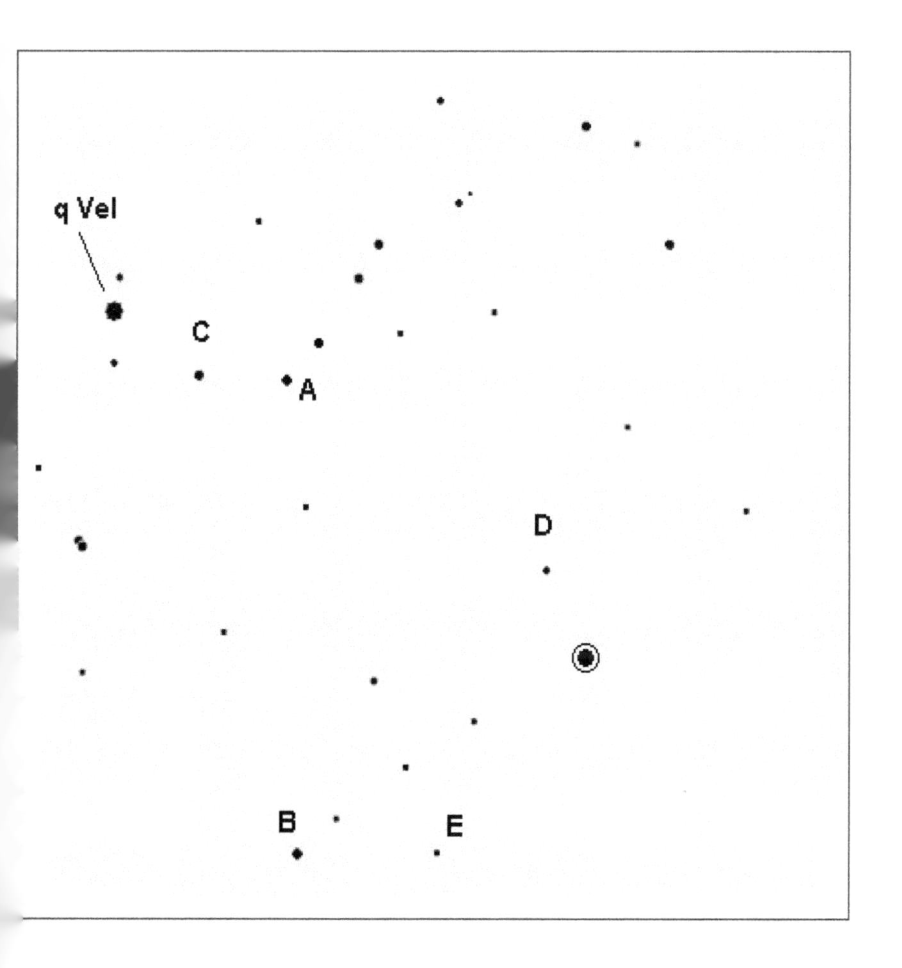

RR Carinae

This red star has a period of 109 days. The chart below also shows two Cepheid variables - QX Carinae is brighter with a range of 6.6 to 7.2 and GX is fainter (9th magnitude). Just out of the field to the North is the briliant cluster NGC 3114.

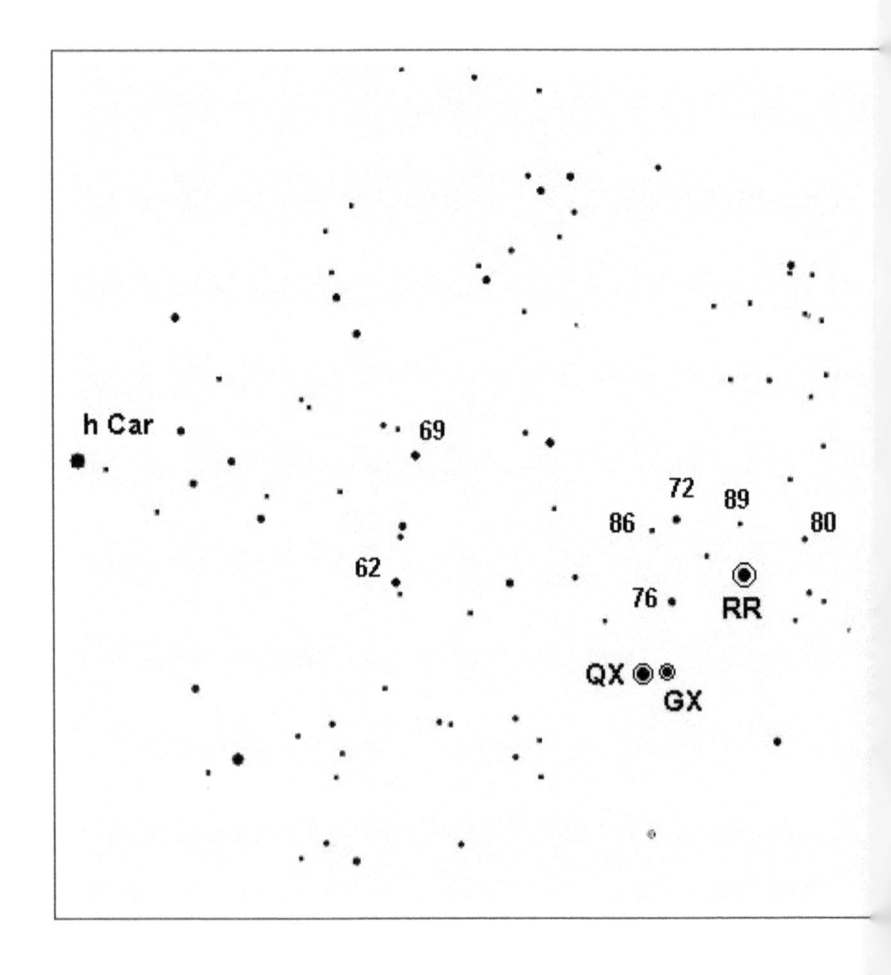

Eta Carinae

The famous eruptive star is fully described in the text, and this close-up view of about 3° also shows two red variables - BO Car, of the 7th magnitude, and RT Car, which normally fluctuates around the eighth.

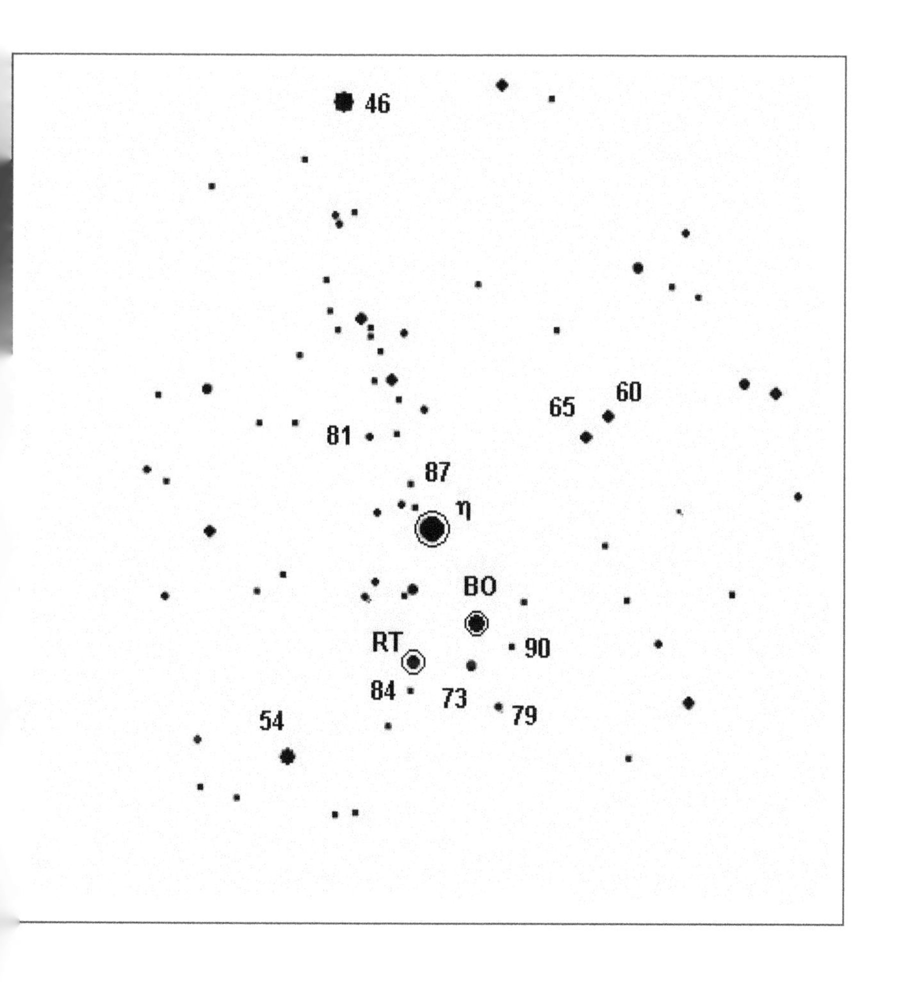

AC Carinae

A red star easily found from the little 7th-mag right angle close by, and which points to the bright star α Pictoris about 4° away.

Comparison Stars:

A = 7.2 B = 7.6 C = 8.0 D = 8.6 E = 9.0

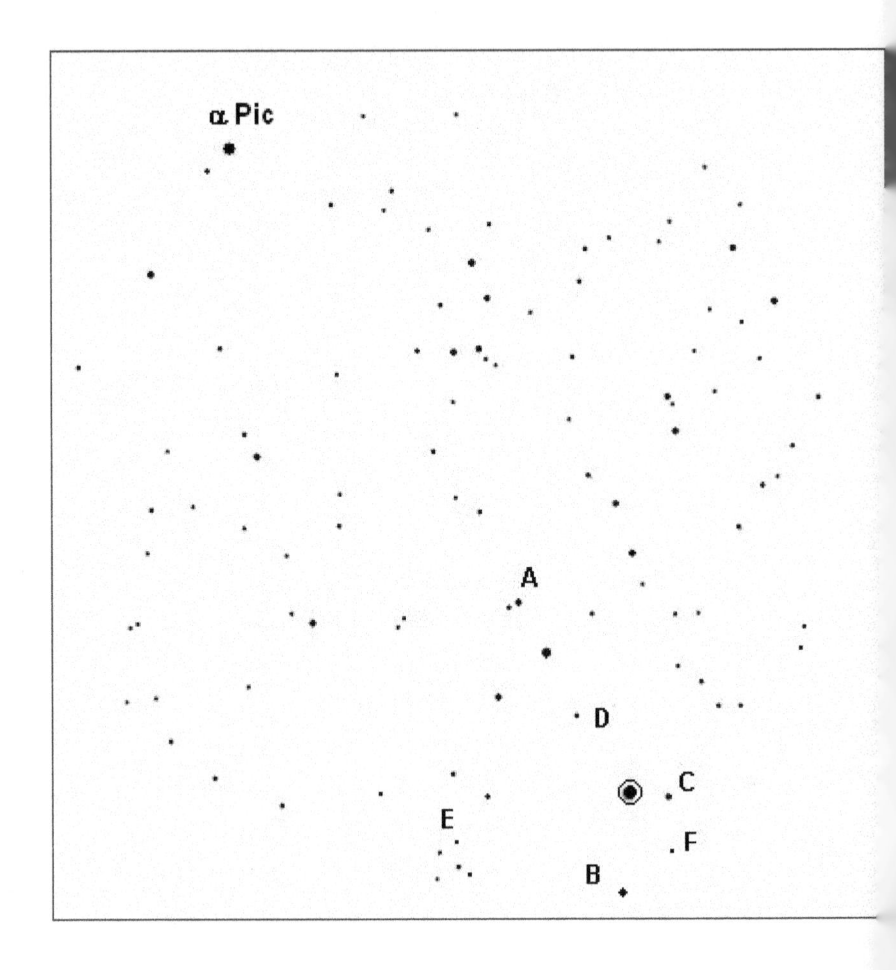

AG Carinae

This highly-luminous eruptive star is in an area of several other variables suitable for binoculars, one of which is BZ Car, described in the text. The star T Carinae was once thought to be variable but is actually constant - at least as far as visual observers are concerned.

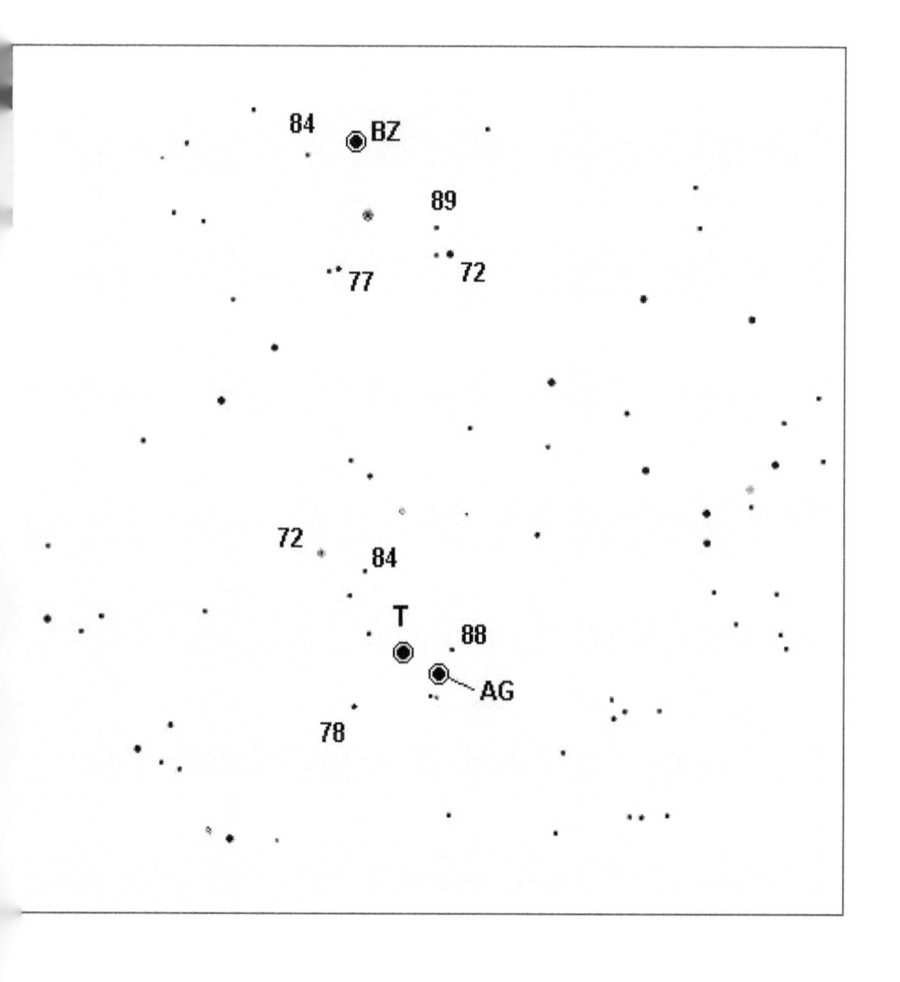

IW Carinae

This 4° field shows IW as well as the nearby bright Mira star R Carinae. Observe IW once a week if you can.

Comparison Stars:

A = 4.0 B = 4.8 C = 5.8 D = 6.5 E = 6.9

F = 7.4 G = 8.0 H = 8.5 K = 9.0 L = 9.3

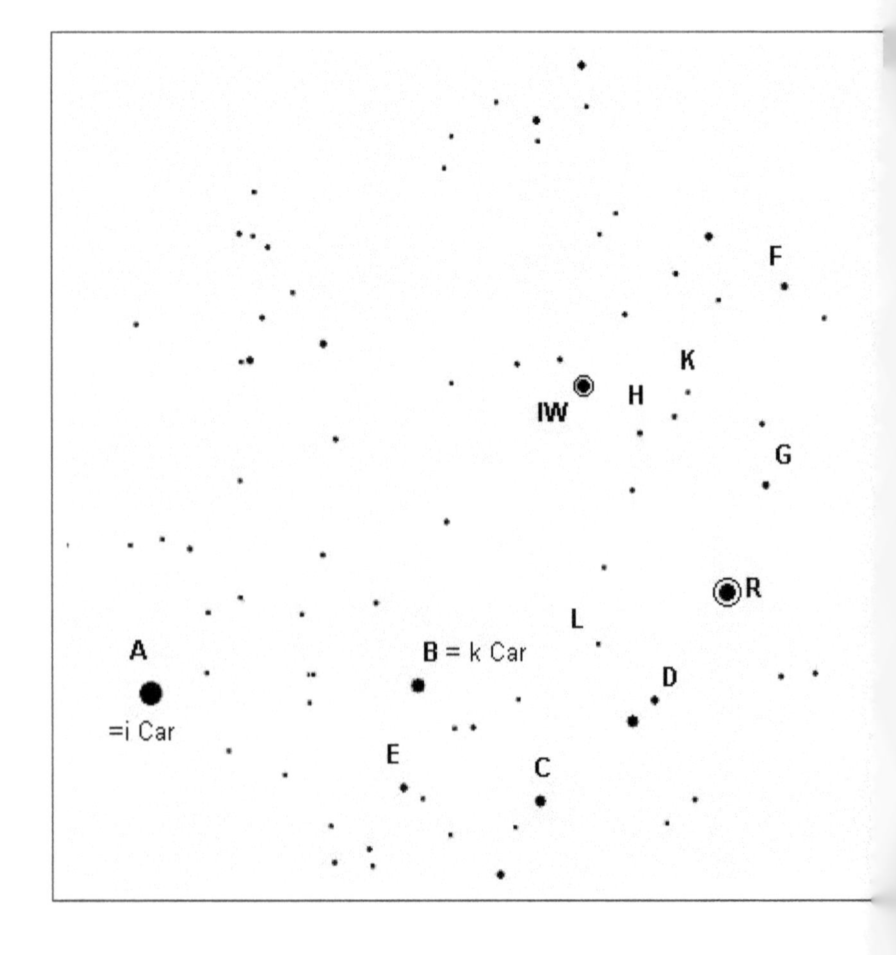

T Centauri

Even some Northern-hemisphere observers may be able to find this bright variable, which with a short period of 90 days should be observed about every 10 days or so.

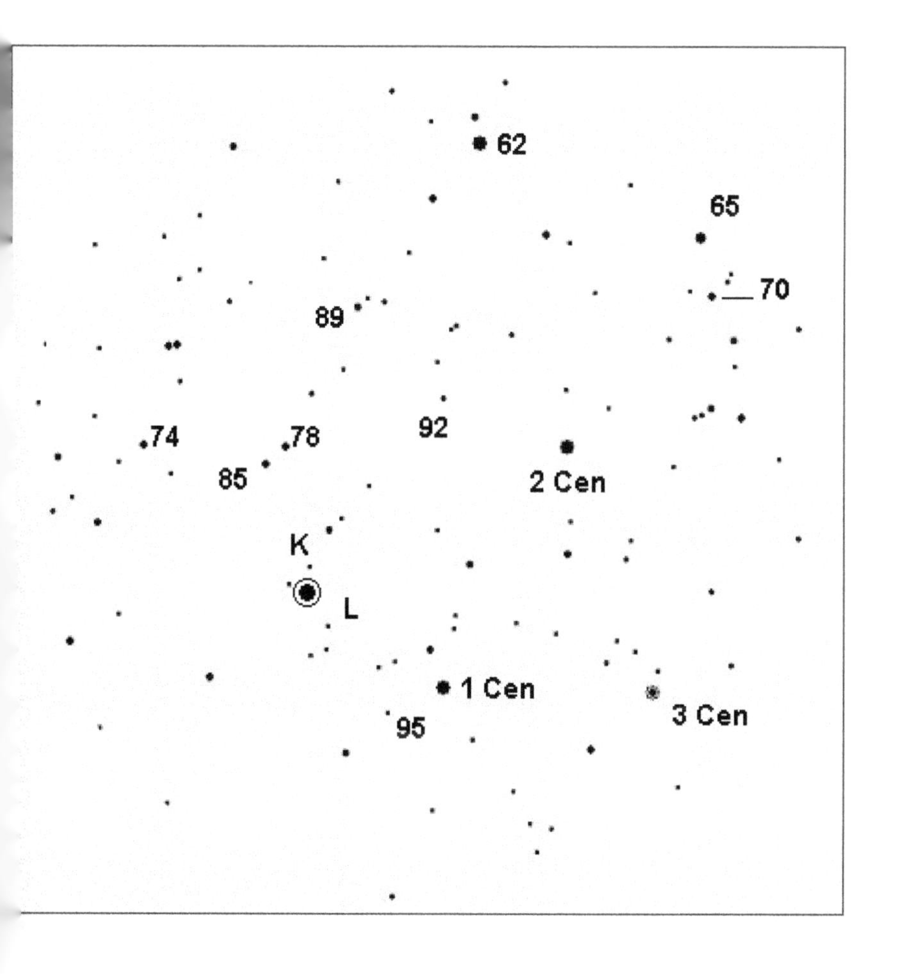

Y Centauri

Even farther North than the previous variable, this has a smaller range of variation as well as being rather less predictable.

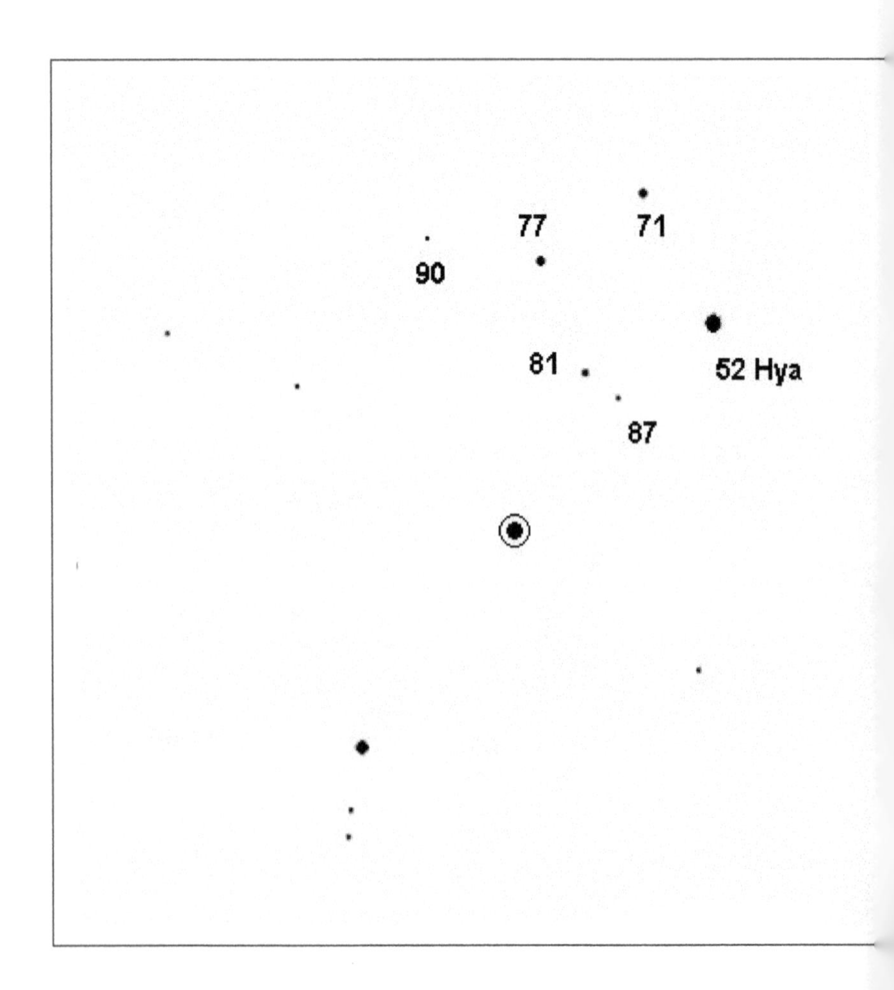

V412 Centauri

In a field rich in stars, including variables, this is a red star with quite a good range, easily found from the brilliant β Centauri. V381 Cen is a Cepheid that varies between 7.3 and 8.0, and SZ is an eclipsing binary with the same range but one magnitude fainter, and it fades once every 4 days.

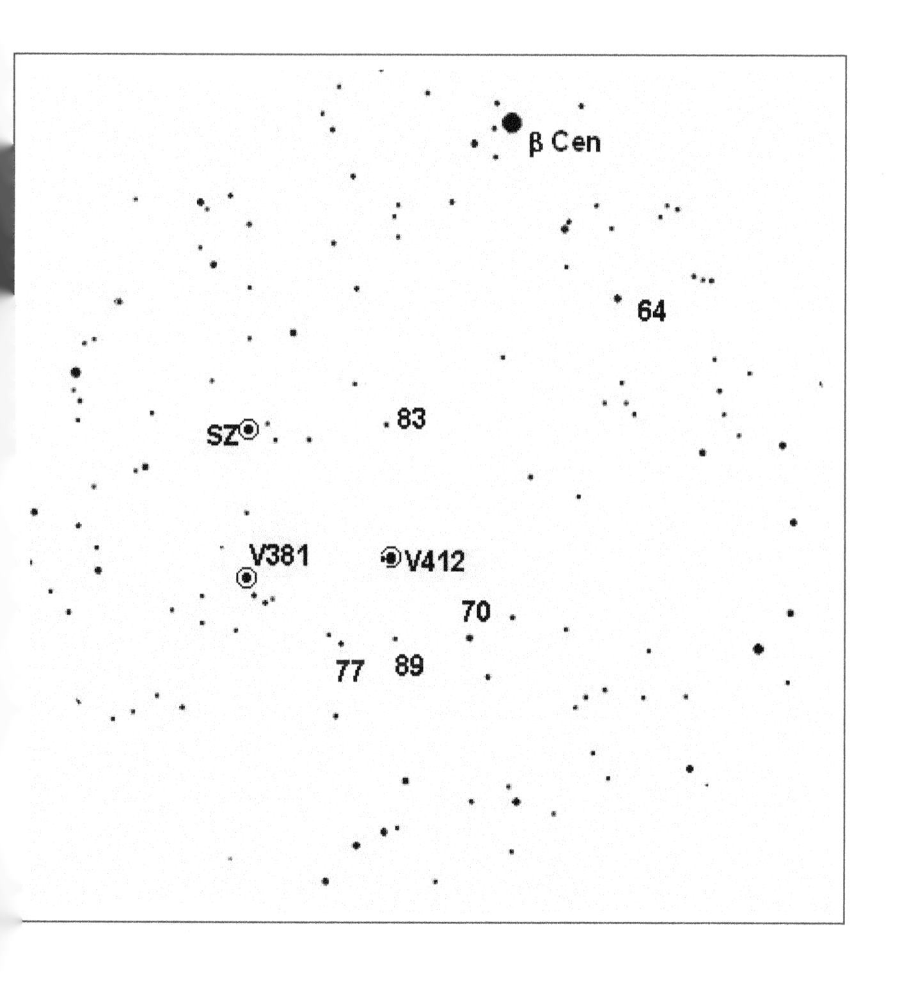

V854 Centauri

A peculiar, and quite active, star of the R Coronae Borealis type, watch this one on every possible occasion, since the fades are unpredictable.

Comparison Stars:

1 = 6.3 2 = 6.9 3 = 7.6 4 = 8.2 5 = 9.0

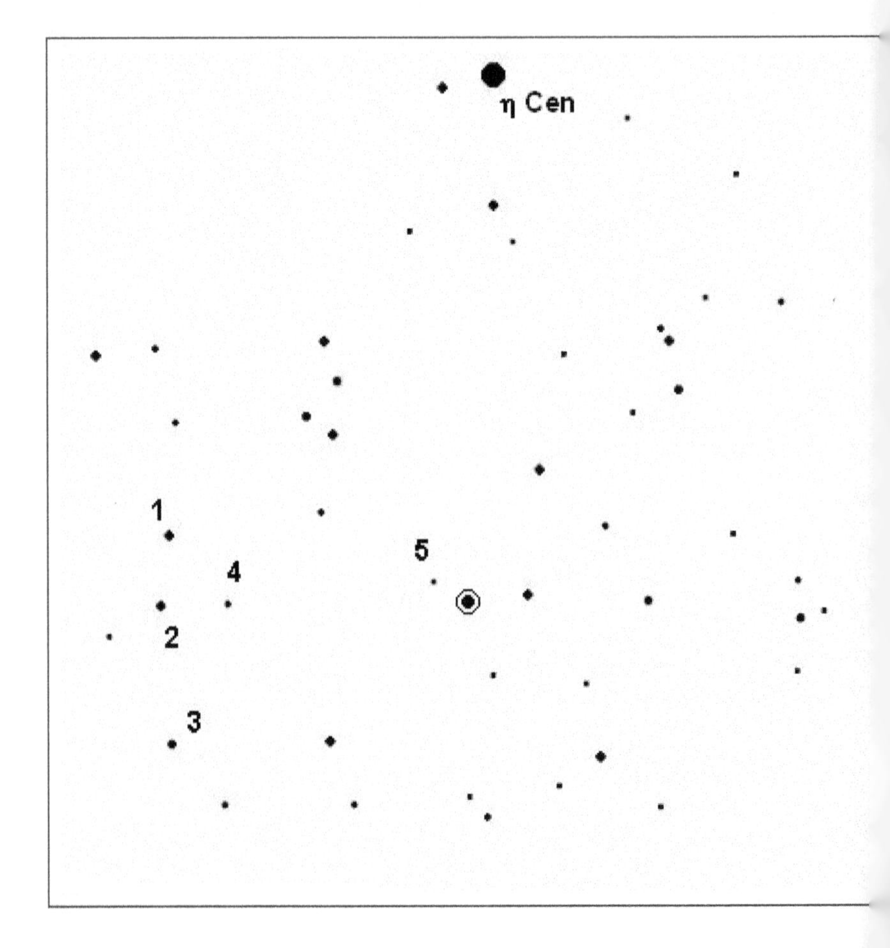

AO Crucis

A red star easily found about three-quarters of a degree south of ζ Crucis.

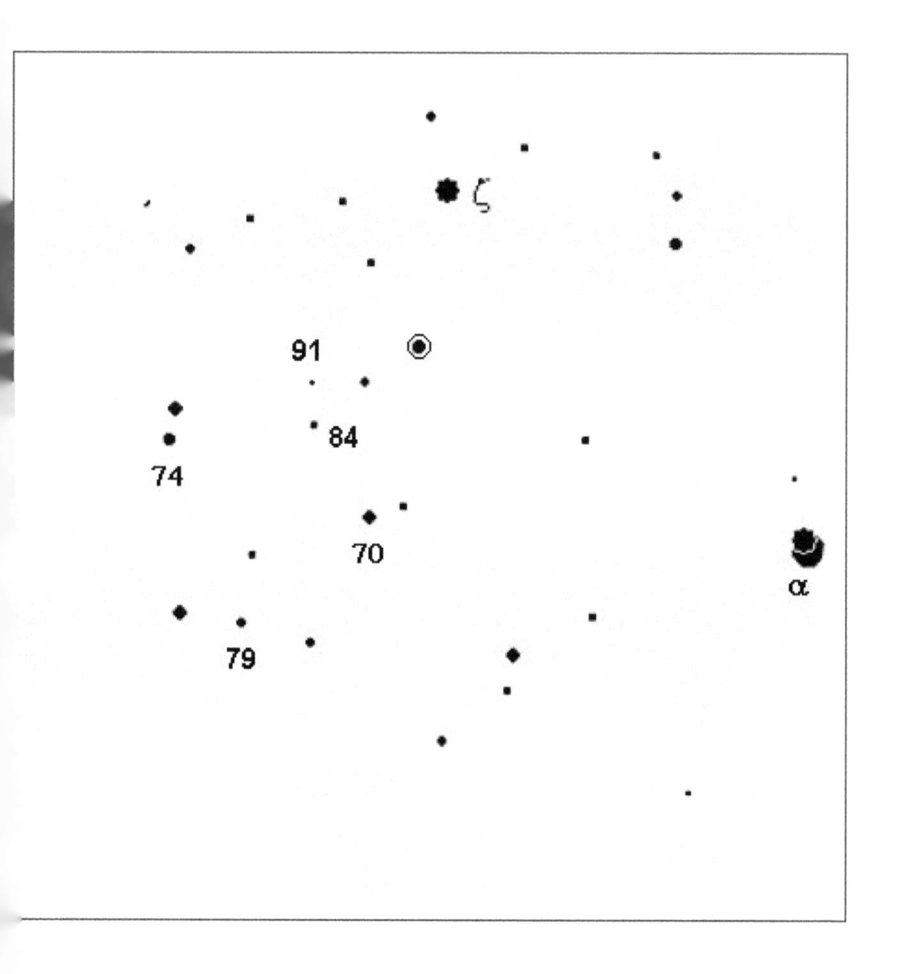

BH Crucis

This peculiar double-maxima star is of the rare spectral type S, which indicates the presence of the element Zirconium in its atmosphere. A star much studied by professional astronomers. Size of this chart is 3 degrees.

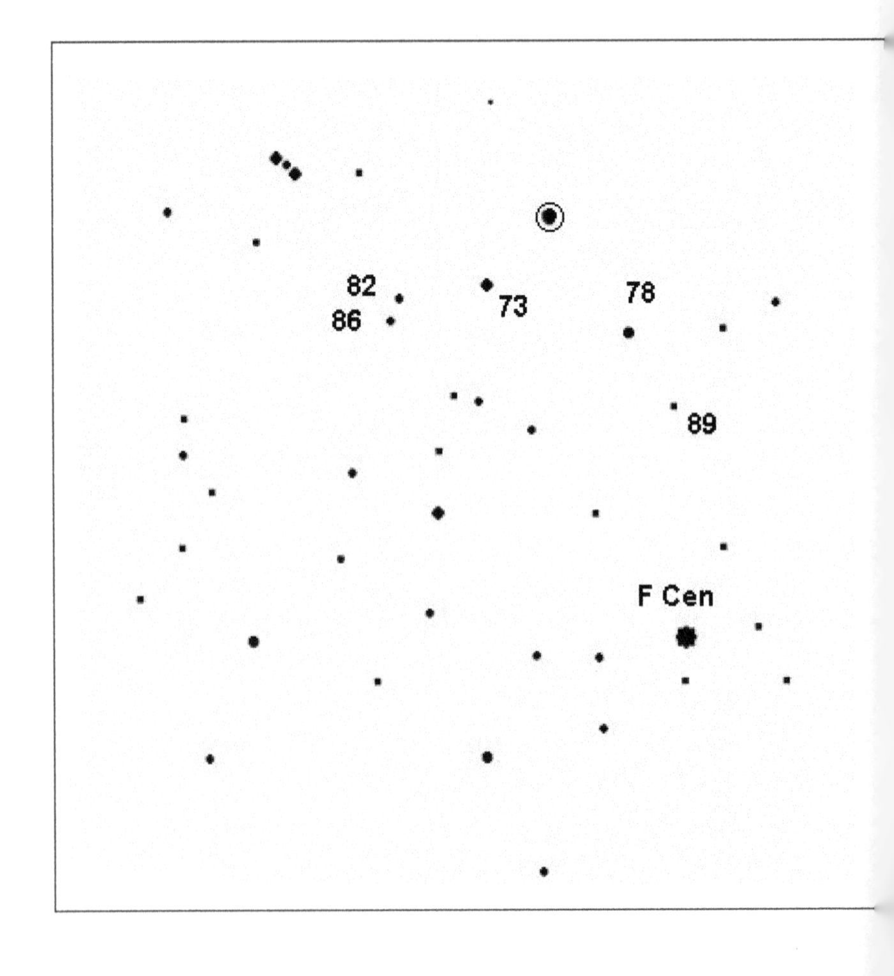

V Horologii

You can find this deep red star a degree South of the 5th magnitude

μ Horologii.

Comparison Stars:

A = 6.6 B = 7.3 C = 7.6 D = 8.0 E = 8.4 F = 8.9

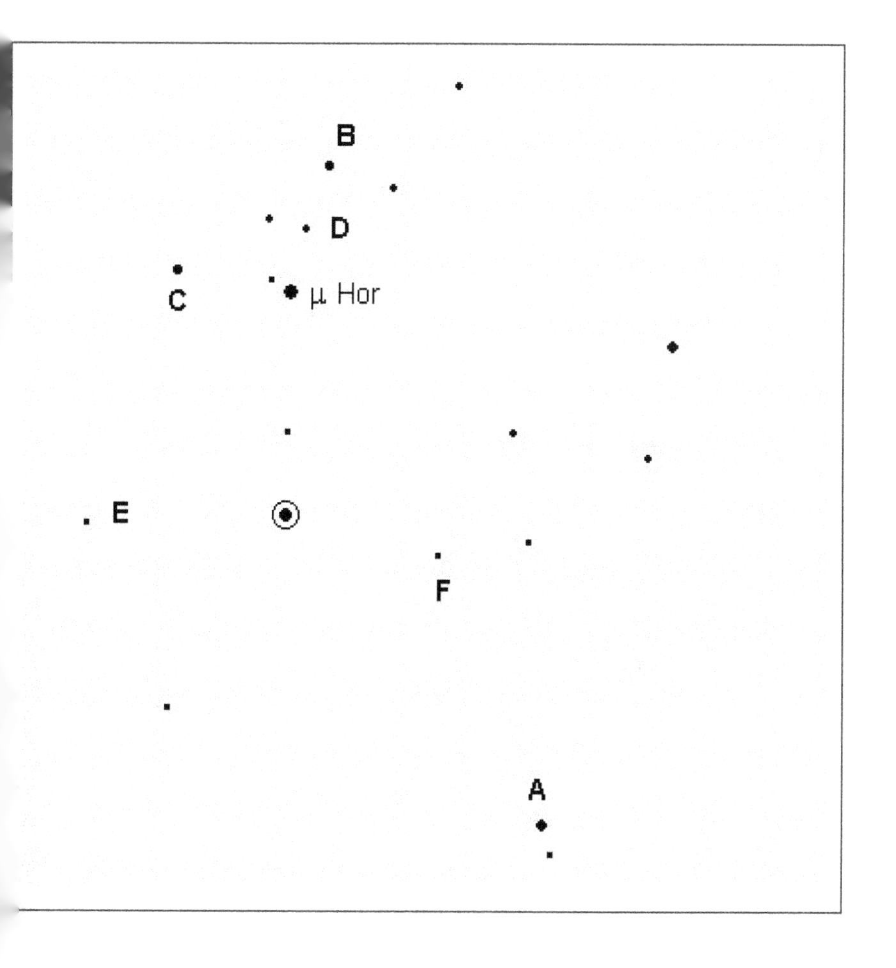

T Indi

Another deep red star, but rather brighter, and in a good field too. Size of this field is 7 degrees.

Comparison Stars:

A = 5.5 B = 6.3 C = 6.9 D = 7.4

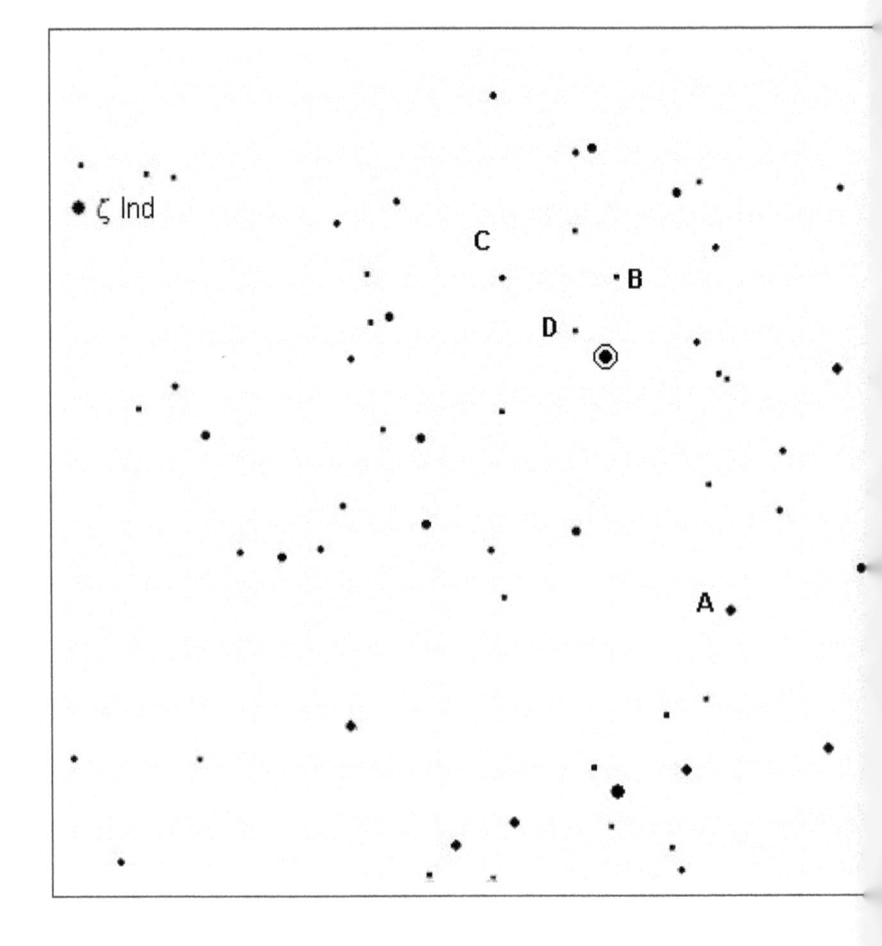

T Microscopii

Observe this red star twice a month. Rather isolated, but star 64 makes a long isosceles triangle with the 4th-magnitude stars ψ and ω Capricorni.

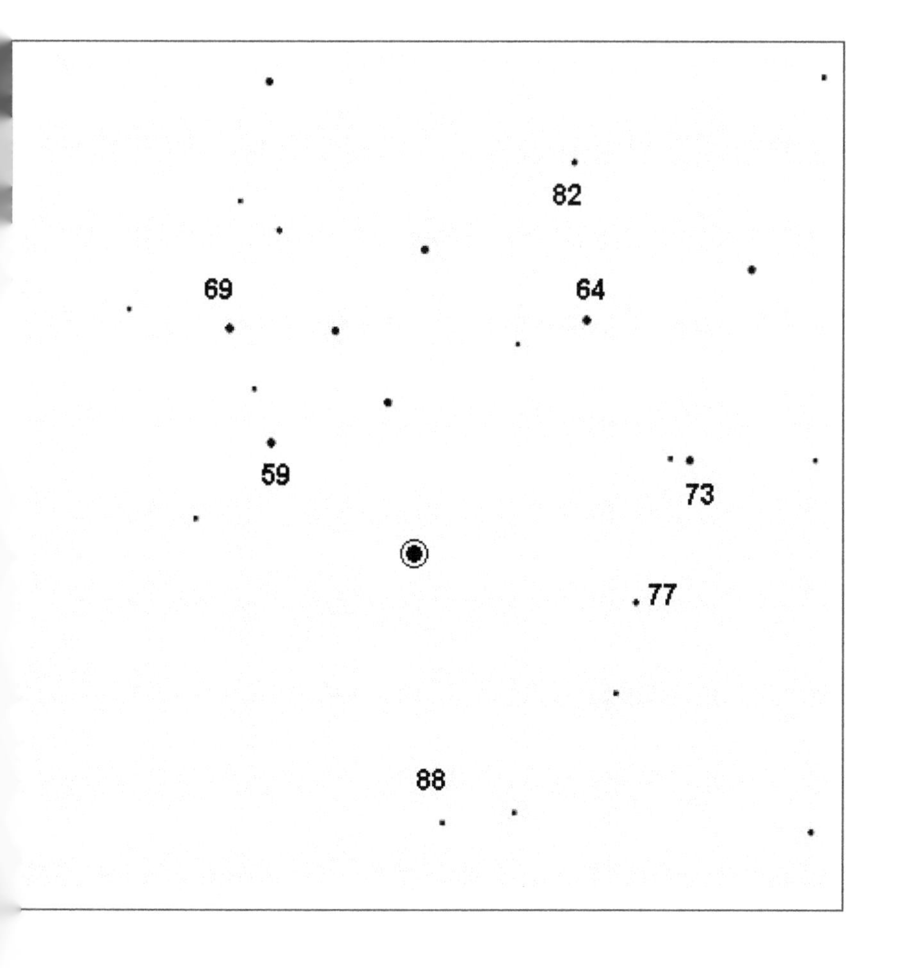

BO Muscae

A star for the smallest of binoculars - and you get another variable into the bargain! R Mus nearby is a Cepheid variable that changes between magnitudes 6.1 and 7.1 in the space of a week.

Comparison Stars:

A = 5.7 B = 6.2 C = 6.4 D = 7.2 E = 7.5

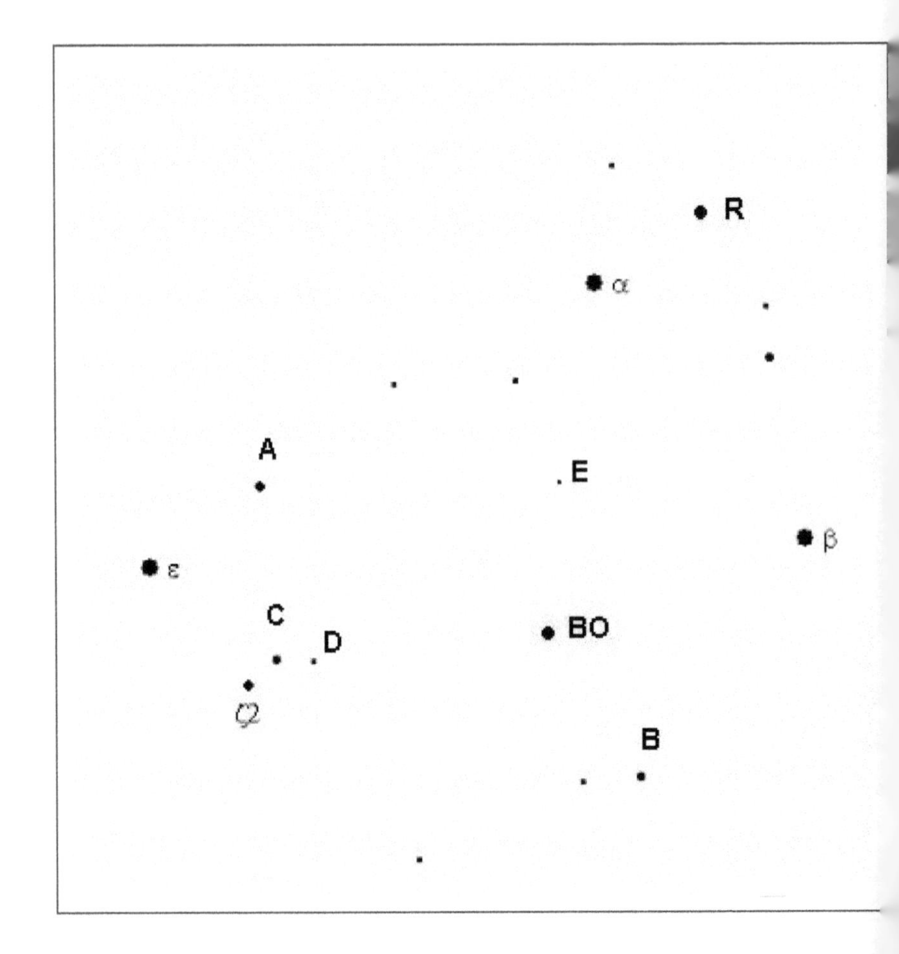

S Pavonis

Owners of large glasses should be able to manage the minima of this good variable, but lesser ones can see most of its changes, along with all of those of the less-regular X Pavonis nearby.

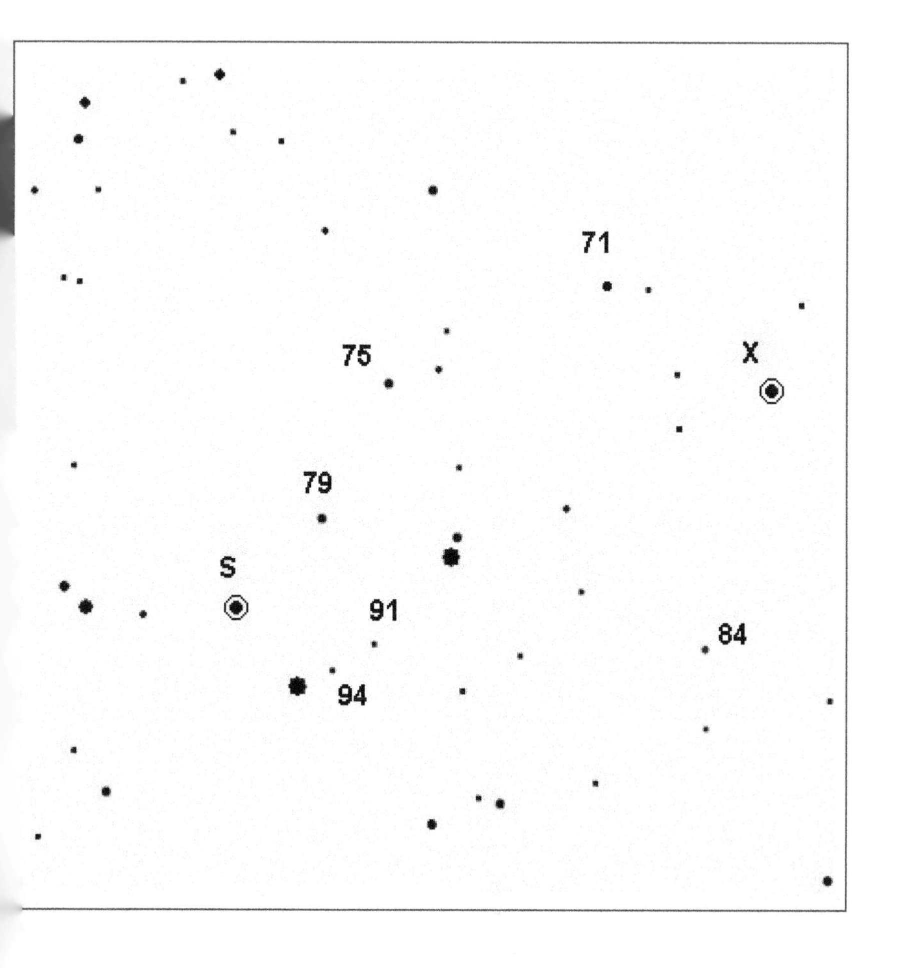

Y Pavonis

Use the 5th-magnitude omicron to find this deep red variable, again with another red variable close by. This one, SX Pav, is brighter but with less-noticeable light changes. Field size is 5°.

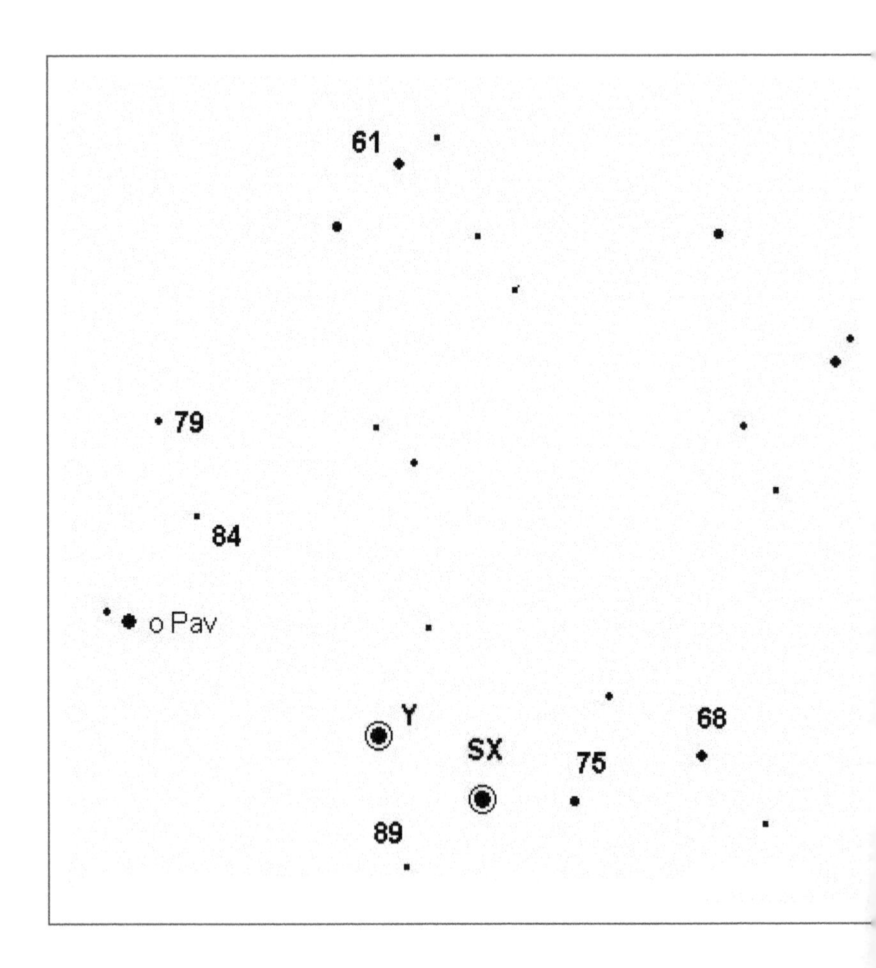

SX Phoenicis

The chief interest here is the extremely short period of only 1 hour and 17 minutes. Take a look at it between observations of other stars and you may see it brightening before your eyes!

Comparison Stars:

A = 6.3 B = 6.7 C = 7.1 D = 7.5 E = 7.9

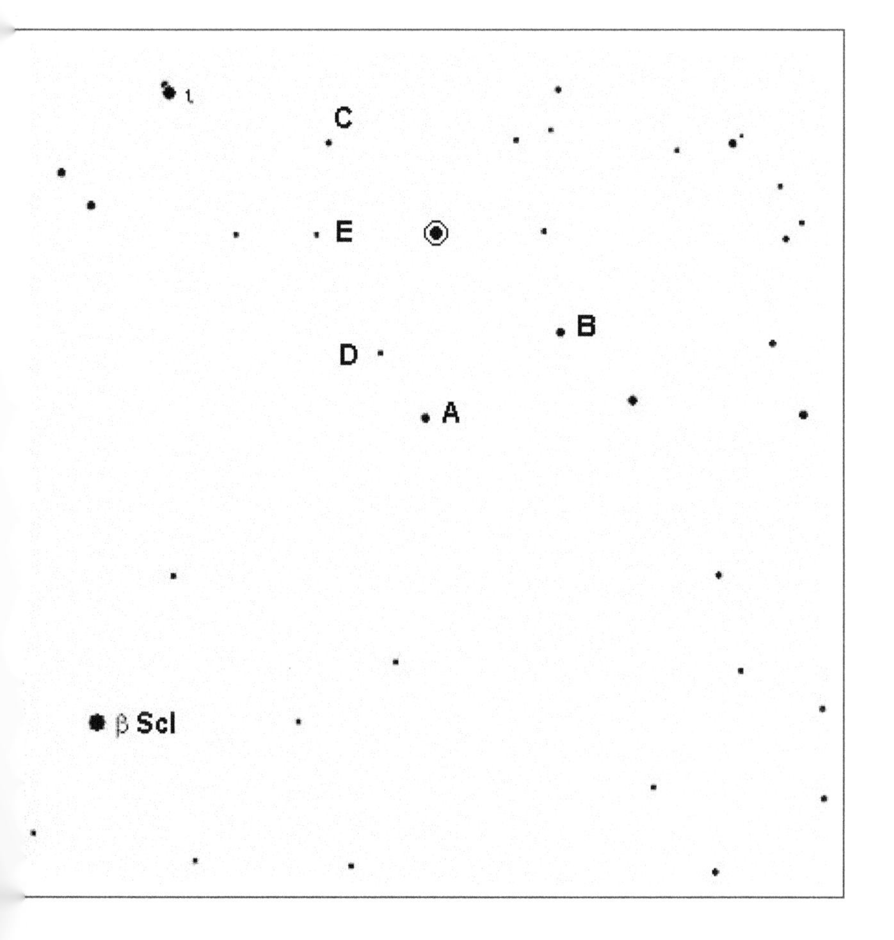

W Pictoris

A deep red variable in a rather obscure area, though you can find the field 5 degrees directly North of the naked-eye star β Pictoris. Owners of large glasses may be able to spot a faint companion to star 80 on the chart. Field size here is just 2 degrees.

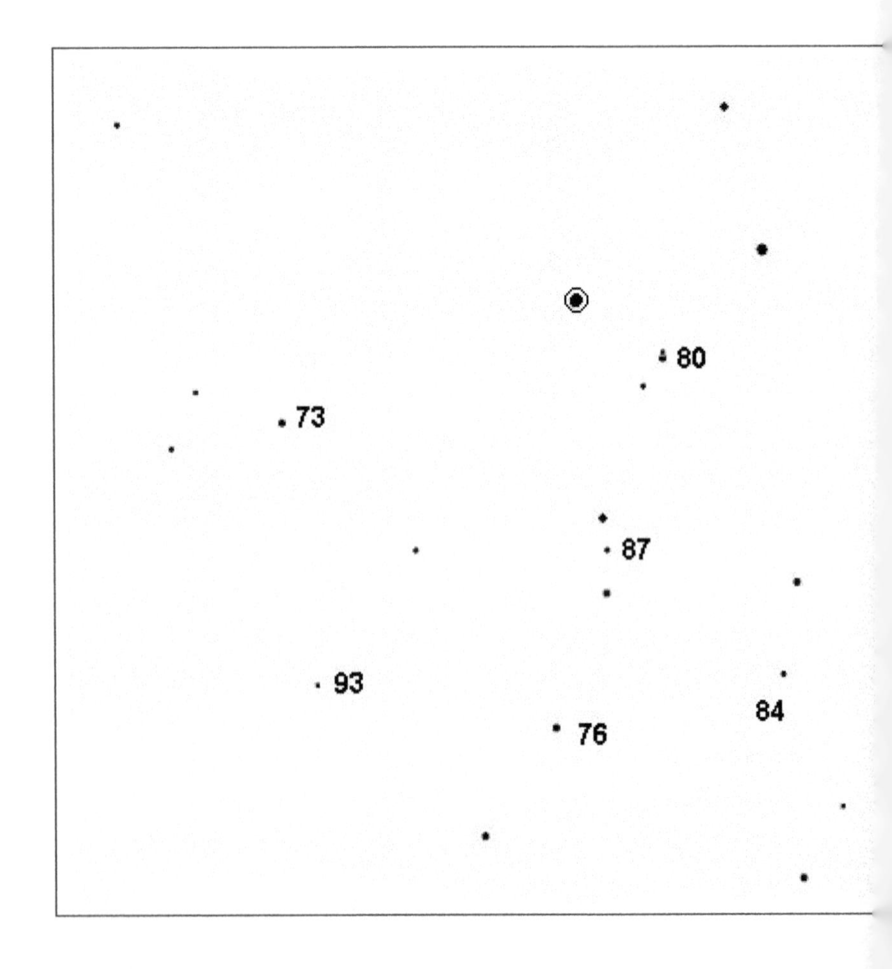

RY Puppis

There is some doubt as to whether this star is actually variable or not, so you may want to keep an eye on it. The bright star on the chart here, f Puppis, is easy to find halfway between two hot, remote and powerful stars - η Canis Majoris and ζ Puppis. Field size 3°.

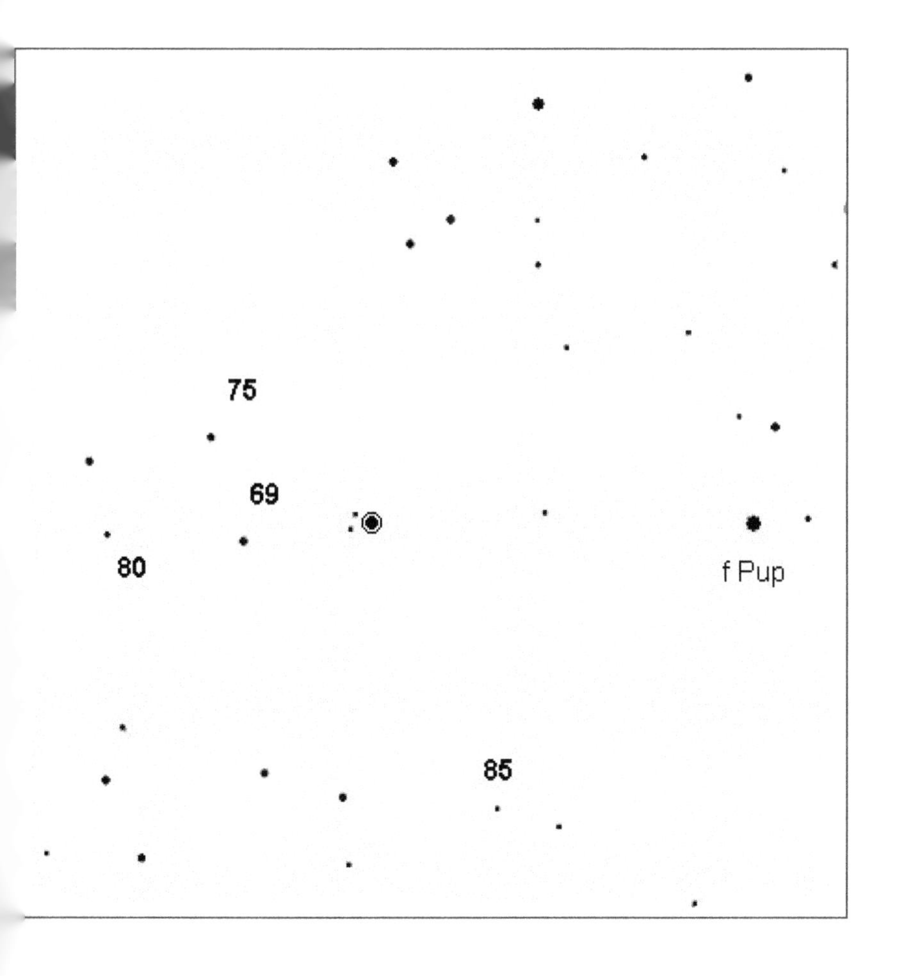

R Sculptoris

A good star for binoculars, with a decent range in light and a lovely red colour. The field forms a long parallelogram with α Sculptoris and α and β Phoenicis. Star 52 is π Scl. Field size here is 6°

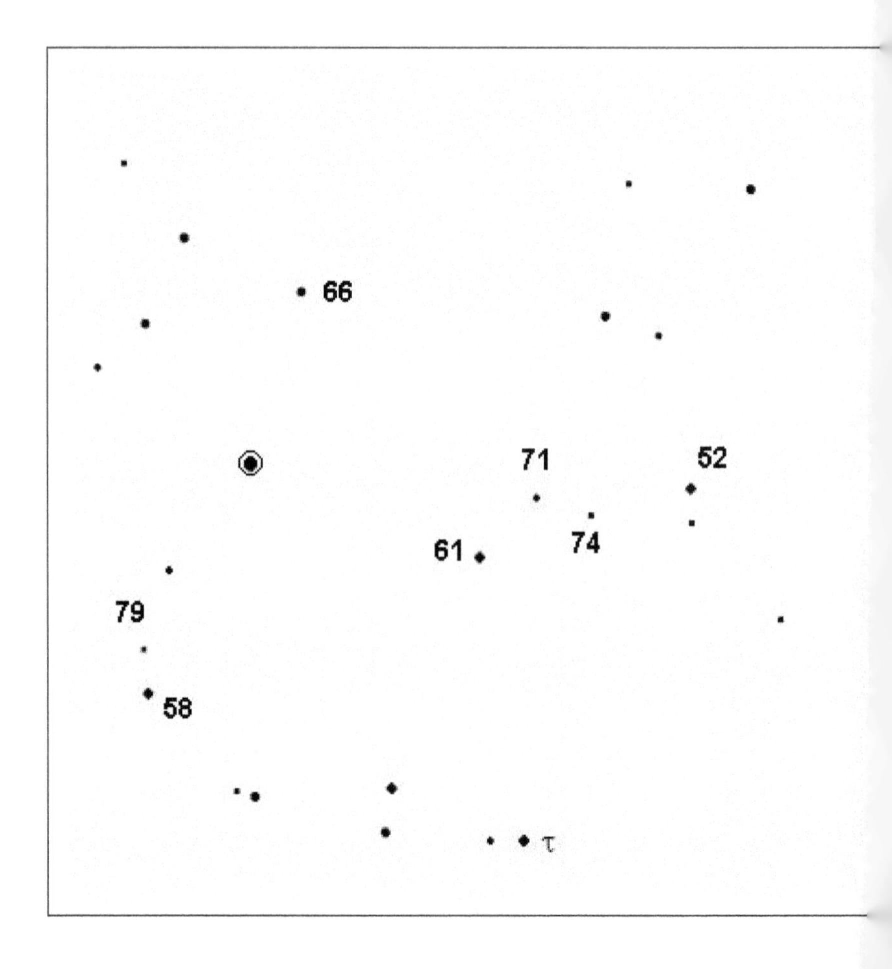

X Trianguli Australis

The slightest optical aid will show this red star, and there is a slightly fainter variable nearby. This is R TrA, a Cepheid with a period of 3.4 days and a range of 6.6 to 7.5.

Comparison Stars:

A = 5.9 B = 6.3 C = 6.7 D = 7.0

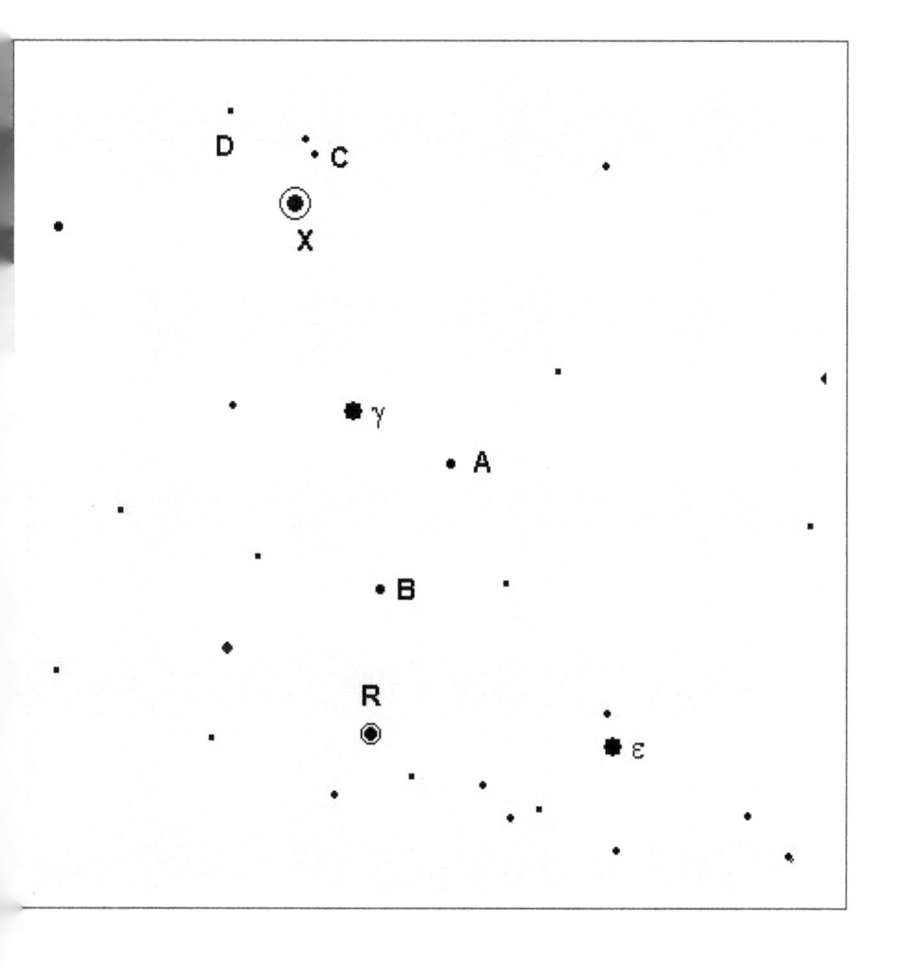

SY Velorum

A red variable in a brilliant field that includes the bright naked-eye star λ Velorum.

Comparison Stars:

A = 6.8 B = 7.3 C = 7.8 D = 8.2 E = 8.5 F = 9.0

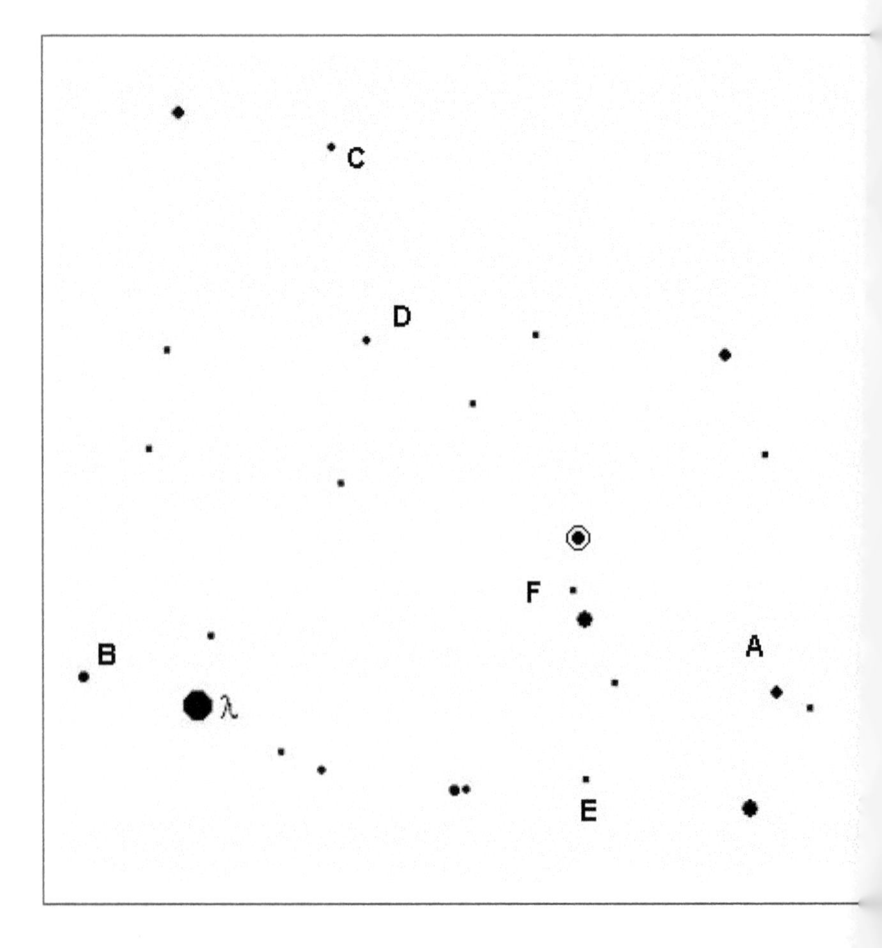

WY Velorum

Again, easily found from a naked-eye star, this time kappa Velorum, this is a red star with a range of about a magnitude and a half. In the same field is a similar star, GL Vel, but whose range is only about half a magnitude. Field size 5 degrees.

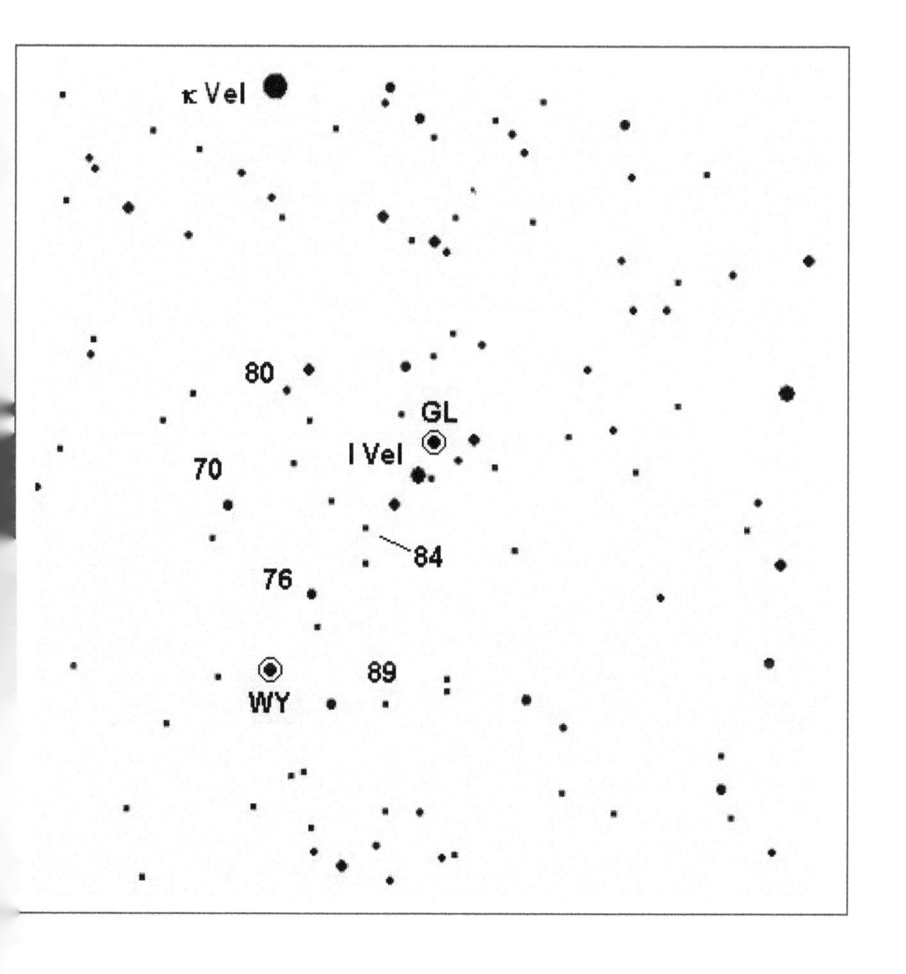

Appendix 6
Predicted Maxima of Long-Period Variables 2010-2019

The table below gives approximate dates for maxima of certain Long-Period variables mentioned in the text. The format is simple; "5" for instance means that in the given year, maximum for that star occurs in the month of May. More accurate predictions are not necessary, since these stars are not regular in any case, so a ball-park figure is actually more useful than an apparently more precise value, which may give a false sense of accuracy.

	2010	**2011**	**2012**	**2013**	**2014**	**2015**	**2016**	**2017**	**2018**	**2019**
R And	10	11		1	2	3	5	6	8	9
o Cet	10	9	8	7	5	4	3	2	1,12	11
R Aqr		2	2	3	4	5	5	6	7	7
R Aql	6	4	1, 10	8	5	2, 12	9	6	4	1, 10
R Cas		5	7	9	11		1	3	6	8
T Cep		2	3	4	4	5	6	7	7	8
χ Cyg		2	3	5	6	7	9	10	12	
R Hya	11	12		1	2	3	3	4	5	6
R Leo	7	5	3	2, 12	10	8	6	5	3	1, 11
X Oph	10	9	8	7	5	4	3	2	1, 12	10
U Ori	2	2	2	2	3	3	3	3	3	3
R Sgr	9	6	3, 11	3	5	11	8	5	1, 10	7
RR Sgr	10	9	8	7	6	5	4	3	3	2
RT	5	3	1, 11	9	7	5	3	1, 11	9	7
RU	6	2, 10	5	1, 9	5	1, 9	5	1, 9	4, 12	8

	2010	2011	2012	2013	2014	2015	2016	2017	2018	2019
RV Sgr	9	7	5	4	2, 12	11	9	8	6	4
R Ser	9	8	8	8	8	7	7	7	6	6
R Tri	6	2, 11	8	5	1, 10	7	4, 12	9	6	3, 12
R Vir		12	5, 10	3, 8, 12	5, 10	3, 8, 12	5, 10	3, 8, 12	5, 10	3, 8, 12
R Hor		3	5	6	7	9	10	12		1
R Nor*		3	7	12		5	9		2	7
R Lep		3	5	7	9	11		1	3	5

Note: the asterisk after R Normae indicates the main, brighter maximum, since this star exhibits paired maxima, a bright one followed by a fainter one.

INDEX

(Page numbers in bold indicate the page on which the chart for that constellation may be found)

Milton Keynes UK
Ingram Content Group UK Ltd.
UKHW041026260124
436746UK00001B/50